Solar Water Heating Systems
Fundamentals and Installation

The United Association of Journeymen and Apprentices of the Plumbing and Pipe Fitting Industry of the United States and Canada

Three Park Place
Annapolis, MD 21401
Phone: 410-269-2000
Fax: 410-267-0261

International Pipe Trades Joint Training Committee, Inc.

AMERICAN TECHNICAL PUBLISHERS
ORLAND PARK, ILLINOIS 60467-5756

International Pipe Trades Joint Training Committee, Inc.

Composed of representatives of the

United Association of Journeymen and Apprentices of the Plumbing and Pipe Fitting Industry of the United States and Canada (UA), the Mechanical Contractors Association of America, Inc. (MCAA), the Union-Affiliated Contractors (UAC) affiliated with the Plumbing-Heating-Cooling Contractors–National Association (PHCC-NA) and the National Fire Sprinkler Association Inc. (NFSA)

Under the Direction
of the
United Association Training Department

This training manual has been prepared for use by United Association journeyworkers and apprentices. It was compiled from information sources which the International Pipe Trades Joint Training Committee, Inc. believes to be competent. However, recognizing that each type of piping system and mechanical equipment installation must be designed to meet particular needs and code requirements, the Committee nor American Technical Publishers, Inc. assumes no responsibility or liability of any kind in connection with this training manual or associated materials, or its use by any person or organization, and makes no representations or warranties of any kind hereby.

This manual may contain instructions for the safe use of various types of tools and equipment that are used in the installation, testing, repair, maintenance, and servicing of equipment related to the subject matter. Those instructions, along with the tables, charts, drawings, or photographs which appear in this training manual, are not intended to set standards for their use and application and manufacturers' type. They are designed to familiarize journeyworkers and apprentices with some of the factors and considerations involved.

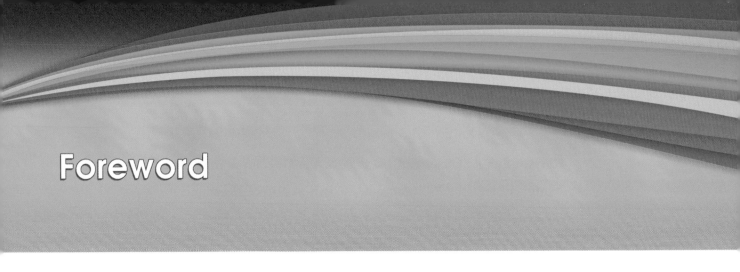

Foreword

The installation, maintenance, and servicing of solar water heating systems is work that is performed by United Association (UA) pipe trades journeyworkers and apprentices. These systems also represent a significant part of the national effort for sustainability, energy conservation, and energy independence.

UA General President William P. Hite initiated an effort to help meet the need for trained and qualified green systems aware workers. The program provides trained, experienced, certified, and immediately available green systems aware workers throughout the United States and Canada at no cost to contractors or their clients (owners and users). With the assistance of UA Director of Training Christopher A. Haslinger, the UA Green Systems Awareness Program was formed. This program encompasses this manual, *Solar Water Heating Systems,* along with other UA manuals such as the Green Awareness Manual, Energy Audit Manual, and Geo-Thermal Certification Manuals, among others. The program also includes the UA's two Sustainable Technologies Demonstration Training Trailers, which demonstrate sustainable systems that would be installed in residential and commercial buildings with actual working systems installed in the trailers.

This manual has been developed by the International Pipe Trades Joint Training Committee, Inc., under the direction of Christopher A. Haslinger, UA Director of Training, and James G. Pavesic, Assistant Director of Training, in cooperation with Philip Campbell, UA Training Specialist.

The International Pipe Trades Joint Training Committee wishes to acknowledge the Solar Water Heating Systems Manual Subcommittee for its contribution. This committee includes Philip Campbell, UA Training Specialist (subcommittee liaison); Carl Cimino, Training Director, Local 393 San Jose, CA; Frank DaCato, Training Director, Local 777, Meriden, CT; Arthur Klock, Training Director, Local 1, New York City, NY; John Sullivan, UA ITP Instructor, Local 1, New York City, NY; Donald J. Berger, UA ITP Instructor, Local 60, New Orleans, LA; Scott Hamilton, Training Director, Local 75, Milwaukee, WI; and James Pruyn, Instructor, Local 597, Chicago IL.

Solar Water Heating Systems was written in partnership with American Technical Publishers. This illustrated manual covers twelve major solar water heating topics and includes service and repair content, troubleshooting tables, and Internet resources. It also includes a DVD containing Quick Quizzes®, an Illustrated Glossary, Solar Radiation Data Sets, Sun Path Charts, Forms and Documents, Flash Cards, Media, and links to videos and other related materials.

The United Association wishes to thank all of those who contributed to this effort. Without all of your dedication and hard work, the development of this manual would not have been possible.

ACKNOWLEDGMENTS

The following companies and organizations have provided photographs, prints, and technical assistance. The inclusion of these companies and organizations does not imply an endorsement of any products and are included only as representative examples.

AAA Solar Supply Inc.
Alternate Energy Technology
American Beauty Tools
American Solar Living LLC
A. O. Smith
Ballymore Company, Inc.
Blue-White Industries
Bosch Thermotechnology Corp.
Butler Sun Solutions Inc.
Cooper Wiring Devices
Copper Development Association, Inc.
Database of State Incentives for Renewable Energy (DSIRE)
DOE/NREL, Gen-Con, Inc.
DOE/NREL, Warren Gretz
Dwyer Instruments, Inc.
Energy Works US, LLC
FAFCO
Florida State Energy Center (FSEC)
Fluke Corporation
Heat Transfer Products, Inc.
Hydroflex Systems, Inc.
IAPMO
Intermatic, Inc.

Interstate Renewable Energy Council (IREC)
Johns Manville
Lab Safety Supply, Inc.
Laura Slivka
Lochinvar, LLC
L P International Inc.
Miller Fall Protection/Honeywell Safety Products
Milwaukee Electric Tool Corporation
MISCO
National Oceanic and Atmospheric Administration (NOAA)
Noble Company
North Carolina Solar Center
Oatey
Panduit Corp.
Plastic Pipe and Fittings Association
Quick Mount PV
RESOL
Rheem Manufacturing Company
RIDGID®
Schott Solar
Sioux Chief Manufacturing Company, Inc.

Solar H$_2$Ot, Ltd.
Solar Pathfinder
Solar Rating Certification Corporation (SRCC)
Solar Service Inc.
SolarWorld Industries America
Solar Works
Solarzentrum North America
Solmetric Corporation
Stiebel Eltron, Inc.
SunMaxx Solar
Sunnovations, Inc.
Taco, Inc.
TCT Solar
ThermoTech
Uponor Wirsbo
U.S. Green Building Council
Vanguard Piping Systems, Inc.
VELUX America Inc.
Viessmann Manufacturing Company, Inc.
Watts Water Technologies, Inc.
Werner Ladder Co.
Xylem Inc.
Zodiac Pool Systems, Inc.

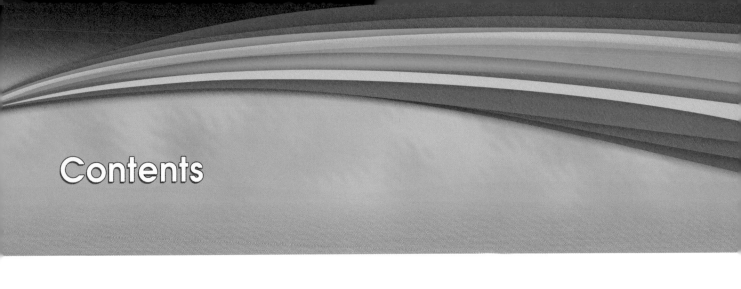

Contents

Contents

DVD Contents

Quick Quizzes®
Illustrated Glossary
Solar Radiation Data Sets
Sun Path Charts

Forms and Documents
Flash Cards
Media
ATPeResources.com

UA WEBSITES

United Association (UA) Information:

For more information about the UA and its training programs in plumbing, pipefitting, sprinklerfitting, welding, and HVAC service, please visit http://www.ua.org.

Ordering UA Training Materials:

UA local unions, their JATCs, and individual members may order this manual and other educational materials from the online bookstore by visiting the members-only section at http://www.ua.org.

Introduction

Solar Water Heating Systems: Fundamentals and Installation was developed in conjunction with the United Association of Journeyman and Apprentices of the Plumbing and Pipefitting Industry for use in solar water heating system training. *Solar Water Heating Systems* covers residential and light commercial solar water heating systems for domestic hot water and swimming pool applications. The textbook covers topics such as solar fundamentals, site assessment, system design and sizing, the installation and service of system components, system startup and maintenance, and worker safety. The book may also be used in preparation for the North American Board of Certified Energy Practitioners (NABCEP) Solar Heating Installer Certification or other solar water heating system installer exams.

Chapter introductions provide an overview of content covered in the chapter.

Troubleshooting charts help to diagnose common solar water heating installation problems.

Facts provide additional technical and historical information.

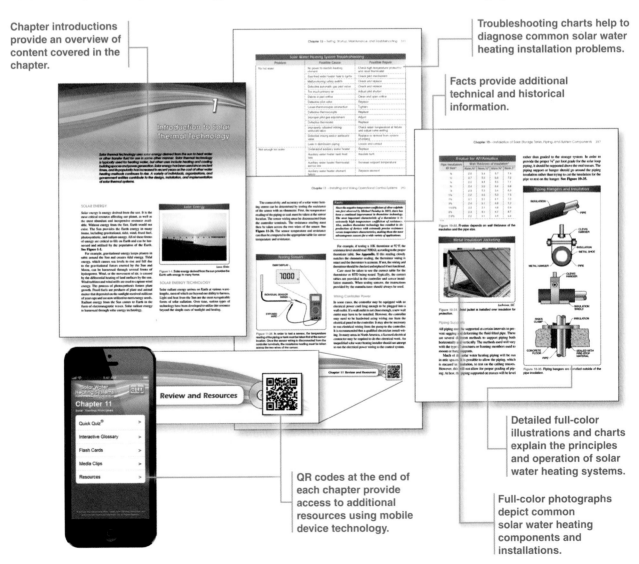

Detailed full-color illustrations and charts explain the principles and operation of solar water heating systems.

Full-color photographs depict common solar water heating components and installations.

QR codes at the end of each chapter provide access to additional resources using mobile device technology.

The *Solar Water Heating Systems: Fundamentals and Installation* Interactive DVD is a self-study aid designed to supplement content and learning activities in the book. The Interactive DVD includes Quick Quizzes®, an Illustrated Glossary, Solar Radiation Data Sets, Sun Path Charts, Forms and Documents, Flash Cards, Media, and a link to ATPeResources.com.

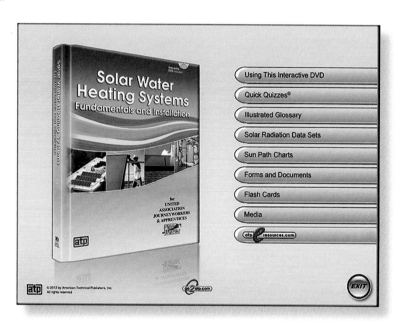

The Interactive DVD enhances content in the textbook with the following features:

- Quick Quizzes® that provide an interactive review of key topics
- An Illustrated Glossary that defines terms, with links to select illustrations and media clips
- Solar Radiation Data Sets that provide solar resource statistics for 239 different locations
- Sun Path Charts that plot the sun's position throughout the year
- Forms and Documents that provide customizable PDF reports, lists, and documents for use in the solar water heating field
- Flash Cards that reinforce understanding of common industry terms and definitions
- Media that reinforces and expands upon book content with videos and animated graphics
- ATPeResources.com, which provides a comprehensive array of instructional resources

Introduction to Solar Thermal Technology

Solar thermal technology uses solar energy derived from the sun to heat water or other transfer fluid for use in some other manner. Solar thermal technology is typically used for heating water, but other uses include heating and cooling building spaces and power generation. Solar energy has been used since ancient times, and its popularity has increased in recent years as the cost of other water heating methods continues to rise. A variety of individuals, organizations, and government entities contribute to the design, installation, and implementation of solar thermal systems.

SOLAR ENERGY

Solar energy is energy derived from the sun. It is the most critical resource affecting our planet, as well as the most abundant and inexpensive resource available. Without energy from the Sun, Earth would not exist. The Sun provides the Earth energy in many forms, including gravitational, tidal, wind, fossil fuel, photosynthetic, and radiant energy. All of these forms of energy are critical to life on Earth and can be harnessed and utilized by the population of the Earth. **See Figure 1-1.**

For example, gravitational energy keeps planets in orbit around the Sun and creates tidal energy. Tidal energy, which causes sea levels to rise and fall due to the gravitational forces exerted by the Sun and Moon, can be harnessed through several forms of hydropower. Wind, or the movement of air, is caused by the differential heating of land surfaces by the sun. Wind turbines and wind mills are used to capture wind energy. The process of photosynthesis fosters plant growth. Fossil fuels are products of plant and animal matter that depended on the sunlight received millions of years ago and are now utilized to meet energy needs. Radiant energy from the Sun comes to Earth in the form of electromagnetic waves. Solar radiant energy is harnessed through solar energy technology.

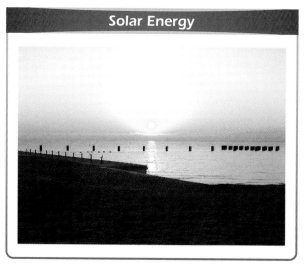

Laura Slivka

Figure 1-1. Solar energy derived from the sun provides the Earth with energy in many forms.

SOLAR ENERGY TECHNOLOGY

Solar radiant energy arrives on Earth at various wavelengths, most of which are beyond our ability to harness. Light and heat from the Sun are the most recognizable forms of solar radiation. Over time, various types of technology have been developed to utilize this resource beyond the simple uses of sunlight and heating.

Solar radiant energy can now be harnessed to produce electricity and light and for heating and cooling purposes. Most solar energy technologies involve pieces of equipment that are added to or incorporated into building structures to collect, convert, and distribute the sun's energy.

Earth receives an abundance of energy from the sun. For example, all of the energy stored in fossil fuel reserves, such as coal, natural gas, and oil, is equivalent to only 20 days of sunlight energy. **See Figure 1-2.** Energy arrives from the Sun at a rate of 500 billion tons (t) of coal/day. This represents more than 1500 times the energy requirements of the United States and Canada. It would take the entire population of the planet one year to consume the solar energy received during only 40 min of daylight. Only 1% of available solar energy is harnessed through the use of existing applications and technologies such as solar architecture, solar lighting, photovoltaics, and solar thermal systems.

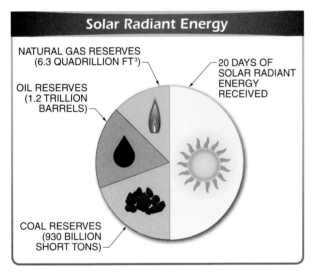

Figure 1-2. The amount of solar radiation received on Earth in 20 days is equal to the entire fossil fuel reserves of coal, natural gas, and oil.

Solar Architectural Design

Many buildings and homes were built with the conservation of energy and natural resources in mind. Incorporating passive solar design techniques into architectural designs reduces the amount of energy needed for lighting, heating, cooling, and other uses. The most influential design method is to thoroughly insulate and seal the building envelope. This method involves the use of features such as extra thick walls, extra insulation, joint and transition sealing, and double-pane windows. Walls and roofs made with stone, concrete, or similar materials absorb and hold heat. Insulating and sealing the building envelope can significantly reduce the heating and cooling loads necessary to condition indoor spaces by allowing less energy to escape outside.

Positioning windows to face the Sun allows maximum sunlight and heat into buildings. If there is too much sunlight during the summer, deciduous trees can be planted to shade windows in the warm weather. Cooling loads are thereby reduced, but sunlight is allowed through in cool weather when the leaves fall. Awnings can also be used to create shade in the summer and allow exposure to the Sun in the winter.

Many of these passive solar energy strategies are included in the Leadership in Energy and Environmental Design (LEED®) Green Building Rating System™ to promote reduced energy use. **See Figure 1-3.** The LEED rating system is an independent rating system that certifies different levels of energy efficiency and environmental protection efforts. The LEED rating system also evaluates building location, construction materials, operations, and occupant comfort.

Solar Lighting

Solar lighting, or daylighting, is a method of capturing and redirecting natural light for use in the interior of a building using special equipment and techniques. Increasing the amount of natural light indoors is good for the environment because it reduces the amount of electricity used. Architectural design, such as window size and placement, or special materials or components may be used for solar lighting to bring additional light indoors. For example, fiber-optic cables are used in hybrid daylighting units to bring sunlight into special hybrid luminaires that use both conventional and solar illuminates in the same fixture.

Photovoltaics

Photovoltaics is solar energy technology that directly converts solar radiation into electricity using crystalline silicon wafers that are sensitive to sunlight. The photovoltaic wafers, or cells, are grouped to form a module. A group of modules, called a photovoltaic (PV) array, produces an appreciable amount of electrical power with no moving parts, noise, or emissions. PV arrays are usually mounted on roofs or nearby on the ground and may include battery storage.

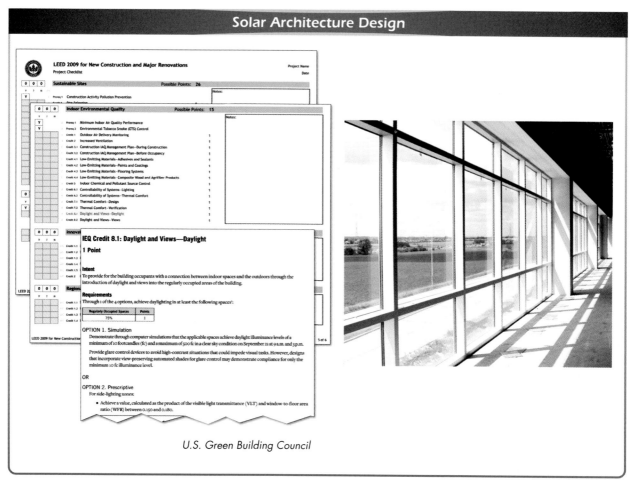

Figure 1-3. A LEED®-certified building may have elements of solar architectural design, such as floor-to-ceiling windows, to promote reduced energy use and increased occupant comfort.

Facts

The LEED® Green Building Rating System™ was created in 2002 and continues to be updated to increase building efficiency throughout building lifecycles. LEED-certified buildings are designed to reduce operating costs, increase asset values, conserve energy and water, promote occupant health and safety, and reduce harmful greenhouse gas emissions.

A *photovoltaic (PV) system* is an electrical system consisting of a PV array and other electrical components needed to convert solar energy into usable electricity. PV systems can be used in almost any application where electricity is needed. They can be as simple as portable systems used for temporary power production. They can also be as complex as multicollector systems used to supplement or replace the needs of a building or city for electricity. The most common type of residential PV system is the utility-connected system, which operates a PV array in parallel with and connected to an electric utility grid. **See Figure 1-4.**

Solar Thermal Technology

Solar thermal technology, sometimes referred to as solar heating and cooling, collects the portion of the solar radiation spectrum referred to as sunlight, which contains visible light and "near-infrared" radiation, and converts it into heat energy for heating and cooling purposes. Heat energy is created when solar radiation is absorbed by fluid in a solar collector, just as a heat lamp heats the skin. The heat energy in the fluid is then distributed through piping to a reservoir or tank to be stored or to other parts of the system to be used immediately. **See Figure 1-5.**

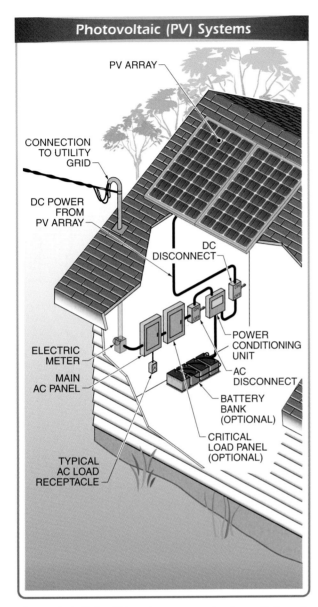

Photovoltaic (PV) Systems

PV ARRAY

CONNECTION TO UTILITY GRID

DC POWER FROM PV ARRAY

DC DISCONNECT

ELECTRIC METER

MAIN AC PANEL

POWER CONDITIONING UNIT

AC DISCONNECT

BATTERY BANK (OPTIONAL)

CRITICAL LOAD PANEL (OPTIONAL)

TYPICAL AC LOAD RECEPTACLE

Figure 1-4. Solar PV systems convert solar radiation directly into electrical energy.

SOLAR THERMAL TECHNOLOGY APPLICATIONS

Solar thermal technology can be used for a wide variety of applications. Currently, solar thermal systems are used primarily in North America for water heating. Solar domestic hot water (SDHW) and swimming pool heating systems are specifically discussed in this manual. However, solar thermal systems can be designed for many different applications such as heating, cooling, generating power, cooking, and harnessing solar chemical energy.

Heating Applications

Solar thermal energy can be used in hydronic and radiant systems to heat building spaces. In hydronic and radiant systems, heated fluid may be used to directly warm a building through radiant floor heating, or the heat may be transferred to water or air through heat exchangers. Often, an individual solar thermal system is used for both DHW and hydronic heating.

Cooling Applications

Solar thermal energy can be used to power a refrigeration cycle for cooling purposes. In the refrigeration cycle the heat energy is used to compress a refrigerant. The compressed refrigerant moves to another part of the system where it expands and cools a set of coils. Air or water is run past the coils, and the cool air or water is used to cool the building.

Power Generation Applications

Solar thermal energy can be used to produce electricity indirectly through the processing of heat transfer fluids. The most common types of systems rely on concentrated solar power. *Concentrated solar power (CSP) technology* is solar thermal technology that uses mirrors and lenses to reflect solar radiation from a large area onto a smaller target area. The concentrated energy heats a transfer fluid such as synthetic oil or molten salt. The heated transfer fluid is then used to heat water and produce steam to run a traditional turbine. These systems are complex and expensive to build, so they are only feasible for utility-scale power plants.

Cooking Applications

Solar cookers use reflected energy from the Sun to heat food and beverages. The food or liquid container essentially becomes the solar thermal collector. The solar cooker reflectors direct and concentrate light onto the inside of the dish- or box-shaped cooking vessel. The cooking vessel is usually dark in color to absorb the maximum amount of heat.

Chemical Energy Harnessing Applications

Directly harnessing solar energy for chemical processes involves the use of less-developed technology but is of great interest to the scientific community. Scientists expect to harness solar energy in chemical reactions, in the future, in a manner similar to the way plants use solar energy through photosynthesis. The energy might then be released through reverse reactions.

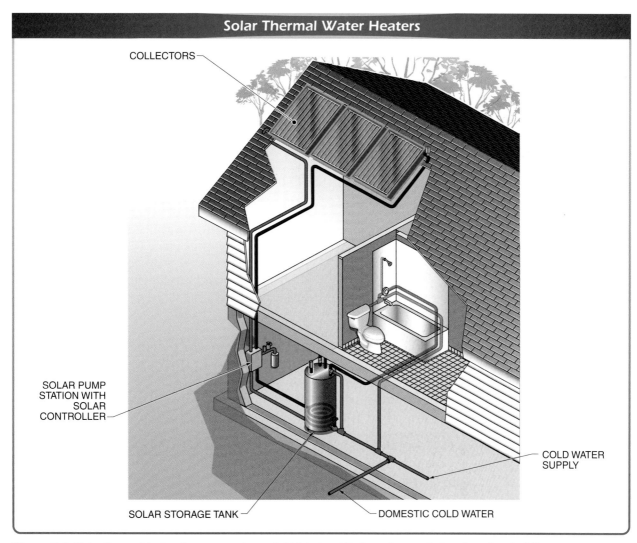

Solar Thermal Water Heaters

COLLECTORS

SOLAR PUMP STATION WITH SOLAR CONTROLLER

COLD WATER SUPPLY

SOLAR STORAGE TANK

DOMESTIC COLD WATER

Figure 1-5. Solar thermal technology converts sunlight into heat that can be used elsewhere in a building.

Residential Uses of Solar Thermal Technology

Solar thermal technology can also be used for residential applications. Residential uses for this technology include domestic hot water, swimming pool and spa heating, and space heating and cooling. Residential use can also include less common applications such as water purification, distillation, and air conditioning.

Commercial Uses of Solar Thermal Technology

Solar thermal technology can be used in commercial facilities such as hotels, schools, apartment and condominium complexes, hospitals, restaurants, laundries, car washes, meat packing facilities, and light commercial businesses. There are agricultural uses for solar thermal technologies such as crop drying, cane sugar refining, dairy processing, meat processing, and food processing.

SOLAR WATER HEATING TEMPERATURE CATEGORIES

Solar water heating technologies may be categorized by the temperature requirements needed for the specific hot water use. Low temperatures up to 100°F (38°C) are used for swimming pool heating and ventilation air preheating. Medium temperatures from 100°F to 212°F (38°C to 100°C) are used for domestic water, space, and industrial process heating. These temperatures are also used in commercial cafeterias, laundries, and hotels. High temperatures in excess of 212°F (100°C) are used for industrial process heating and electricity generation.

SOLAR WATER HEATING TECHNOLOGY LOCATIONS

Solar water heating technology can be used all across North America and worldwide, even in areas with long, cold winters. The amount of heat produced is directly related to the amount and intensity of sunlight, or solar irradiance, a site receives. **See Figure 1-6.**

Solar water heating production depends on sunlight conditions, rather than ambient temperatures. Sun energy also varies seasonally. During the summer months, with extended daylight hours, solar water heating technology may produce more hot water than required. The opposite is true during the winter months. Solar thermal technology may not always be reliable due to daily changes in the weather. If it is cloudy for an extended period of time, solar thermal technology may not produce enough hot water. However, solar thermal energy can be stored and used when the Sun is not shining.

For example, a solar water heating system will not produce 100% of the hot water needs of a home or building year-round, even when significant storage is provided. As a result, solar water heating systems need to be mated to a back-up water heater. A solar water heating system will not replace the space heating system of a building, but it may be able to supplement the heating system and help it to use less energy. Most system owners can meet 40% to 85% of their total DHW energy requirements from pre-engineered solar equipment, depending on the location. Solar thermal technology is powerful and simple technology that can help building owners save money on heating and reduce their carbon footprint.

HISTORY OF SOLAR THERMAL TECHNOLOGY IN NORTH AMERICA

Some techniques used to harness solar energy have been used for thousands of years. For example, early cave dwellers preferred caves with openings facing southeast to allow the morning Sun to warm them. They used the Sun to bake bricks into strong building materials and to dry food for preservation.

Solar energy has also been used to heat water since the dawn of humankind. Using the Sun to heat water was, and still is, fairly simple. A container is filled with water, set in the sunlight, and the water is heated by the solar energy after a short period of time.

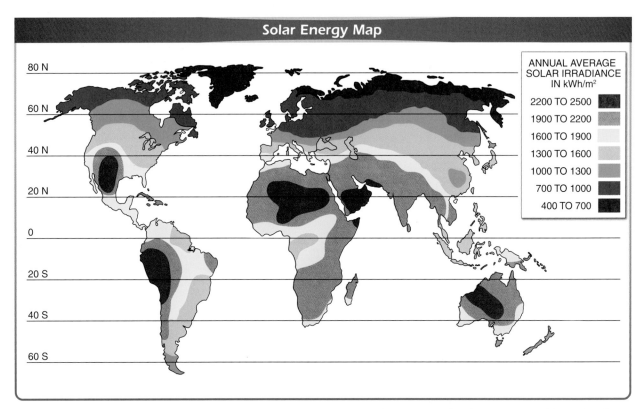

Figure 1-6. The amount of heat produced by a solar thermal system is directly related to the amount and intensity of sunlight, or solar irradiance, a site receives.

Ancient Greeks built houses that took advantage of sunrays during moderately cool winters and avoided the heat of the Sun during the summer. The Romans advanced solar energy by adapting home building design to different climates, using glass to enhance the effectiveness of solar heating and including solar design to green houses and public baths. For example, solar energy was used to heat Roman baths to increase the temperature in the caldarium (room with hot water) without warming the frigidarium (room with cold water). Solar architecture became such a part of Roman life that sun-rights guarantees were eventually enacted into Roman law.

Early Solar Thermal Technology

One of the earliest instances of modern day solar thermal use occurred in the 1700s. A leading naturalist of the time, Horace de Sassure, began to experiment with solar hot boxes. These precursors to modern solar collectors were simple insulated boxes painted black on the inside and with one side covered with glass. They were very similar to today's solar cookers. Indeed, many modern solar principles were identified during these early experiments.

Some of the earliest documented cases of solar thermal energy use in North America involve the pioneers that moved west after the Civil War. They left black pots in the Sun all day in order to have heated water in the evening. The first solar water heater that operated under a similar concept was a metal tank that was painted black and placed on a roof where it was tilted toward the sun. The concept worked, but it usually took all day for the water to heat. Then, as soon as the Sun went down, it cooled off quickly because the tank was not insulated.

Solar Water Heating Technology—1891 to 1941

Clarence Kemp of Baltimore patented the first commercial solar water heater in 1891. The solar water heater was equipped with several cylindrical water tanks of galvanized iron that were painted black. Kemp insulated the tanks in felt paper and placed them in a glass-covered wooden box for better heat retention. This invention earned Kemp the title "father of solar energy in the United States."

The solar water heater improved the lives of homeowners. During the summer, it eliminated the need to heat water on the stove. Firing up the stove to heat water warmed the entire house. In the winter, the solar water heater was drained to protect it from freezing, and homeowners resumed heating water on the stove. Kemp claimed the solar water heater could be used from early April through late October in Maryland. In Southern California, it could be used year-round. High energy costs in California made using free solar energy even more logical. By 1900, 1600 of Kemp's solar water heaters were installed in Southern California.

In 1909, William Bailey revolutionized the industry by designing the first flat-plate collector. The most visible difference with his design was a separate collector and storage tank. The collector had a grid of copper pipes and was covered with glass. A metal absorber plate was added beneath the pipes to transmit the solar heat in the box to the water in the pipes. The storage tank was insulated. Since these improvements kept water warm throughout the day and night, this solar hot water heater was called the Day and Night Collector. The system could be connected to a back-up gas heater, wood stove, or coal furnace. An electric heater could be placed inside the storage tank to heat the water automatically if it dropped below a preset temperature.

Bailey's business grew until a cold spell hit Southern California in 1913. Copper pipes in the collectors froze and burst when the temperature dropped to 19°F. Seeing that this could be a significant problem, he placed nonfreezing liquid in the collector pipes. The nonfreezing liquid traveled through a coil in the storage tank to heat the water. More than 4000 Day and Night Collectors were sold by the end of World War I. The peak year was 1920 when more than 1000 were sold.

Bosch Thermotechnology Corp.
Large commercial solar projects often include both solar water heating and photovoltaic systems.

Solar hot water heater sales began to decrease when natural gas prices dropped and gas companies offered incentives, including free installation, to switch to gas. Bailey recognized the trend and used his experience to produce gas water heaters. His company made its last solar water heater in 1941.

Gas discoveries nearly put an end to solar water heating in California, but this was not the case in Florida, where solar energy was the only way to heat water inexpensively. The Solar Water Heater Company was established in Florida in 1923. By 1925, the population of Miami had increased to more than 75,000. Business flourished in Miami until the building boom subsided in early 1926 and a hurricane struck the area in September. The plant closed shortly thereafter.

In 1931, the plant reopened with an improved collector. Charles Ewald changed the wooden box to metal so that the collector would last longer in humid environments. He also insulated the box and replaced the steel tubing with more durable and better conducting soft copper. He discovered that soft copper withstood temperatures as low as 10°F. Ewald added more piping and placed it strategically for optimum efficiency. His design produced hotter water in greater volume. He called it the Duplex.

He also developed a method of matching the needs of the homeowner with an appropriately sized collector and storage tank. This revived the industry in 1934. The following year, New Deal legislation boosted home building and, in turn, the solar heating business. Inexpensive FHA home improvement loans stimulated the market. By 1941, nearly 60,000 solar water heaters had been sold in Florida. About 80% of new homes in Miami had solar water heaters, and they were used in more than 50% of the buildings within the city. In addition to Florida, solar water heaters were used in Louisiana, Georgia, and other parts of the world, such as Japan.

Solar Water Heating Technology—1942 to 1985

The solar water heating boom of the early 1900s did not last. At the start of World War II, the government placed a freeze on the nonmilitary use of copper, stalling the solar water heating market.

After the war, the rise of skilled labor and copper prices made the collectors less affordable. Electricity prices also dropped in the 1950s, making electric water heaters more appealing. The installation and initial costs of electric water heaters were also lower than solar water heaters and they were automatically controlled providing a new convenience to the water heater. Solar water heating was not the same

bargain anymore in the United States, especially when oil import limits were allowed to surpass 50% of total domestic output. A similar scenario happened later in Japan when it began to import oil in the 1960s. The peak year for solar water heating sales in Japan was 1966.

Solar energy remained popular until abundant sources of fossil fuel became available. Interest in solar energy surged during the oil embargoes of 1973 and 1979. Federal and state tax incentives led to rising sales in the early 1980s. Sales flourished, but the industry paid a high price for this brief period of prosperity. Many new companies entered the solar field with little knowledge about the technology and were not concerned about long-term relationships with customers. This led to poor installations and gave the solar industry a bad reputation. In 1985, oil prices began to stabilize and then drop. Most of the newer companies began to leave the solar industry as the price of oil continued to decline over the next few years.

Solar Water Heating Technology in Canada

Solar water heating technology in Canada, along with other regions in North America, was a viable and emerging energy sector by the 1980s. Canada was a world leader with many innovative technologies and patents, which are still widely used today.

The Canadian government implemented the National Energy Program, but when the government changed in 1986, all support for solar water heating technology disappeared. Solar water heating technology was just beginning to become commonly used, but no one was available to repair and maintain the systems. Within the next two years, 85% of the solar industry in Canada went bankrupt.

Modern Solar Water Heating Technology

Solar water heating technology in North America has improved since the 1980s. Improvements have been precipitated by both certification design review and the increase of experienced installers. There are now more safeguards available to ensure proper system design and installation. The national Solar Rating & Certification Corporation (SRCC) provides solar collector and solar system certification and rating services. Training is a critical element in the industry. As with any mechanical system, these systems must be installed and serviced properly for optimum operation.

Today, solar water heating systems are in use around the world. There are nearly 1.5 million buildings with solar water heating systems in Tokyo, Japan alone. In Israel, 30% of the buildings use solar DHW. Greece and Australia are also leading users of solar thermal technology. More than 1.2 million buildings in the United States have solar water heating systems, and there are an additional 250,000 solar-heated swimming pools.

However, there is a lot of room for expansion in the solar energy industry in North America. Using solar energy is the easiest way to save energy and money. Wherever the Sun shines, solar water heating systems can work. Modern designs may be different from early solar designs, but the concept used is the same.

SOLAR WATER HEATING INDUSTRY STRUCTURE

Many individuals, organizations, and government entities impact the solar water heating industry. Various levels of knowledge and skill are required to design, install, inspect, and commission solar water heating systems. The process involves a number of qualified individuals and organizations, each with important roles in ensuring the safe, reliable, and long-term performance of solar thermal systems.

Like most industries, the structure of the solar water heating industry is composed of several levels. **See Figure 1-7.** These levels may include a variety of enterprises: large and small, domestic and foreign, and public and private. The industry consists of component manufacturers, distributors, contractors, installers, credentialing and trade organizations, consumers, support and standards organizations, and code and regulating organizations.

Solar Water Heating Manufacturers

There are thousands of manufacturers of solar water heating system components in North America. Solar water heating systems include components, such as solar collectors, storage tanks, valves, packaged controls, pumps, and piping, and many additional types of accessories and products. **See Figure 1-8.** Most component manufacturers are already established as plumbing or mechanical product manufacturers since 80% of the solar water heating system is composed of components used in conventional plumbing and mechanical systems. However, when components unique to solar water heating systems, specifically solar thermal collectors, are examined, there are significantly less manufacturers involved.

According to a 2004 industry survey, there are only seven manufacturers of solar thermal collectors in Canada. In 2009, the United States Energy Information Administration reported that there were 88 manufacturers and/or importers active in manufacturing, importing, and/or exporting solar thermal collectors. These companies shipped 13.8 million sq ft of solar thermal collectors in 2009.

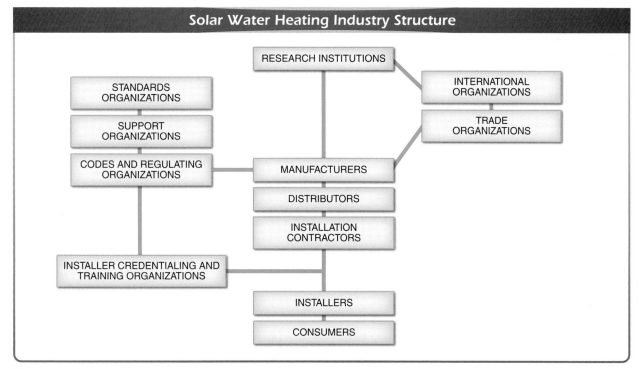

Figure 1-7. The solar water heating industry has several levels in its structure with many interconnected individuals, organizations, and government entities that all impact the industry.

Solar Water Heating Manufacturers

Bosch Thermotechnology Corp.

Figure 1-8. Some components of solar water heating systems are solar collectors, storage tanks, valves, packaged controls, pumps, and piping.

In 2009, the solar thermal industry remained highly concentrated, with the five largest companies accounting for 79% of total shipments. This concentration, however, was the lowest recorded in the past 10 years. The decrease is likely due to the start-up companies that have entered the market over the last 3 years.

The number of solar thermal collector manufacturers increases every year as the number of installations increases. As more government funding for the solar thermal industry is distributed and made available to the end users, these numbers are expected to steadily increase.

Solar Water Heating Distributors

Solar water heating systems are distributed or can be purchased in several different ways. Solar water heating system components other than collectors are purchased through plumbing and mechanical supply outlets. In general, installation contractors may purchase collectors from wholesale distributors who purchase them directly from the manufacturers. Solar water heating system distributors are typically existing plumbing and mechanical supply distributors. In some instances, contractors may purchase collectors directly from the manufacturer. Some manufacturers have a preferred network of installers to whom they distribute their collectors in order to have greater control of the quality of the installation.

Collectors may also be purchased over the Internet from a variety of sources. Homeowners may order collectors directly from manufacturers and then outsource the labor or install them on their own.

Solar Water Heating Installation Contractors

Solar water heating systems are installed by either installation contractors or homeowners. In most areas of North America, homeowners may install these systems without licensing or certification requirements. However, inspection requirements must still be followed. A homeowner should not attempt this type of installation without knowledge of solar water heating technology or skills in plumbing or mechanical system installation.

In many areas of the country, solar water heating systems are required to be installed by licensed or certified installation contractors. Contractors may be required to be licensed by the state, province, county, or city where the installations will take place. Therefore, contractors are usually established, licensed plumbing or mechanical contractors.

Sometimes installation contractors must be certified by the solar collector manufacturer in order to install their product. These contractors usually specialize in solar thermal installations. In most instances, the installation contractors are small contractors with 1 to 10 employees. This is especially true in the residential solar thermal technology market.

> **Facts**
>
> *Information on licensed plumbing contractors and solar installation professionals can be found at the website www.findsolar.com.*

Solar Water Heating Installers

A solar water heating installer may be directly employed by an installation contractor who specializes in solar thermal installations or by a plumbing and mechanical contractor. Safe and high-quality solar water heating system installations are essential for the success and acceptance of this technology. **See Figure 1-9.** Both the contractor and installer have important roles in ensuring quality installations. They should not only be trained and skilled in solar water heating technology installation but also plumbing installation; building construction; and electronic controls installation, maintenance, troubleshooting, and repair. Quality craftsmanship should be exhibited in all installation parameters.

Solar Water Heating Installers

Figure 1-9. Properly trained installers/technicians are responsible for safe and high-quality solar thermal installations.

In many areas of North America, the installer may be required to be a licensed journeyman plumber. This is due to the fact that a solar water heating system is required to be connected to a potable water supply. Many state, provincial, county, and city administrative codes require that any person who performs work on or makes connections to the potable water supply or sanitary drainage system of a building be a licensed journeyman plumber or licensed apprentice. This requirement is even more critical if the solar water heating system is a direct system because the system itself is the potable hot water supply.

Another requirement that is becoming more prevalent in the solar water heating industry is for the installer to be a certified solar water heating installer. This is to ensure that the system is installed in a proper manner. Solar water heating installer training, licensing, and certification requirements vary depending on local codes and other regulations. Requirements may be established at the national, regional, state, provincial, or local level.

Subsidies are often used to fund solar water heating installations. Funding agencies wish to ensure that the systems function as they are designed and contribute to the reduction of fossil energy use. To accomplish this, they require that the installers or at least the installation contractor be certified in solar water heating installations. The criteria for solar water heating system installer certification may include the following:
- previous experience in solar water heating installations and plumbing system installations
- knowledge of building, plumbing, and mechanical codes and standards

- familiarity with applying for permits and approvals from local building authorities
- awareness of individual capabilities and limitations
- willingness to seek outside expertise as required

In addition to the certification criteria, solar water heating system installer certification may require the ability to perform the following tasks:
- recommend well-engineered, quality components
- select and size systems and equipment to meet performance expectations
- ensure equipment is properly labeled and safety hazards are identified
- locate and orient solar panels to maximize performance and accessibility
- mount panels with strong, weather-sealed attachments
- complete work in a timely manner while practicing safe and orderly work habits
- employ safe and accepted methods of solar thermal equipment installation
- complete inspections, commissioning, and acceptance tests
- provide owners/operators with appropriate documentation, instructions, and training
- provide follow-up service for completed work as required

Experienced installers should have considerable design knowledge and familiarity with many types of solar water heating systems and components. They should be able to diagnose and troubleshoot even complex systems effectively. Installers are perhaps the most visible members of the solar thermal industry to the consumer, making it vital that they be professional and qualified individuals.

Credentialing and Training Organizations

Across North America, there are many methods used to regulate solar installations such as a variety of license classifications and practitioner certifications. These methods allow for the management of the market through incentive requirements. Currently, 14 states have defined specific solar license classifications, while 17 states have included language regarding practitioner certification in the state solar incentive programs. Canadian installer certification is mostly being driven by local incentive programs that require some form of credentials. Establishing regulations to protect the consumer ensures that solar thermal installations are designed and implemented with quality and safety in mind.

Licensure and certification are two approaches that can be used to regulate the industry. *Licensure* is a mandatory system of standards, usually controlled by state government, to which a practitioner must conform in order to practice a given profession. *Certification* is a voluntary system of standards, usually set by key stakeholders, that practitioners can choose to meet in order to demonstrate accomplishment or ability in their profession.

The advantages of licensure include state control over the selection and enforcement of a specific standard and the understanding and acceptance by the consumer of the concept of state regulation. The advantages of certification include standardization for a specific job based on common measures of competence to verify skills and portability across regions, states, and employers. The use of both licensure and certification balances the enforcement of standards through regulation and ensures quality through measured competency.

Training programs for the installation and operation of solar water heating systems are operated by trade associations, manufacturers, schools, and trade unions. These programs promote quality installations but can vary widely in length and scope, making it difficult for a consumer to choose a skilled installation contractor or installer for a desired system.

The length of a training course depends on the competencies and prerequisites the learner brings to the class and the kind of job the learner will be qualified for upon completion. A course should be long enough to expose learners to all critical tasks involved in the job they are being trained to do.

North American Board of Certified Energy Practitioners. An organization seeking to standardize installer qualifications is the North American Board of Certified Energy Practitioners (NABCEP). **See Figure 1-10.** NABCEP is composed of representatives from throughout the renewable energy industry. NABCEP works with renewable energy and energy efficiency industries and professionals to develop and implement quality credentialing and certification programs for practitioners.

NABCEP certification for solar water heating installers requires both experience and education, ensuring that certified professionals meet minimum levels of expertise in solar water heating systems installation. Certification is awarded upon successful completion of an exam and evidence of installation experience. NABCEP certification is voluntary and

does not replace local training and licensing requirements, but it provides a measure of quality for solar energy system installers.

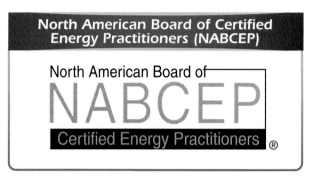

Figure 1-10. The NABCEP works with renewable energy and energy efficiency industries and professionals to develop and implement quality credentialing and certification programs for practitioners.

Interstate Renewable Energy Council. An organization that has been working to clarify the confusing training element in solar thermal technology is the Interstate Renewable Energy Council (IREC). The IREC, in an effort to ensure continuity, consistency, and quality in the delivery of training, has implemented the Institute for Sustainable Power Quality (ISPQ). Through the institute, a framework of standards and metrics have been created to provide a means to compare content, quality, and resources across a broad range of training programs covering renewable energy, energy efficiency, and distributed generation technologies. This international framework ensures the legitimacy of program instruction.

The IREC Solar Licensing Database provides state licensing requirements along with links to assist users in finding state-specific resources. The database also includes several summary tables, which provide a high-level perspective of the solar licensing landscape. The database is designed as a resource that brings together the varying requirements for installation.

The Renewable Energy Training Catalog lists practitioner training courses, entry-level classes, workshops, and related training programs. The database of accredited training programs can be found at http://irecusa.org/irec-programs/ispq-training-accreditation/ispq-awardees/.

United Association. The United Association (UA) provides the only North American standardized solar thermal systems training for union plumbing, pipe-fitting, and HVAC apprentices and journey workers. The training curriculum is created at an international level and presented in local training centers throughout the United States and Canada. **See Figure 1-11.** These training centers are staffed by field-experienced and professionally trained instructors. The training centers have solar thermal systems for training and supplemental domestic hot water purposes. The solar thermal training program includes a minimum of 40 hours of classroom instruction and individualized hands-on installation experience.

Solar Water Heating Consumers

The consumer, or end user, is the owner of the building or structure that uses the solar water heating system. Consumers include homeowners, developers, and business owners. Consumers drive the growth of the solar water heating industry but may not be knowledgeable about the systems. This system knowledge may include the types of system features available, the advantages and disadvantages of solar water heating, and how to finance a system. Installation contractors and installers work with the consumer or agents of the consumer, such as architects, engineers, general contractors, or managers, to educate the consumer about the many options available.

Much of the industry marketing effort is aimed toward educating potential consumers about various options and incentives in the solar water heating system marketplace. Efforts to educate consumers about solar water heating systems come from various sources including local and national government entities, nonprofit organizations, and trade and manufacturing associations. Organizations that aim to educate consumers about the solar water heating industry include, but are not limited to, the following:

- U.S. Department of Energy (DOE) and U.S. Environmental Protection Agency (EPA)—The DOE and EPA jointly developed the Energy Star Program, which is used to not only rate the energy efficiency of various products but also assist in educating consumers.

- Solar Energy Industries Association (SEIA)—SEIA is a national trade association that works to make solar energy a mainstream and significant energy source by expanding markets, removing market barriers, strengthening the industry, and educating the public on the benefits of solar energy.

- SunMaxx Solar—SunMaxx Solar is a provider of solar water heating products, installation services, information, and training. They provide consumers with information on solar water heating basics, solar collectors, certifications and incentives, and system payback. Consumers can use the information to make educated choices regarding the value of solar thermal energy. SunMaxx Solar also provides online training videos and webinars.

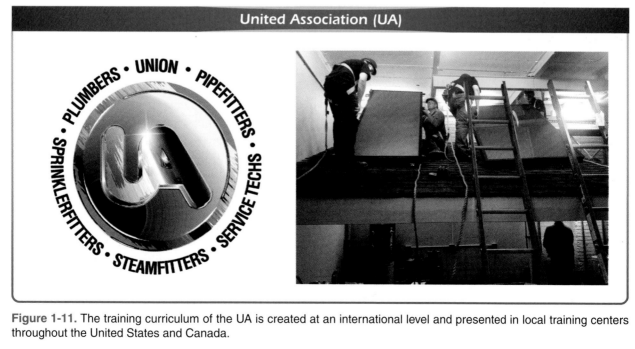

United Association (UA)

PLUMBERS • UNION • PIPEFITTERS • SERVICE TECHS • STEAMFITTERS • SPRINKLERFITTERS

Figure 1-11. The training curriculum of the UA is created at an international level and presented in local training centers throughout the United States and Canada.

Solar Thermal Support Organizations

Numerous nonprofit organizations work to promote and further develop the solar thermal industry, including organizations involved in research, marketing, installer training, standards development, and code regulations. These organizations also aim to make systems safer for the installer and the owner/operator.

Solar Thermal Research Institutions

Around the world, numerous universities and national laboratories are working to further develop the solar thermal industry. Research institutions are the underlying force for improving existing solar thermal manufacturing techniques and developing the next generation of technologies. These institutions also educate and train the leaders and high-tech workforce of the future.

National Renewable Energy Laboratory. One of the most well-known research organizations is the National Renewable Energy Laboratory (NREL). The NREL seeks to advance the energy goals of the DOE and the nation.

Laboratory scientists and researchers support critical market objectives to accelerate research from scientific innovations to market-viable alternative energy solutions. At the core of this strategic direction are research and technology development competencies. These areas span the understanding of renewable resources for energy, the conversion of these resources to renewable electricity and fuels, and ultimately the use of renewable electricity and fuels in homes, commercial buildings, and vehicles. The laboratory thereby directly contributes to the goal of finding new renewable ways to power homes, businesses, and cars.

Florida Solar Energy Center. Another research organization that has great impact on the solar thermal industry is the Florida Solar Energy Center (FSEC). The mission of the FSEC is to research and develop energy technologies that enhance the economy and environment of Florida and the nation and to educate the public, students, and practitioners on the results of the research. The FSEC was created by the Florida legislature in 1975 to serve as the energy research institute for the state. The main responsibilities of the center are to conduct research, test and certify solar water heating systems, and develop education programs.

National Research Council. The National Research Council (NRC) is the premier organization of Canada for research and development. The NRC has been Canada's leading federal resource for science- and technology-based research for more than 90 years. It operates the Institute for Research in Construction, which produces the *National Building Code of Canada*, and tests solar products for conformance to the standards of the Canadian Standards Association (CSA®).

Facts

The NREL works with the solar industry to lower the cost of solar water heating systems. Researchers assist with prototype development of new materials to improve material durability and performance of these systems.

International Organizations

Many international organizations impact the solar thermal industry in North America. The purpose of these international organizations is to promote cooperation in the development of technologies that use renewable resources and energy efficiently. International organizations for the solar thermal industry are as follows:

- International Energy Agency (IEA)
- International Renewable Energy Agency (IRENA)
- Renewable Energy & Energy Efficiency Partnership
- Solar Energy International (SEI)
- United Nations Development Program—Environment and Energy
- United Nations Statistics Division—Energy Statistics
- World Bank—Energy Division

International Energy Agency. The International Energy Agency (IEA) is an autonomous organization that works to ensure reliable, affordable, and clean energy for its 28 member countries and beyond. Founded in response to the 1973–1974 oil crisis, the initial role of the IEA was to help countries coordinate a collective response to major disruptions in oil supply through the release of emergency oil stocks to the markets. While this continues to be a key aspect of its work, the IEA has evolved and expanded. It is at the heart of global dialogue on energy, providing authoritative and unbiased research, statistics, analyses, and recommendations.

An important part of the program involves collaboration in the research, development, and demonstration of new energy technologies to reduce excessive reliance on imported oil, increase long-term energy security, and reduce greenhouse gas emissions. The IEA focuses on the following main areas:

- energy security—promoting diversity, efficiency, and flexibility within all energy sectors

- economic development—ensuring the stable supply of energy to IEA member countries and promoting free markets to foster economic growth and eliminate energy poverty
- environmental awareness—enhancing international knowledge of available options for tackling climate change
- engagement worldwide—working closely with non-member countries, especially major producers and consumers, to find solutions to shared energy and environmental concerns

The Solar Heating and Cooling Program, established in 1977, is one of the first programs of the IEA. The program is unique in that work is accomplished through the international collaborative effort of experts from member countries and the European Union. It is enhanced through collaboration with other IEA programs and solar trade associations in Europe, North America, and Australia. Some of the programs involved include the Energy Conservation in Buildings and Community Systems Program, Photovoltaic Power Systems Program, and SolarPACES. This international approach accelerates the pace of technology development, promotes standardization, enhances research and development (R&D) programs, permits national specialization, and saves time and money.

Trade Associations

Businesses within an industry, even competitors, often join together to form trade associations. The members of trade associations support each other and promote the industry to consumers in an effort to increase the market for all involved. Manufacturers, installation contractors, and other groups in the solar thermal industry have formed national and regional alliances. These organizations host trade shows and conferences to facilitate industry contacts and educate the public. Not all businesses within the industry belong to trade associations, but most find them to be mutually beneficial alliances.

Solar Energy Industries Association. The Solar Energy Industries Association (SEIA) is the U.S. trade association of solar energy manufacturers, dealers, distributors, contractors, installers, architects, consultants, and marketers. SEIA works to expand the use of solar technologies in the global marketplace. Through advocacy and education, SEIA works to build a strong solar industry. SEIA works with its 1000 member companies to make solar energy a mainstream and significant energy source by expanding markets, removing market barriers, strengthening the industry, and educating the public on the benefits of solar energy.

Canadian Solar Industries Association. The Canadian Solar Industries Association (CanSIA) represents the interests of its members by working to increase the use of solar energy in Canada. In addition, the association recognizes that the development of solar energy technologies carries obligations to the purchasers and users of these technologies. Therefore, the association, through programs and activities, seeks to ensure that the solar industry in Canada provides systems and services that meet expectations of value, performance, and safety.

State Solar Energy Trade Associations. There are a number of state trade organizations. Two of the largest are the New York Solar Energy Industries Association (NYSEIA) and the California Solar Energy Industries Association (CALSEIA).

NYSEIA is an association of solar energy companies and other individuals in New York State interested in solar energy. NYSEIA works to help commercial and residential energy users understand how the use of renewable energy can help lower energy costs and how to receive energy through environmentally friendly means such as solar thermal technology. NYSEIA actively works with the government and utility companies to find ways to increase the use of renewable energy.

CALSEIA is an association representing solar photovoltaic and solar water heating contractors, installers, manufacturers, distributors, consultants, engineers, and designers. CALSEIA promotes the widespread use of solar water heating and photovoltaic systems in California and nationwide by educating consumers and supporting solar legislation and the conducting of business in a professional and ethical manner. CALSEIA accomplishments include developing the Solar Rights Act, Solar Shade Control Act, Feed-In Tariff (FIT) Law, and Net Metering Law.

Standards Organizations

Many organizations that develop standards and safety guidelines for the electrical, plumbing, and mechanical industry are also involved, directly or indirectly, in the solar thermal industry. Several organizations also publish standards and guidelines that are specifically related to solar thermal systems.

Product listing organizations test products for safety and conformity to standard requirements. Products that pass are certified, or listed, as matching the manufacturer specifications and fulfilling the requirements for safety. These products may then bear the mark, which is considered a symbol of quality, of the listing organization.

Safety Standards. Safety standards have been written by various state, provincial, and federal organizations. The organization that regulates worker safety requirements at the national level in the United States is the Occupational Safety and Health Administration (OSHA). Regulations related to solar thermal installations and more specifically construction work, fall protection, material handling, outdoor work, and general workplace safety in the United States are contained in OSHA 29 Code of Federal Regulations (CFR) 1910 and 1926. Each state and province has its own OSHA standards that must be followed.

Miller® Fall Protection/Honeywell Safety Products
Proper fall arrest systems must be used when working at heights greater than 10′.

Solar Rating and Certification Corporation. The Solar Rating and Certification Corporation (SRCC) was incorporated in 1980 as a nonprofit organization. The primary purpose of the SRCC is to develop certification programs and national rating standards for solar energy equipment. The SRCC is an independent third-party certification entity. It is the only national certification organization established solely for solar energy products.

The SRCC currently administers a certification, rating, and labeling program for solar collectors and a similar program for complete solar water heating systems. The operating guidelines, test methods, minimum standards, and rating methodologies of the certification program are based on nationally accepted equipment tests on solar equipment by independent laboratories, which are accredited by SRCC.

The scope of the program includes swimming pool and recreational heating, space heating and cooling, and water heating. The program is administered according to SRCC Document OG-100, *Operating Guidelines for Certifying Solar Collectors*, and its companion document SRCC Standard 100, *Test Methods and Minimum Standards for Certifying Solar Collectors*. Participation in the SRCC solar collector certification program is voluntary, and all manufacturers of applicable products are eligible to participate. SRCC testing of collectors is a test of durability and performance, with the test procedures for performance being specified by another standard writing group, which is the American Society of Heating, Refrigerating, and Air Conditioning Engineers.

American Society of Heating, Refrigerating, and Air Conditioning Engineers. The American Society of Heating, Refrigerating, and Air Conditioning Engineers (ASHRAE), founded in 1894, is an international organization of 51,000 persons. ASHRAE fulfills its mission of advancing heating, ventilation, air conditioning, and refrigeration to serve humanity and promoting a sustainable world through research, standards writing, publishing, and continuing education. ASHRAE Standard 93, *Methods of Testing to Determine the Thermal Performance of Solar Collectors,* and ASHRAE Standard 96, *Methods of Testing to Determine the Thermal Performance of Unglazed Flat-Plate Liquid-Type Solar Collectors,* are specific standards used to test solar collectors.

Canadian Standards Association. The Canadian Standards Association (CSA®) is a nonprofit organization with the aim of producing product standards and certification. The CSA also provides advisory services, training materials, and published standards documents. There are a number of performance, safety, and installation standards that pertain to solar energy products in Canada. The standards applicable to solar thermal systems within Canada include the following:

- CSA Standard F378-87, *Solar Collectors* (similar to the SRCC OG-100 program)
- CSA Standard F379-09, *Packaged Solar Domestic Hot Water Systems (Liquid-to-Liquid Heat Transfer)* (similar to the SRCC OG-300 program)
- CSA Standard F383-08, *Installation Code for Solar Domestic Hot Water Systems*

Plumbing codes within Canada require that systems be certified according to CSA Standard F379 and installed according to CSA Standard F383.

Other Standards Organizations. As this text goes to press, the International Plumbing and Mechanical Officials (IAPMO) is partnering with ASTM International, previously known as the American Society for Testing and Materials (ASTM), to develop a family of consensus ANSI standards related to solar water heating, combination solar water and space heating, solar space cooling, solar system performance prediction, and heat metering. The consensus-based standards development approach by IAPMO and ASTM promises open standards that are accessible to all. The IAPMO/ASTM ANSI solar standards family will be introduced beginning in early 2013.

IAPMO is a nonprofit organization dedicated to the plumbing and mechanical industry. IAPMO develops standards, presents educational seminars, tests journeymen plumbers, and certifies plumbing inspectors.

ASTM International is a technical society and primary developer of voluntary standards, technical information, and services that promotes public health and safety and contributes to the reliability of products, materials, and services. ASTM International publishes standard specifications, test methods and procedures, and guides and definitions for materials, products, systems, and services.

The American National Standards Institute (ANSI) is a national organization that helps identify industrial and public needs for national standards. ANSI standards are commonly produced and copublished by ANSI and member technical societies, trade associations, and United States government departments.

Code and Regulating Organizations

Solar thermal systems are regulated by several codes including building, plumbing, and mechanical codes. The specific code enforced for an installation depends on the codes adopted by the authority having jurisdiction (AHJ) in the area where the installation is located. In some situations, there may not be any codes adopted or even a regulating agency. This is especially true in remote rural areas. In any case, manufacturer installation instructions must always be followed. It is also a good rule to ensure that the installation conforms to the plumbing and mechanical codes adopted by the nearest AHJ.

Building Codes. Building codes are developed by several organizations. All codes are not written alike. Therefore, it is necessary to establish which codes are used at the installation location. Structural requirements, including roof penetrations and roof supports, are regulated by building codes. The most prevalent building code in the United States is the *International Building Code®*, published by the International Code Council. For one- and two-family dwelling units, the *International Residential Code®* may be the governing document.

The model building codes in Canada are the *National Building Code of Canada* and the *National Plumbing Code of Canada*. Some provinces within Canada have their own building codes, which are modeled after the national codes. All plumbing codes within Canada require systems to be certified according to CSA Standard F379 and installed according to CSA Standard F383.

Plumbing and Mechanical Codes. Solar water heating DHW systems are part of plumbing or mechanical systems. Plumbing and mechanical systems are regulated by several model codes in the United States through either state/local codes or national model codes.

The two dominant model plumbing and mechanical codes adopted in the United States are published by the International Association of Plumbing and Mechanical Officials (IAPMO) and the International Code Council (ICC). IAPMO publishes the *Uniform Plumbing Code®* and the *Uniform Mechanical Code®*. The ICC publishes the *International Plumbing Code®*, the *International Mechanical Code®*, and the *International Energy Conservation Code®*. It should be noted that individual states, provinces, or other public entities may adopt measures or provisions that regulate solar water heating systems. The local AHJ should be consulted regarding the code and code edition used in the area where a system is to be installed.

There are other codes that may be used in various locations. For example, IAPMO also publishes the *Uniform Solar Energy Code®* and the *Green Plumbing and Mechanical Code Supplement*. These codes offer additional information on solar thermal systems that can be beneficial for the installer even if the codes are not specifically adopted in the area.

Other more specific codes may also apply to solar thermal installations. Water heaters are regulated by plumbing codes and, if using gas energy systems, they may also be regulated in the United States by National Fire Protection Association (NFPA) 54, *National Fuel Gas Code,* or in Canada by CSA Standard B149, *Natural Gas and Propane Installation Code*. Electrical wiring and connections are regulated by NFPA 70®, *National Electrical Code®*, or by CSA Standard C22.1, *Canadian Electrical Code*.

GOVERNMENTAL ENERGY POLICIES

Federal, state, provincial, and local government energy policies are primarily aimed at reducing the dependency on fossil fuel energy, mandating efficient energy use, and encouraging the development of renewable energy resources. These policies revolve around establishing an energy portfolio that sustains future growth and energy independence. Government entities primarily use incentive programs to encourage the use of renewable energy, including solar thermal technology. Government incentive programs may include any of the following:

• grants
• rebates
• renewable energy credits
• low-interest or no-interest loans
• sales and property tax exemptions
• income tax credits
• tax deductions for individuals and corporations
• cash payments based on energy production

Facts

The DOE's Office of Energy Efficiency and Renewable Energy (EERE) provides various grants and cooperative agreements to promote increased development of solar projects. Financial assistance is provided to business, industry, and local governments to fund the demonstration of solar thermal projects.

United States Energy Policies

There are many federal programs available throughout the United States. They include programs for corporate depreciation, corporate exemptions, corporate tax credits, federal grants, federal loans, industry recruitment/support, performance based incentives, personal exemptions, and personal tax credits.

Through March 2010, approximately $232 million was targeted for investment in grant and rebate funding for PV and solar water heating installations through the State Energy Program (SEP) as part of the American Recovery and Reinvestment Act (ARRA). An additional $219 million is targeted for renewable energy programs for which PV and/or solar water heating technologies are eligible. It is possible that a portion of the $567 million targeting energy efficiency could be used to support solar water heating installations. In addition, individual states have invested $550 million in loan programs for which solar technologies may be eligible.

The SEP, authorized by Congress and funded by the DOE, has existed for approximately 15 years. This program provides funding to state energy offices, based on formulas linked to state population, for the implementation of programs and policies that support renewable energy and energy efficiency. In 2009, the SEP received the most funding in its history. The budget rose from approximately $33 million in 2008 to over $3.1 billion in 2009. This unprecedented 96-fold increase in funding was provided by the federal stimulus bill. The SEP was one of several clean-energy-related programs whose funding was enhanced dramatically within the DOE's Office of Energy Efficiency and Renewable Energy (EERE).

Another undertaking funded by the DOE is the Energy Policy Act of 2005. The Energy Policy Act of 2005 established a 30% tax credit (up to $2000) for the purchase and installation of residential solar electric and solar water heating property and a 30% tax credit (up to $500 per 0.5 kW) for fuel cells. Initially scheduled to expire at the end of 2007, the tax credits were extended through December 31, 2008, by the Tax Relief and Health Care Act of 2006.

In October 2008, the Energy Improvement and Extension Act of 2008 extended the tax credits once again (until December 31, 2016), and a new tax credit for small wind-energy systems and geothermal heat pump systems was created. In February 2009, the maximum credit amount for all eligible technologies (except fuel cells) placed in service after December 31, 2008, was removed through ARRA.

State programs vary widely. For example, California has exempted solar hot water from property taxes since 1999. Several other states have similar exemptions or allow municipalities to treat solar hot water differently for property tax purposes. In 1995, Arizona began to offer individuals a personal investment tax credit of 25% of the cost of solar hot water installations up to a maximum of $1000. Currently, Hawaii and 13 other states have similar credits ranging from 10% to 50% of the cost, up to a maximum dollar amount, depending on the state. Ten states offer direct rebates for residential solar hot water systems, ranging from $500 per system (Maryland) to $5000 per system (Pennsylvania and Delaware).

Facts

Canadian residential and commercial buildings use 70% of their energy for heating.

Canadian Energy Policies

Under the income tax regulations in Canada, certain capital expenditures are eligible for accelerated capital cost write-offs. These tax regulations apply to systems that produce heat and/or electricity from fossil fuels or alternative renewable energy sources, at 30% and 50% respectively, on a declining balance basis. Expenses incurred during the development and implementation of renewable energy and energy conservation projects also may be fully deducted.

Many provinces offer tax exemptions and/or grants to offset system costs for residential, commercial, institutional, and industrial customers. These incentives may reduce total system costs by 15% to 35%.

Database of State Incentives for Renewables and Efficiency

States, municipalities, and utilities offer incentive programs for solar thermal and other alternative energy systems. These may include rebates, loans, or service charge reductions. For information on the various incentive programs that are available, the Database of State Incentives for Renewables and Efficiency (DSIRE) should be consulted. DSIRE is composed of information regarding federal, state, local, and utility incentives and policies that promote renewable energy and energy efficiency. The website can be found at http://www.dsireusa.org. **See Figure 1-12.** Established in 1995 and funded by the U.S. DOE, DSIRE is an ongoing project of the North Carolina Solar Center and the IREC.

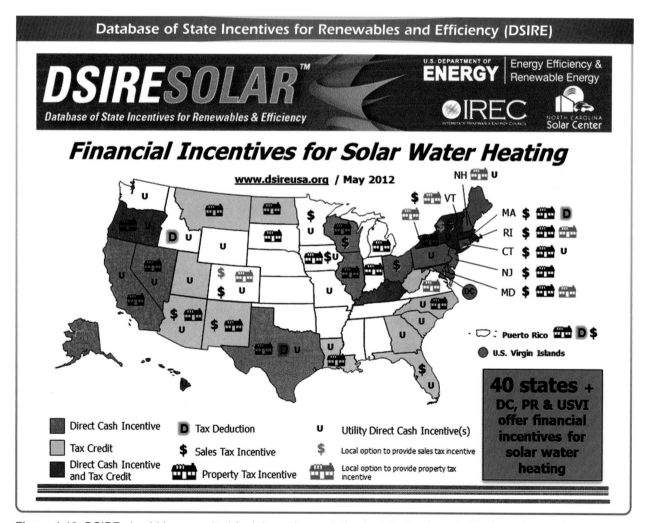

Figure 1-12. DSIRE should be consulted for information on federal, state, local, and utility incentives and policies that promote renewable energy and energy efficiency.

SOLAR WATER HEATING SYSTEM ECONOMICS

There are 110 million households in the United States that require energy for water heating. Of the U.S. households that require energy for water heating, 39% use electricity to heat water, 54% use natural gas, and only 1% use solar energy. **See Figure 1-13.** This limited use of solar water heating systems is directly related to the historically low cost of energy in the United States, particularly natural gas. However, as the price of fossil fuel continues to rise, supplies continue to deplete, and concerns over climate change increase, solar thermal technology is becoming more appealing.

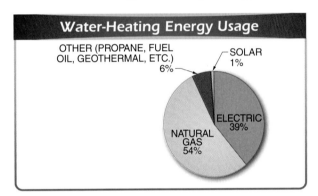

Figure 1-13. Of the U.S. households that require energy for water heating, 39% use electricity, 54% use natural gas, and only 1% use solar energy.

Solar Fraction

While it may seem that an individual uses relatively little hot water in the course of a day, an average of 18% of residential and 7% of commercial building energy is used to heat water. However, a solar water heating system cannot satisfy all hot water needs due to weather and seasonal solar radiation variation. Therefore, the performance of a solar DWH system may be measured by its solar fraction.

Solar fraction is the amount of solar thermal energy produced divided by the water heating energy demand of a building and is used to represent the annual amount of water heating energy produced by the solar water heating system. A system with a 60% solar fraction reduces water heating demand (and also water heating energy costs) by 60% and is given a rating of 0.6. Typical annual solar fractions in the United States are in the range of 40% to 80%. **See Figure 1-14.**

Solar Water Heating System Costs

Currently, solar water heating systems are significantly more expensive to purchase and install than conventional water heaters. In some cases, solar water heating systems in retrofit situations can be up to ten times more expensive than conventional water heaters. The average cost of an installed solar water heating system that uses a natural gas water heater for backup is approximately $8000. The average cost of an installed conventional natural gas water heater is approximately $700. The average energy savings over the course of a year with the use of a solar DHW system are estimated to be approximately $250/yr. Without incentives, it would take almost 30 yr for the consumer to receive the payback, or return on investment (ROI), for the initial cost of the solar system.

Driving down the initial cost of solar thermal technology is essential to improving its economic appeal, and in turn, marketability. This is precisely why the various incentive programs are in place. Depending on the incentive, the initial cost of the solar water heating system can be significantly reduced, and the payback period can be shortened. **See Figure 1-15.**

One other item that can be considered in the cost analysis of solar versus fossil fuel water heating is the increase in home or building equity. The increased value of the property can also be subtracted from the cost of the solar system, making the investment even more attractive.

Greenhouse Gas Reduction

Cost savings is not the only reason for a consumer to choose to install a solar water heating system. With the use of sustainable technologies, like solar water heating, the amount of greenhouse gases emitted into the atmosphere is reduced.

The United States government has set specific targets for the reduction of energy-related greenhouse gas emissions. The goal of the government is to reduce greenhouse gas emissions by 17% by 2020 and 83% by 2050, from a 2005 baseline. Solar thermal technology can contribute significantly to meet this goal. Solar thermal systems generate clean primary energy that replaces both natural gas and fossil-fueled electricity. The potential amount of energy saved in the United States through solar water heating alone is more than 1 quadrillion Btu. This amount corresponds to an emissions reduction potential of approximately 1% of the total U.S. annual carbon dioxide (CO_2) emissions.

The goal of the U.S. DOE Solar Heating and Cooling (SHC) Program is to reduce U.S. CO_2 emissions by at least 2.5% by the year 2020 and 10% to 15% by 2050. The program promotes various forms of solar heating and cooling such as solar water heating, solar cooling, industrial and agricultural solar process heating, and solar swimming pool heating.

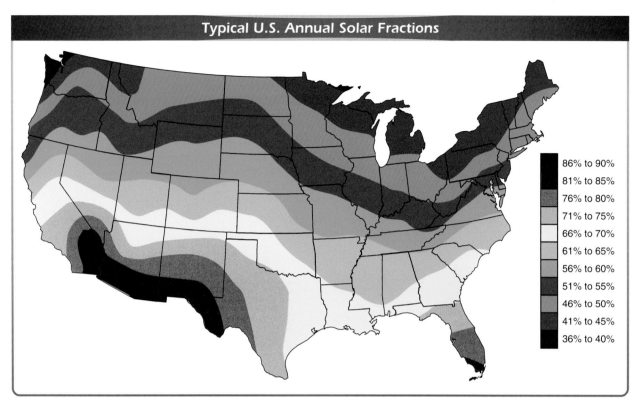

Figure 1-14. Typical annual solar fractions in the United States are in the range of 40% to 80%.

State	Standard Solar Water Heating System‡	30% Federal Tax Credit	State Personal Tax Credit	State Rebate†	Utility Rebate*	Net Cost of Solar Water Heating System	Payback Period‖
New York	$8000	$2400	$2000	$1.50/kWh displaced annually	$1500	$2100§	8.4
Rhode Island	$8000	$2400	$2000			$3600	14.4
Hawaii	$8000	$2400	$2800	$750	$800	$1250	5.0
Maryland	$8000	$2400		$500		$5100	20.4
Massachusetts	$8000	$2400	$1000	$2000	$0	$2600	10.4
Pennsylvania	$8000	$2400		$2400	$500	$2700	10.8
Louisiana	$8000	$2400	$4000	$150	$750	$700	2.8
New Hampshire	$8000	$2400		$1500	$750	$3350	13.4
Connecticut	$8000	$2400		$2000		$3600	14.4
California	$8000	$2400		$1250	$1875	$2475	9.9
North Carolina	$8000	$2400	$1400		$1000	$3200	12.8
Georgia	$8000	$2400	$2500		$500	$2600	10.4
Oregon	$8000	$2400	$1500	$1500	$1000	$1600	6.4

Solar Water Heating System Payback

* may vary depending on locality
† local rebates may also be available
‡ includes components and installation
§ does not include state rebate
‖ in years (net cost divided by $250 in annual energy savings)

Figure 1-15. Multiple incentives can significantly reduce the initial cost of the solar water heating system and make the payback period much more reasonable.

Chapter 1 Review and Resources

Solar Thermal Principles

The fundamentals of harvesting radiant energy for solar water heating applications are relatively straight forward. A solar collector captures sun energy and transfers that energy in the form of heat to a fluid that circulates through the collector. The heated collector fluid is then used for domestic hot water (DHW) purposes or used as a transfer fluid to heat the DHW in a storage tank. The storage tank saves the unused DHW to be used at a later time.

FUNDAMENTALS OF SOLAR RADIATION

Solar water heating system operation is simple at first glance. However, closer inspection reveals the complexity of system design required to efficiently heat domestic hot water (DHW) daily for an extended period of time (20 yr to 30 yr on average). The collector, storage tank, and pump (if used), must be designed for extended use and constructed of the proper materials. The piping system and components must also be designed for the most efficient method of moving the transfer fluid and DHW to the point of use. They must also consist of the proper materials to provide the desired system longevity.

The collector, tank, piping, and other components of the solar water heating system can be controlled and adjusted by engineers through system design and by installers upon installation. However, the Sun, which is crucial for system operation, is unreliable and cannot be controlled.

The Sun moves constantly across Earth's sky longitudinally and latitudinally. This occurs throughout the day and over the course of a year. The atmosphere and weather of Earth also cause radiant energy from the Sun to be unreliable since clouds, dust, smoke, etc., block or absorb the energy available to the system. The geographical location of the installation site also impacts the availability of sun energy. **See Figure 2-1.**

Solar Thermal Principles

Figure 2-1. Clouds, dust, smoke, the time of day, and the time of year affect the efficiency of solar collectors.

The Sun and weather cannot be controlled, but solar water heating technology can be adjusted to most efficiently capture solar radiation. To accomplish this, the sun path through the sky throughout the year must be examined at each individual installation site. The solar collector must then be installed based on this information and on the efficiency needs of the system and its application.

An installer of solar water heating systems need not be an astronomer to understand the basic principles of solar radiation. In fact, most of these principles are taught in elementary school science classes. Solar radiation principles, such as Earth and Sun orientations, change of seasons, daily sun path, and sun angle, need to be reviewed in reference to their impact on solar thermal systems.

Solar Radiation

Solar radiation is energy from the Sun in the form of electromagnetic radiation, which consists of waves with electric and magnetic properties. These waves are represented in the electromagnetic spectrum of radiation. The *electromagnetic spectrum* is a continuum of all electromagnetic waves arranged by frequency and wavelength. **See Figure 2-2.** The portion of the electromagnetic spectrum to which solar collectors respond is visible light and near-infrared radiation.

Solar Irradiance. Solar radiation can be measured, and its measured quantity is solar irradiance. *Solar irradiance* is the power of solar radiation per unit area and is commonly expressed in watts per square meter (W/m^2) or kilowatts per square meter (kW/m^2). Solar irradiance is calculated as if the electromagnetic waves, generally referred to as sunlight, were striking an imaginary surface.

Facts

Sunlight takes about 8 min to reach Earth. The light travels at about 186,000 mi/sec and about 93 million mi. This distance from the sun to the Earth is also known as an astronomical unit (AU) and is used to measure distances between planets and other celestial bodies.

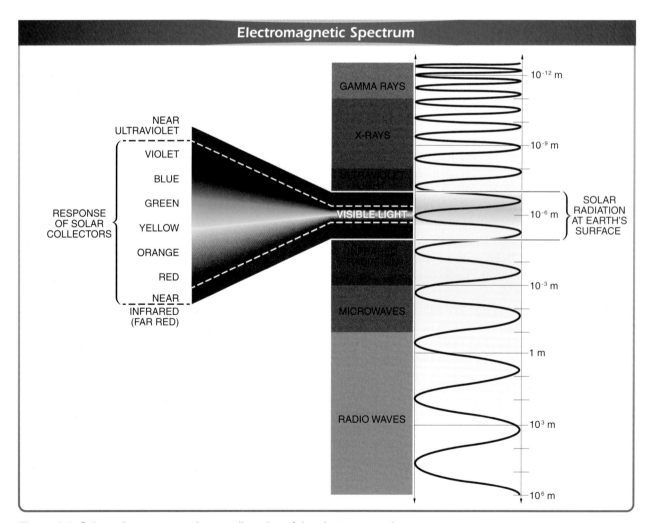

Figure 2-2. Solar collectors use only a small portion of the electromagnetic spectrum.

Solar irradiance at the edge of Earth's atmosphere is measured at an average of 1366 W/m². This is referred to as the Earth's solar constant. **See Figure 2-3.** As solar radiation travels farther from the Sun, the solar constant, or the amount of solar radiation, decreases. For example, the solar constant at Mars (154.8 million miles from the Sun versus 93 million miles for Earth) is 591 W/m².

Wind turbines transform wind energy into electrical energy and are another form of renewable energy.

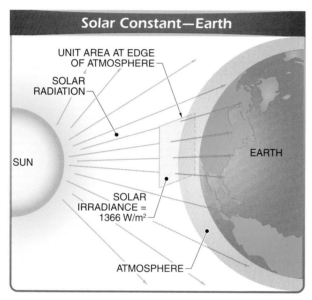

Figure 2-3. Earth's solar constant, 1366 W/m², is the solar irradiance measured at the edge of the atmosphere.

Solar irradiation is a measure of solar irradiance over a period of time. Solar irradiation is expressed as units of watts per square meter (W/m²) or kilowatts per square meter (kW/m²) per element of time (e.g., hour, day, or year). Solar irradiation quantifies the amount of energy received on a surface and is one of the key factors used in sizing and estimating the performance of solar collectors.

Insolation. *Insolation* is the amount of solar irradiation reaching the surface of Earth. Therefore, when discussing solar irradiation, insolation is also being referenced. Insolation is commonly used to rate the solar energy potential of a location by calculating the average energy received on a surface per day. It is typically expressed as kWh/m²/day or equivalent peak sun hours. Color-coded maps can be used to show how insolation varies across different regions of North America and at different times of the year. **See Figure 2-4.**

Atmospheric Effects on Solar Radiation

Various elements within the atmosphere affect solar radiation as it travels to the surface of the Earth. These elements cause the solar irradiance to be lower than the solar constant. Before solar radiation even enters Earth's atmosphere, 6% is reflected back into space by the atmosphere. Various elements in the atmosphere also absorb approximately 15% of the average solar radiation, while another 20% is reflected back into space by cloud cover. As a result, depending on the atmospheric conditions at any individual collector site, solar radiation reaches the surface of collectors as direct radiation and diffused radiation. **See Figure 2-5.**

Direct Radiation. *Direct radiation* is solar radiation that reaches Earth in a direct line or beam from the Sun without any change in direction. It can be easily recognized as sunlight that, when blocked, creates shadows. Only 51% of the direct radiation at the edge of Earth's atmosphere reaches the Earth's surface. Solar concentrating collectors, which track the path of the Sun, use only direct radiation to produce electrical or thermal energy.

Diffuse Radiation. Solar radiation can also be scattered toward Earth by elements, such as water vapor, clouds, smoke, and dust, in the atmosphere. *Diffuse radiation* is the scattered solar radiation that eventually reaches the surface of a solar collector. Solar collectors can also use diffuse radiation to create thermal energy. Approximately 10% to 20% of the total amount of usable solar radiation consists of diffuse radiation.

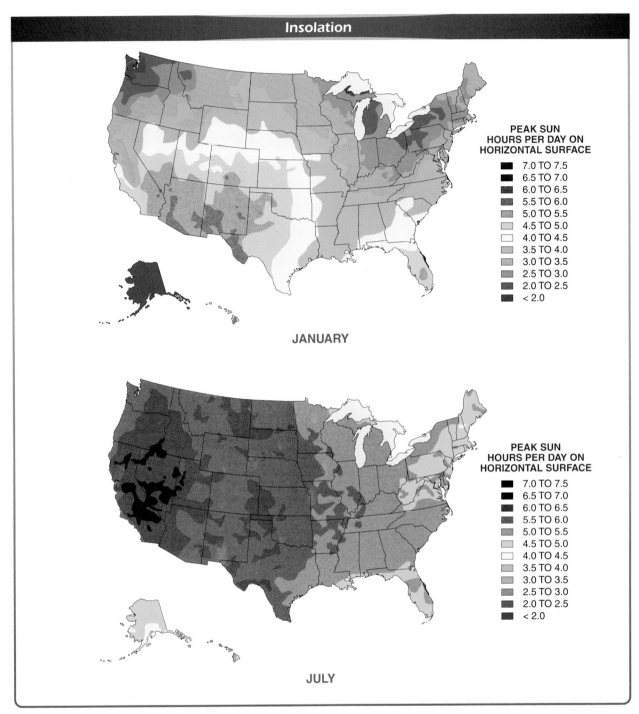

Figure 2-4. Insolation is the amount of solar irradiation that reaches the surface of Earth and varies depending on geographic location and time of year.

Air Mass. Another characteristic of Earth's atmosphere that affects the total amount of solar radiation available for use in a solar collector is air mass. *Air mass* is the amount of atmosphere that solar radiation must penetrate to reach Earth's surface.

Facts

The atmosphere of Earth is more than 300 mi thick. However, 99% of the atmosphere is within 19 mi of Earth's surface, and 99.9999% is within 62 mi.

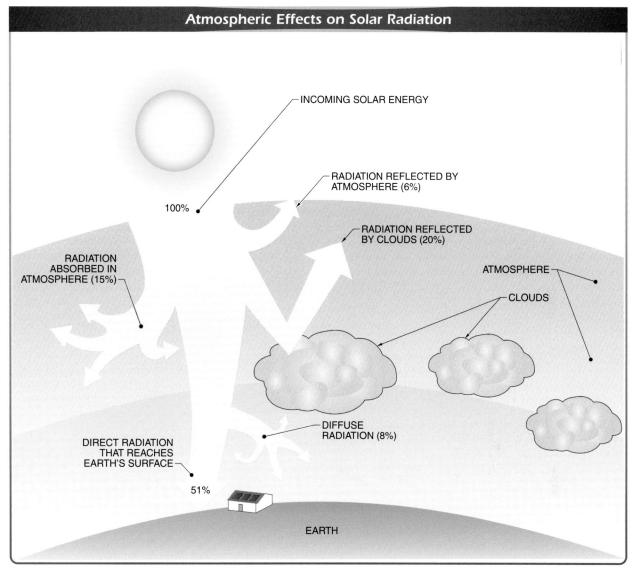

Figure 2-5. About half of the solar radiation emitted from the Sun is reflected, absorbed, or diffused by various elements in the atmosphere.

When the Sun is at its zenith, the amount of atmosphere solar radiation has to travel through is at its minimum. *Zenith* is the highest point in the sky reached by the sun directly overhead at a particular location. **See Figure 2-6.** As the angle of the Sun deviates from its zenith, the amount of atmosphere, or air mass, increases. This increased air mass increases the effects of atmospheric conditions on solar radiation, thereby contributing an additional decrease in the amount of solar radiation available for use. The angle of the Sun at any location depends on the time of day, time of year, and latitude of that location.

Total Global Radiation. *Total global radiation* is the total amount of solar radiation that eventually reaches a collector installation site and is available for use in solar thermal systems. It is essentially the sum of direct and diffuse radiation after all of the atmospheric effects are taken into consideration. It is also referred to as insolation.

It is not necessary for the installer to perform all the calculations necessary to determine the amount of insolation available for a collector. Design engineers will need to perform these calculations for large industrial or commercial installations to gain peak efficiency. For normal residential installation purposes, the average insolation for most locations and climate conditions in North America has been calculated at an average of 1000 W/m². This figure is approximately two-thirds of the solar constant (1366 W/m²).

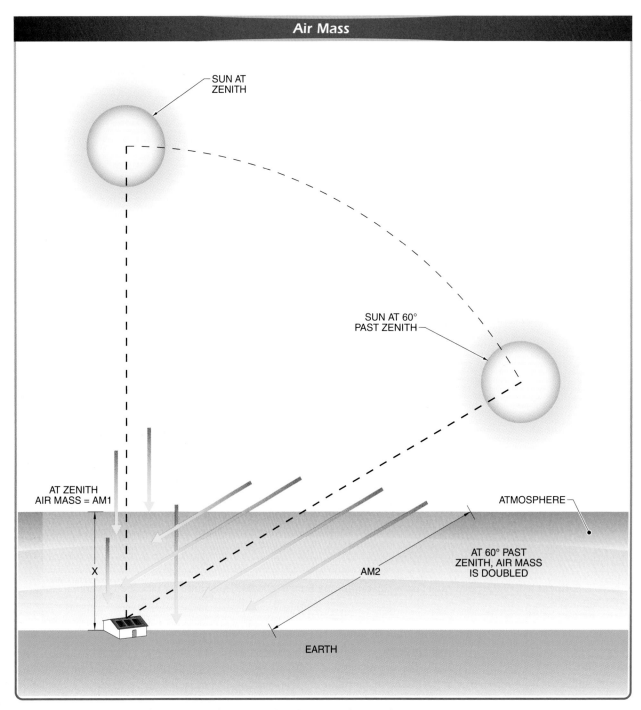

Air Mass

SUN AT
ZENITH

SUN AT 60°
PAST ZENITH

ATMOSPHERE

AT ZENITH
AIR MASS = AM1

AT 60° PAST
ZENITH, AIR MASS
IS DOUBLED

X

AM2

EARTH

Figure 2-6. The more the Sun deviates from its zenith, the more air mass increases.

SOLAR PATH

The insolation on the surface of Earth varies over time because of the Earth-Sun geometric relationship. Earth's daily rotation, tilted axis, and elliptical orbit around the Sun constantly change the angle of the Sun relative to a fixed location such as a solar collector. These daily changes in the sun angle will increase and decrease insolation. These changes can be charted by creating a solar path diagram through the sky. **See Figure 2-7.**

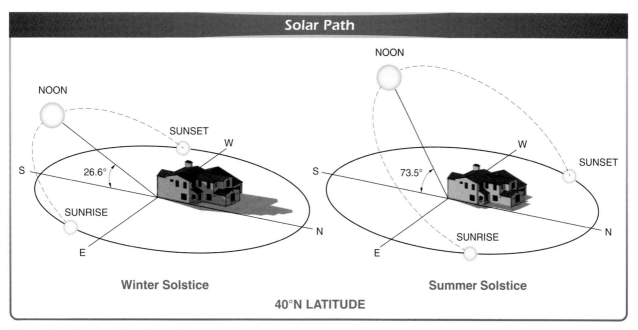

Figure 2-7. The position of the Sun relative to a point on Earth changes due to Earth's daily rotation, tilted axis, and elliptical orbit around the Sun.

Earth Rotation

Due to the daily rotation of the Earth, the Sun appears to travel across the sky in an arc pattern from dawn to dusk. At noon, the Sun is directly overhead of the collector location and thus is at the minimum distance that solar radiation must travel to reach the collector. At sunrise and sunset, the Sun is at its most extreme angle from the collector, which is the maximum distance solar radiation must travel. At noon, the collector receives the maximum amount of insolation. At sunrise and sunset, it receives the least amount of insolation. At noon local time, the Sun is not always at its zenith at a specific location due to the creation of modern time zones.

Solar Time. *Solar time* is a timescale based on the apparent motion of the Sun crossing overhead. Before the modern era, time was kept at each city according to solar time. The speed of travel and communication in the modern era caused problems with determining corresponding times between one city and the next city, especially with regards to train schedules. In 1879, worldwide standard time zones were proposed. By 1900, most of the world was setting clocks to standard time rather than solar time.

Time zones were created to keep daylight hours in the same time frame everywhere in the world. A time zone straddles a standard meridian line at each 15° increment in longitude (15°, 30°, 45°, etc.). Solar noon

corresponds to standard time noon at the time zone standard meridian line only. Areas within the same time zone but not on the standard meridian line will not be at solar noon when it is standard time noon. The time difference between solar noon and standard time noon can be up to ±45 minutes. **See Figure 2-8.**

Solar Noon. *Solar noon* is the moment when the Sun crosses a point along a north-south longitudinal line. The proper orientation of the solar collector should be calculated for solar noon if the highest efficiency in solar energy harvesting is required.

Solmetric Corporation
Electronic site analysis tools can be used to determine the sun path and shading for a given location.

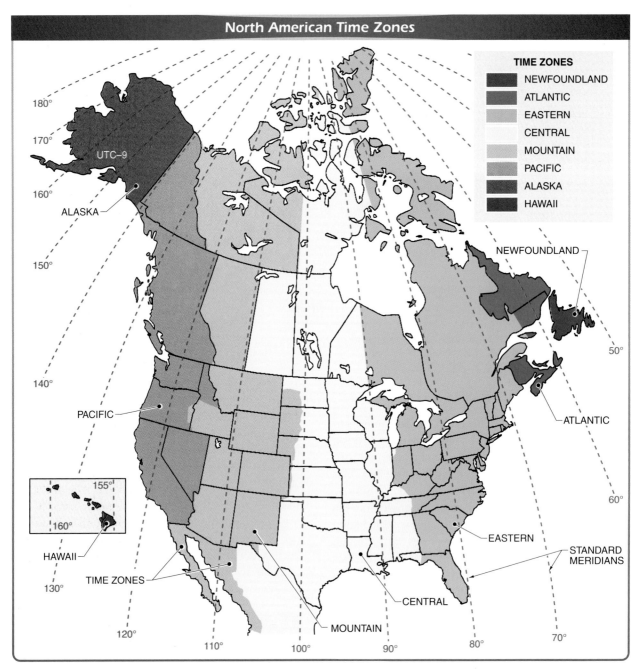

Figure 2-8. One region of North America may fall under a different time zone than another.

Earth Tilt

Yearly changes in the angle of the solar radiation reaching Earth occurs because of the tilt of Earth's axis of rotation as it makes its yearly elliptical orbit around the Sun. Earth's axis is tilted at an angle of 23.5° from vertical. As Earth travels around the Sun, the northern hemisphere tilts away from the Sun during winter and toward the Sun during summer. The opposite is true for the southern hemisphere. **See Figure 2-9.**

These seasonal changes cause a distinct variation in the amount of insolation available for solar thermal use. During summer, daylight lasts longer, allowing more time for solar radiation to be collected. Since the northern hemisphere is angled closer to the Sun, solar radiation travels through less atmosphere to reach a collector and the angle of solar radiation is more direct. This increases the amount of insolation available that can be harvested.

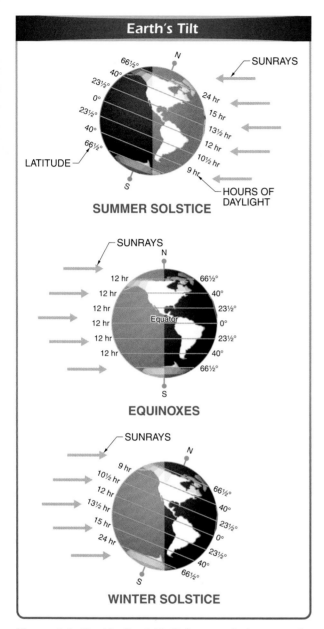

Figure 2-9. Earth's tilt of 23.5° from vertical causes the northern hemisphere to tilt away from the Sun during winter and toward the Sun during summer as Earth travels around the Sun.

During winter, daylight is shorter and the northern hemisphere is angled away from the Sun. There is less time for solar thermal collection, solar radiation travels through more atmosphere to the collector, and the angle of solar radiation is less direct. These properties of the winter months result in a decrease in the amount of insolation available for solar thermal harvesting.

These yearly changes in the sun angle relative to the location of a solar collector can also be charted as a solar window. **See Figure 2-10.** A *solar window* is the depicted elevation and path of the Sun at winter solstice (December 21) and at summer solstice (June 21) used to create a visual image of all possible elevations of the Sun throughout the year. Visualizing the solar window at the location of a solar collector is essential for determining the proper orientation of the solar collector.

The solar path and solar window for any location can be also plotted graphically. Site assessment software allows the installer to graph the solar window at any latitude. Included on the graph are the Sun's altitude and azimuth angles, time of day, and paths for various dates of the year. **See Figure 2-11.**

The graph can also provide an elevation view of the solar window, which will be beneficial when positioning the collector. It should be noted that 80% to 90% of solar insolation occurs between the hours of 9 AM and 3 PM. Thus, the solar collector must at least be oriented to capture insolation during this time window. **See Figure 2-12.**

COLLECTOR ORIENTATION

An understanding of the solar path and solar window is crucial for the proper installation of the solar collector. The solar collector will not function efficiently if it is not properly oriented (positioned) to collect solar radiation. The solar path and window provide the information needed to determine the two most critical elements of proper collector placement: collector orientation and collector tilt.

In the northern hemisphere, the solar path begins with the Sun rising from an easterly direction and setting in a westerly direction. As the year progresses, the sunrise changes in direction from the northeast to the southeast. In northern latitudes, the southeasterly direction is more pronounced. However, at all times of the year at solar noon, the Sun is at a due south direction. Also, due south is where the most insolation occurs. Therefore, solar collectors should face due, or true, south.

True South. *True south* is the direction directly opposite true north. Therefore, true north must be found in order to find true south. The true north direction may be determined by a local land surveyor or by using a compass, the shadow method (the shadow cast at solar noon is true south), a GPS-supported device, or maps.

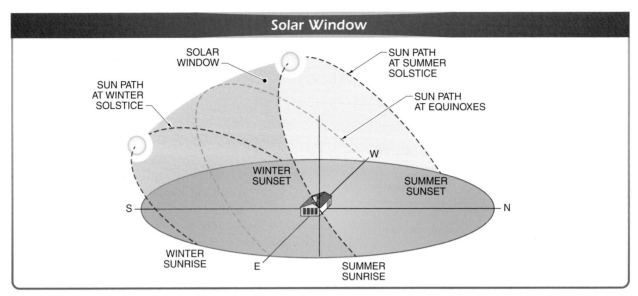

Figure 2-10. A solar window is the visual image of all possible elevations of the Sun throughout the year.

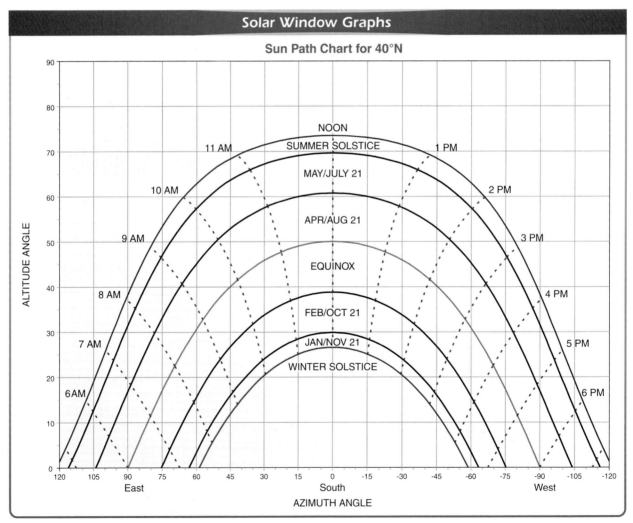

Figure 2-11. A solar window graph includes information on the altitude and azimuth angles of the Sun, time, and sun path for various dates of the year.

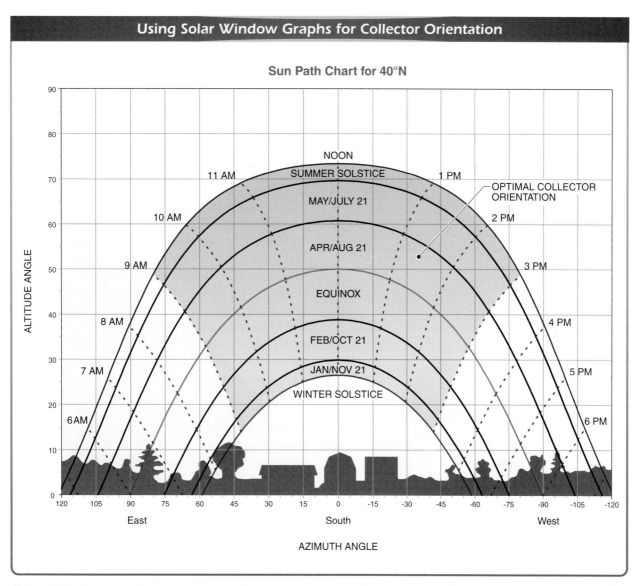

Figure 2-12. A solar window graph can provide an elevation view of the solar window that is beneficial in orienting collectors to capture the most insolation.

A magnetic compass can be used to easily determine magnetic north.

Most mechanical and digital devices use magnetic compass readings to determine the direction of north. The compass needle points to magnetic north rather than geographic true north. *True north* is the direction along Earth's surface toward the geographic North Pole. The magnetic North Pole is close to Canada's Ellesmere Island, about 11° from true north. **See Figure 2-13.**

By traveling along the Earth's surface from north to south, it is evident that the angle and distance from magnetic to geographic north varies. In order to find true north at any location, the compass variation or the magnetic declination must be found.

Magnetic Declination. *Magnetic declination* is the angle differential between magnetic north and geographic, or true, north. The declination is positive when the magnetic north is east of true north. This angle, or degree, must be added to the magnetic north reading to determine true north. The declination is negative when the magnetic north is west of true north. This angle, or degree, must be subtracted from the magnetic north reading to determine true north. There are several ways to find the declination angle such as using isogonic charts, websites, and declination applications for computers and smartphones.

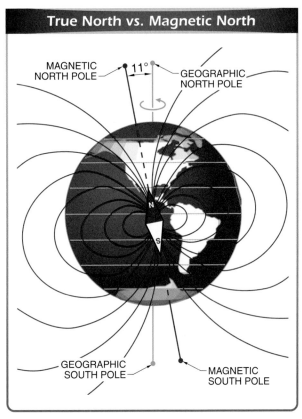

Figure 2-13. True north is the direction along Earth's surface toward the geographic North Pole, which is 11° from the magnetic North Pole.

An isogonic chart has isogonic lines that represent degree declination angles. **See Figure 2-14.** The numbers on the western half of the map indicate the given number of degrees that need to be subtracted from the compass reading to find true north. The numbers on the eastern half of the map need to be added to the compass reading to find true north. For example, in Southern California, between 12° and 14° must be subtracted from what the compass reads as north.

Some government websites, such as http://ngdc. noaa.gov/geomag-web/#declination, provide declination angles using the latitude and longitude, zip code, or city and state as inputs. **See Figure 2-15.** For example, after inputting the city and state New York, New York, the isogonic chart shows a variation of 13°7′36″ west. This means the compass needle in this geographical location actually points 13°7′36″ west of true north. In this case, the angle would be added to find true north. True south is then directly opposite this reading. This is the proper orientation of the solar collector.

Sometimes when using computer or mobile device applications for finding the declination angle, the latitude and longitude of the location may be needed. Mobile device applications, such as the free iOS app "Declination," use the global positioning system (GPS) of the device to calculate the declination at that current location.

Typically, the solar collector should be mounted at an angle to the Sun that maximizes its performance and places the collector in direct insolation from 9 AM to 3 PM solar time. It is between these hours that a fixed point receives 80% to 90% of solar radiation for the entire day. Collectors should face as close as possible to true south. However, a variation of up to 30° is generally acceptable and will not significantly reduce panel performance. This is especially true in southern latitudes where the Sun is more directly overhead for most of the year. Collectors can also be oriented to use either more morning sunlight or afternoon sunlight if critical orientation is needed. Manufacturer installation instructions for the collector should be followed to determine acceptable variations.

Solar Service Inc.
Solar collectors should be mounted at an angle to the Sun that maximizes its performance. This can be accomplished using a frame or rack.

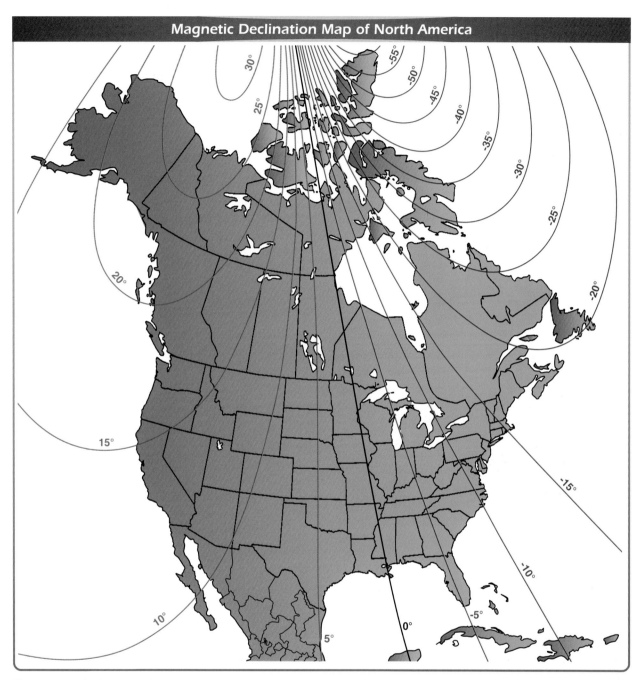

Figure 2-14. An isogonic chart has isogonic lines that represent degree declination angles.

Collector Tilt

Most of the usable solar radiation travels in a straight line from the Sun to the surface of Earth. A flat surface such as a solar collector receives more solar energy when the angle of incidence is closer to zero or perpendicular to its surface. The *angle of incidence* is the angle differential between the angle at which incoming solar radiation strikes the surface of the solar collector and a line perpendicular to that surface. The collector receives less energy when the angle of incidence increases.

The solar path chart reveals that the maximum amount of insolation occurs at solar noon. Significantly less solar radiation is received in the early morning and late evening when the angle of incidence is much greater. Therefore, the collector should be tilted up to meet the solar radiation emitted at solar noon.

National Oceanic and Atmospheric Administration (NOAA) Website

Figure 2-15. Government websites can provide accurate declination data for installers.

It seems logical that to harvest the most insolation the collector should track the Sun and adjust its tilt and orientation as the day progresses. However, this would require very expensive and heavy equipment. Solar tracking is only feasible for industrial concentrating collectors. Most installations are designed as fixed position installations with an orientation to harvest the most solar energy possible throughout the year.

For most residential installations in North America, it has been determined that the solar collector tilt angle for year-round use should be equal to the latitude at the solar collector location. **See Figure 2-16.** For mostly summer energy collection, the tilt angle should be latitude minus 15°. For mostly winter energy collection, the tilt angle should be latitude plus 15°. This additional 15° of tilt will compensate for the lower heat harvest obtainable in the winter months by allowing more collector surface to directly face the Sun in the winter than in the summer.

Essentially, the solar collector should be oriented to face true south. It should also be tilted to meet sun radiation at the degree from horizontal of the collector site latitude. Collector installation may deviate from these parameters, if needed, for customized seasonal energy harvesting.

For example, in southern latitudes, the installer may want to tilt the collector to latitude plus 15° to capture more radiation during winter months rather than summer months when the ambient temperature is already warm. Or, an installer may want to install the collector of a radiant hydronic system at the location's latitude angle for the best year-round insolation capture. In every instance, the manufacturer recommendations and instructions should be followed.

PRINCIPLES THAT AFFECT SOLAR THERMAL TECHNOLOGY

A solar water heating system functions by converting harvested solar energy into heat, transferring that heat to a fluid, storing the heated fluid, and providing the fluid for an intended purpose. The fluid is used as either domestic hot water (DHW) or a transfer fluid to heat DHW. The efficiency of the system depends on how well solar thermal technology principles are applied. System efficiency also depends on the successful adaptation of the principles of heat transfer to solar thermal technology. Therefore, it is crucial that the installer have an understanding of the basic principles of heat transfer.

Absorption and Reflection

Heat transfer initially occurs within the solar collector. Solar radiant energy is absorbed by the collector and converted into heat. The collector absorber, usually a dark, coated metal plate, is the most important part of the collector. The collector manufacturer must design the absorber to absorb the greatest possible amount of radiant energy. Therefore, it must reflect the least possible amount of radiant energy.

Absorption, Absorptance, and Absorbance. *Absorption* is the physical process of absorbing light. It is the transformation of radiant power to another type of energy, in this case heat, by interaction with matter. Within the solar collector, the absorber plate collects solar radiant energy, or sunlight, and converts that energy into heat. *Absorptance* is the rate at which a solar collector absorbs radiation. *Absorbance* is the amount of radiation absorbed by a solar collector.

Reflection, Reflectance, and Reflectivity. *Reflection* is the change in direction of a wave, such as a light or sound wave, away from a barrier the wave encounters. In this case, the wave is solar radiation, or sunlight, and the barrier encountered is the combined absorber material and glazing that covers the collector. *Reflectance* is the ratio of the total amount of radiation reflected by a surface to the total amount of radiation striking the surface. *Reflectivity* is a measure of the ability of a surface, such as an absorber and glazing combined, to reflect radiation.

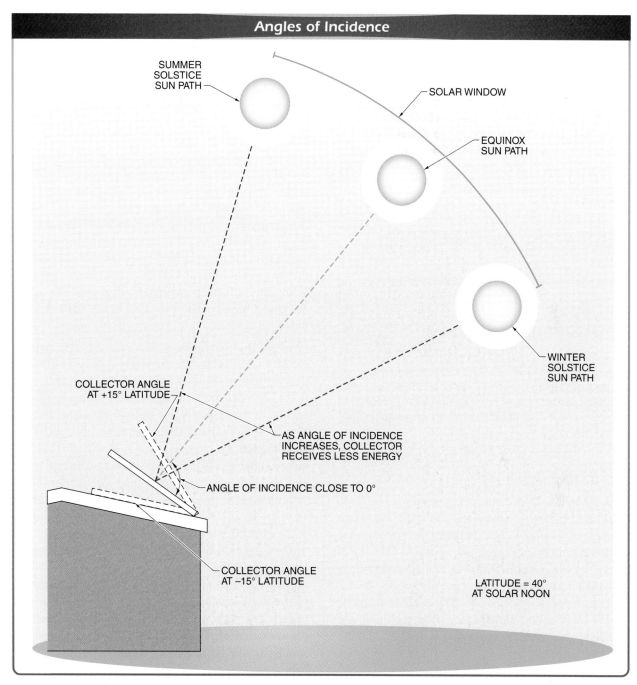

Angles of Incidence

SUMMER
SOLSTICE
SUN PATH

SOLAR WINDOW

EQUINOX
SUN PATH

WINTER
SOLSTICE
SUN PATH

COLLECTOR ANGLE
AT +15° LATITUDE

AS ANGLE OF INCIDENCE
INCREASES, COLLECTOR
RECEIVES LESS ENERGY

ANGLE OF INCIDENCE CLOSE TO 0°

COLLECTOR ANGLE
AT –15° LATITUDE

LATITUDE = 40°
AT SOLAR NOON

Figure 2-16. For most residential installations in North America, it has been determined that the solar collector tilt angle for year-round use should be equal to the latitude at the solar collector location.

Certain materials reflect more insolation than others. Reflective materials are typically light with a smooth or shiny finish, while absorptive materials are typically dark with a matte finish. The materials used to cover collectors and absorb sun energy are selected for their ability to absorb a high percentage and reflect a minimum amount of energy. **See Figure 2-17.** The absorber and absorber-coating efficiency are determined by the rate of absorption versus the rate of reflectance. This, in turn, affects the ability of the absorber and absorber coating to retain heat and minimize emissivity and reradiation. *Emissivity* is the ability of a material to emit energy by radiation. High absorbance and low reflectivity improves the potential for collecting solar energy and transferring that energy to a fluid.

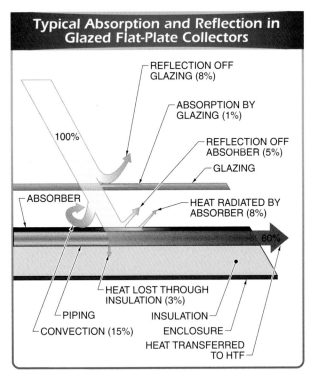

Typical Absorption and Reflection in Glazed Flat-Plate Collectors

- REFLECTION OFF GLAZING (8%)
- ABSORPTION BY GLAZING (1%)
- 100%
- REFLECTION OFF ABSORBER (5%)
- GLAZING
- ABSORBER
- HEAT RADIATED BY ABSORBER (8%)
- 60%
- HEAT LOST THROUGH INSULATION (3%)
- PIPING
- INSULATION
- CONVECTION (15%)
- ENCLOSURE
- HEAT TRANSFERRED TO HTF

Figure 2-17. The materials used to cover the collector and absorb sun energy are selected for their ability to absorb a high percentage and reflect a minimum amount of energy.

Heat Transfer

Heat, a form of kinetic energy, is transferred in three ways: conduction, convection, and radiation. Heat transfer, also called thermal transfer, occurs only if a temperature difference exists and only in the direction of decreasing temperature. Thus, heat is always transferred from higher temperatures to lower temperatures. By using knowledge of the principles governing the methods of heat transfer and through the proper selection and fabrication of materials, the collector designer and the installer can obtain the required heat flow for the solar thermal system. **See Figure 2-18.**

Conduction. *Conduction* is the transfer of heat through a solid material resulting from a difference in temperature between different parts of the material. Heat energy is associated with the motions of the particles making up the material. It is transferred by such motions, shifting from regions of higher temperature, where the particles are more energetic, to regions of lower temperature where they are less energetic. The rate of heat flow between two regions is proportional to the temperature difference between them and the heat conductivity of the substance.

The molecules in solids are bound and contribute to the conduction of heat mainly by vibrating against neighboring molecules. A more important mechanism, however, is the migration of energetic free electrons through the solid. Metals, which have a high free-electron density, such as copper, are good conductors of heat. Nonmetals, such as glass, have few free electrons and do not conduct as well as metals.

Conduction occurs within the absorber itself. As one part of the absorber becomes heated, the heat is conducted to cooler parts of the absorber. Conduction also occurs when the absorber comes in contact with the transfer fluid in the absorber piping. Conduction of heat continues within the heat exchange system. In an internal-coil heat exchanger, the transfer fluid heats the exchange piping that conducts heat to the DHW within the storage tank. With plate exchangers, the transfer fluid conducts heat to the plates, which conduct the heat to the DHW flowing through the plates.

Convection. *Convection* is the mode of heat transfer in fluids (liquids and gases). Convection is based on the fact that, in general, fluids expand when heated and, thus, decrease in density. As a result, the warmer, less dense portion of fluid tends to rise through the cooler fluid. If heat continues to be supplied, the cooler fluid that flows in to replace the rising warmer fluid will also become heated and rise. Thus, a current, called a convection current, becomes established in the fluid. Warmer, less dense fluid continually rises from the point of heat application, and cooler, denser fluid flows outward and downward to replace the warmer fluid.

Convection occurs in the solar thermal system when the heated transfer fluid within the absorber piping rises within the collector and is replaced by cooler fluid downstream. Convection continues within the solar storage tank with heated fluid entering the bottom of the tank and rising as cooler fluid descends to be heated. This process creates the stratification of hot and cooler regions within the storage tank. Convection is thus the principle of heat transfer in which the hottest fluid rises to the top of the storage tank and is transported to the point of use. Convection is also the primary principle of thermosiphoning.

Radiation. *Radiation* is the exchange of thermal radiation energy between two or more objects. Thermal radiation is electromagnetic radiation and arises as a result of a temperature difference between two objects. Unlike conduction and convection, radiation does not require a medium between the two objects for heat transfer to take place. Rather, the transfer agents are photons, which travel at the speed of light.

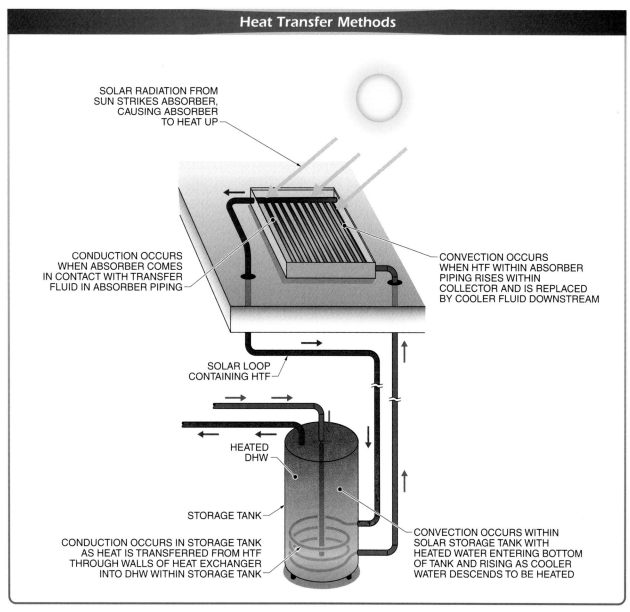

Heat Transfer Methods

SOLAR RADIATION FROM SUN STRIKES ABSORBER, CAUSING ABSORBER TO HEAT UP

CONDUCTION OCCURS WHEN ABSORBER COMES IN CONTACT WITH TRANSFER FLUID IN ABSORBER PIPING

CONVECTION OCCURS WHEN HTF WITHIN ABSORBER PIPING RISES WITHIN COLLECTOR AND IS REPLACED BY COOLER FLUID DOWNSTREAM

SOLAR LOOP CONTAINING HTF

HEATED DHW

STORAGE TANK

CONDUCTION OCCURS IN STORAGE TANK AS HEAT IS TRANSFERRED FROM HTF THROUGH WALLS OF HEAT EXCHANGER INTO DHW WITHIN STORAGE TANK

CONVECTION OCCURS WITHIN SOLAR STORAGE TANK WITH HEATED WATER ENTERING BOTTOM OF TANK AND RISING AS COOLER WATER DESCENDS TO BE HEATED

Figure 2-18. Conduction, convection, and radiation are heat transfer methods that take place in solar collectors.

Heat transfer into or out of an object by thermal radiation is a function of several factors. These factors include surface reflectivity, emissivity, surface area, temperature, and geometric orientation with respect to other thermally participating objects. In turn, the surface reflectivity and emissivity of an object are functions of its surface conditions (e.g., roughness and finish) and composition.

Radiation is the primary method of heat transfer in a solar thermal system. Solar energy is radiated, or transferred, from the Sun to the collector absorber. As each solar collector component and the solar thermal system become heated, some heat is radiated out from the components and out into the atmosphere. Too much radiation

of heat out of the solar thermal system will cause a heat loss within the system. Insulation is used within the collector and on the piping system to retain as much radiant heat as possible within the system, thereby increasing system efficiency.

All three modes of heat transfer must be taken into consideration in all aspects of solar thermal technology. This is critical not just with the design of the solar collector and storage system but with the type of materials used in the piping system. It is also a crucial consideration in the methods used to retain and conserve the heat generated by the system such as insulation and thermal expansion remedies.

Temperature Differential and Delta T

For heat to be transferred, two objects or substances must have different temperatures. This applies whether the heat transfer method is conduction, convection, or radiation. Heat transfers from the hotter object to the cooler object. The greater the temperature difference between two points within an object or between two separate objects, the greater the driving force to move heat from the warmer to the cooler point. Heat transfer occurs at a faster rate when there is a higher temperature differential.

Temperature Differential. *Temperature differential* is the difference in temperature between any two points at a given instant. For example, if the temperature of DHW flowing out of a storage tank is 120°F and cold water flowing into the storage tank is 50°F, the temperature differential (delta T) is 70°F.

Delta T. *Delta T*, also shown as ΔT, is the difference, or change, in temperature either occurring or needed. Delta T is used to calculate the efficiencies of solar collectors or systems. One critical delta T is the temperature differential between the inlet and outlet temperatures of the solar collector heat transfer fluid (HTF).

Thermal Mass

Another important principle in the transfer of heat is the ability of a material to absorb and store thermal energy. *Thermal mass*, also known as heat capacity, is the capacity of a material to store thermal energy for extended periods. The higher the thermal mass, the more efficiently the material can store heat. The better a material can store heat, the more efficient the system that uses the material will be.

Concrete and masonry are two materials that have high thermal mass. These materials are used in the design of solar passive buildings, which store heat during the day and radiate that heat inside the building at night. Water also has high thermal mass, which makes it a good medium for heat storage in solar thermal systems. **See Figure 2-19.**

Facts

A propylene glycol and water solution, which is a common heat transfer fluid (HTF) used for freeze protection in closed systems, has a specific heat capacity 15% to 20% lower than water alone.

Thermosiphoning

Thermosiphoning is a method of passive heat exchange based on natural convection in which liquid circulates without the need for a mechanical pump. Thermosiphoning simplifies the circulation of liquid and thus the heat transfer process by avoiding the cost and complexity of a conventional pump or circulator.

In thermosiphon solar water heating systems, the storage tank is located above the collectors so that as the HTF is heated in the collectors, the HTF rises by convection to the storage tank. **See Figure 2-20.** Thermosiphon systems are considered passive systems because they use the natural action of heated fluid instead of pumps. They can be direct or indirect systems depending on whether heat exchangers are used.

Substances with High Thermal Mass

MASONRY CONCRETE

Alternate Energy Technology
WATER

Figure 2-19. Masonry, concrete, and water have high thermal mass, which allows for the storage of large amounts of heat.

Thermosiphon Water Heating Systems

SunMaxx Solar

Figure 2-20. A thermosiphon water heating system relies on convection to circulate HTF through the system.

A phenomenon related to thermosiphoning is reverse thermosiphoning, which must be avoided in solar water heating installations. *Reverse thermosiphoning* is the movement of warm water up to a collector through convection when ambient temperatures are low enough to cool the HTF in the collector, thereby creating a reversal of intended flow.

Reverse thermosiphoning can occur in any solar water heating installation if proper precautions are not taken. During the night, after sunset, the temperature of the fluid in a rooftop collector can drop to well below the temperature of the fluid in the storage tank. According to the natural laws of convection, cooler, denser fluids will fall and be replaced by warmer, less dense fluids that rise. As a result, the solar loop can essentially reverse its designed flow by allowing cool fluid from the collector to cool the storage tank as the warm fluid in the tank rises to warm the collector. Therefore, all the stored thermal energy that was produced by the system during the day is lost. Reverse thermosiphoning can be easily prevented by practicing good installation techniques such as installing heat traps and check valves.

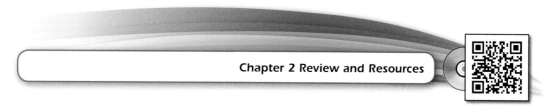

Chapter 2 Review and Resources

3

Collectors

All water heaters have some form of energy source. An electric water heater has one or more electric elements that transform electricity into heat to create hot water. A gas water heater burns natural gas or propane to heat water. A solar water heater harvests the radiant energy from the sun in the solar thermal collector to produce heat for creating hot water. Simply stated, solar collectors are another energy source for creating hot water for domestic use, space heating, and power generation.

SOLAR COLLECTOR DESIGN CHARACTERISTICS

There are several types and designs of solar collectors that can be categorized in different ways depending on their applications. They also exist in various price ranges from relatively inexpensive, simple collectors to expensive, highly complicated industrial collectors. However, all solar collectors share some of the same design characteristics.

Absorbers

Every solar collector has an absorber. An *absorber* is a solar collector component that captures the radiant energy from the sun and converts that energy into thermal energy, or heat, which it transfers to a heat transfer medium. An absorber can be a flat surface or a fin, which is usually copper or aluminum with attached piping to facilitate the conduction of heat to the heat transfer medium inside the piping. **See Figure 3-1.** An absorber is designed to have the highest absorptivity and the least amount of emissivity. For this reason, a selective coating is usually applied to the metal.

Selective Coatings. A *selective coating* is a very thin layer of material applied to an absorber to increase absorptance efficiency. The selective coating can be as simple as flat black paint or as complex as layered proprietary coatings specifically manufactured for use on the absorbers of solar collectors. The difference between selective coatings is the degree of efficiency that the coating brings to the retention and conduction of heat energy in the absorber. Price is also an important factor. A solar collector with black paint will be far less expensive than one with a proprietary coating.

Immediately after striking the surface of the absorber, 5% of the solar radiation will be reflected back into the atmosphere. An absorber coated with black paint will absorb 95% of the solar radiation, but it will only retain 50% of the total thermal energy received. The absorber with black paint will radiate 45% of the total thermal energy back to the atmosphere as heat. **See Figure 3-2.**

Absorbers

Figure 3-1. An absorber captures radiant energy from the sun and converts that energy into heat.

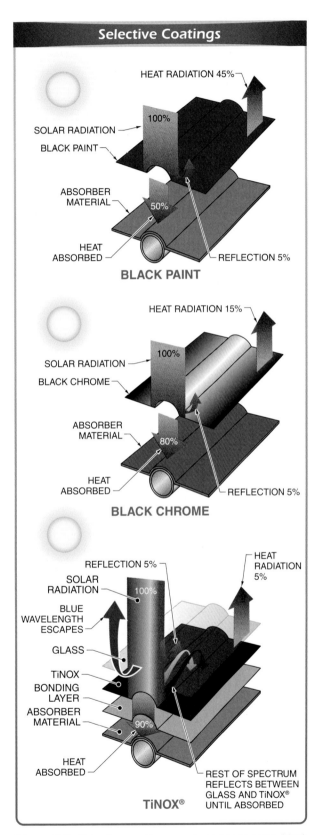

Figure 3-2. Selective coatings, such as black paint, black chrome, and TiNOX®, increase the efficiency of an absorber.

One of the most commonly used selective coatings is black chrome. Black chrome, like black paint, immediately reflects 5% of the solar radiation back to the atmosphere after it strikes the surface of the absorber. An absorber electroplated with black chrome will absorb 95% of the solar radiation, but will only retain about 80% of the total thermal energy received. A black chrome absorber will radiate about 15% of the total thermal energy back to the atmosphere as heat.

One popular proprietary selective coating is called TiNOX®. TiNOX is a complex multilayered coating that absorbs 95% of the solar radiation and retains 90% of the total thermal energy received. Like black paint and black chromium, a TiNOX-coated absorber will reflect 5% of the solar radiation back into the atmosphere.

The advantage TiNOX has over other coatings is that only 5% of the total energy received is radiated back to the atmosphere as heat. The various layers of TiNOX act as energy traps that capture heat energy and allow only a small amount of the spectral wavelength to escape. This escaping wavelength is in the blue color range, which is why TiNOX absorbers look blue. TiNOX is one of the most efficient selective coatings, but it is also one of the most expensive.

Enclosures

An *enclosure* is a collector component that houses an absorber and piping, protects the absorber from ambient temperatures, and captures some of the radiant heat from the absorber. **See Figure 3-3.** Depending on the type of solar collector, the enclosure could be a box frame, such as that in a flat-plate collector, or glass tubing, such as that in an evacuated-tube collector. Some solar collectors, such as swimming pool collectors, do not require an enclosure.

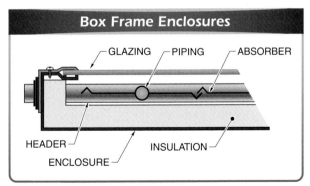

Figure 3-3. Box frame enclosures house the absorber and piping, have insulation and glazing to protect the absorber from ambient temperatures, and capture radiant heat.

Insulation

Insulation is a material used as a barrier to inhibit thermal transmission. Insulation is used to protect the absorber and piping from ambient temperatures and heat loss. Framed enclosures typically have insulation applied to the sides and bottom to keep the absorber and piping from contacting the enclosure, thus reducing conduction heat loss. The insulation also keeps most of the radiant heat released from the absorber and piping in the enclosure to achieve as high efficiency as possible. Evacuated-tube collectors contain a vacuum around the absorber and piping that acts as insulation. Flexible seals are used around piping connections that penetrate the enclosure to make the collector airtight.

Glazing

Glazing is the part of a collector that covers and protects the absorber, allows solar energy to reach the absorber, contains radiant heat from the absorber within the enclosure, and consists of glass or plastic materials. A flat-plate collector that uses a box frame enclosure usually has glazing over the top of the absorber. An evacuated-tube collector uses glass tubing to enclose the absorber.

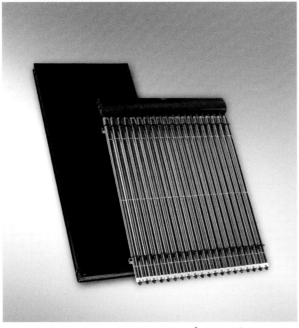

Viessmann Manufacturing Company Inc.
Flat-plate and evacuated-tube collectors have very different appearances but operate under the same solar thermal principles.

Collector Areas

Solar collectors come in various sizes and shapes. In order to make accurate comparisons between the collectors and their different designs, the effective area of a collector must be known. It is important to know the three types of collector surface areas in order to correctly compare, size, and install the desired solar collector. **See Figures 3-4.**

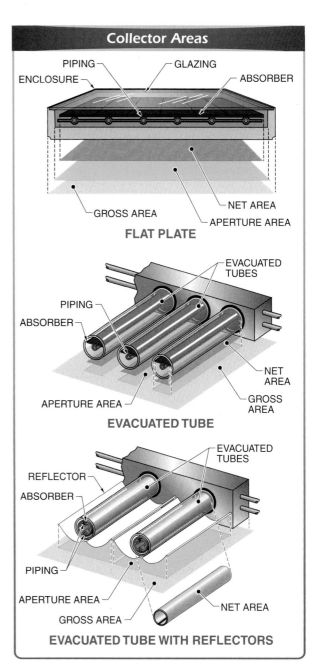

Figure 3-4. The net collector area, also called the effective collector area, varies by the type and design of the solar collector.

The three types of collector areas include the gross collector, aperture, and net area. The *gross collector area* is the overall collector dimension. This is the physical footprint of the collector. The *aperture area* is the unshaded opening of the collector that allows access to solar radiation. The *net area*, also called the effective collector area, is the surface of the absorber that can accept radiation from the sun. The net area varies depending on the type and shape of the absorber.

SOLAR COLLECTOR CLASSIFICATIONS

All solar collectors are classified using three general categories. Solar collectors can be classified by the type of transfer medium used, operating temperature, and glazing.

Heat Transfer Medium Types

A solar collector can be classified as either a dry collector or a wet collector, depending on the type of heat transfer medium used in the absorber. The type of heat transfer medium is determined by the function of the collector and the climate in which the collector is located.

Dry Collectors. A *dry collector,* also known as a solar air collector, is a solar collector that uses air as the heat transfer medium. These collectors are most commonly used for space heating. Dry collectors heat the air directly and can be used for preheating makeup air in HVAC systems. A dry collector typically has a sealed enclosure and an absorber plate that uses a selective coating for higher efficiencies. **See Figure 3-5.**

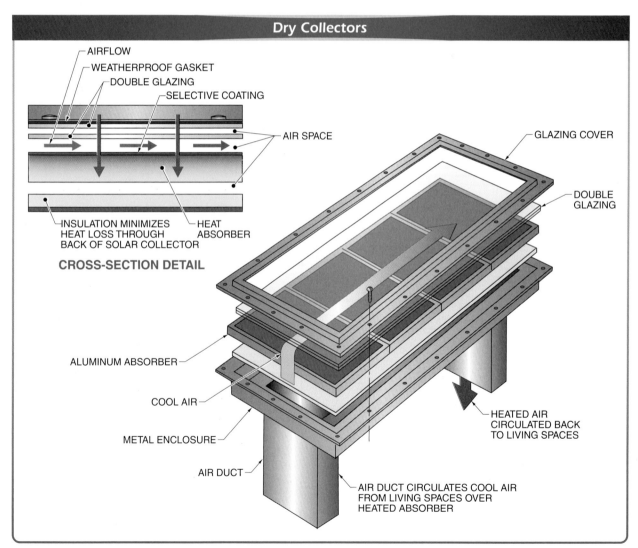

Dry Collectors

AIRFLOW
WEATHERPROOF GASKET
DOUBLE GLAZING
SELECTIVE COATING
AIR SPACE
INSULATION MINIMIZES HEAT LOSS THROUGH BACK OF SOLAR COLLECTOR
HEAT ABSORBER

CROSS-SECTION DETAIL

GLAZING COVER
DOUBLE GLAZING
ALUMINUM ABSORBER
COOL AIR
METAL ENCLOSURE
AIR DUCT
HEATED AIR CIRCULATED BACK TO LIVING SPACES
AIR DUCT CIRCULATES COOL AIR FROM LIVING SPACES OVER HEATED ABSORBER

Figure 3-5. Dry collectors use air as the heat transfer medium and can be used for preheating makeup air for HVAC systems.

Wet Collectors. A *wet collector* is a solar collector that uses a fluid as the heat transfer medium. This fluid is called the heat transfer fluid (HTF) and can be potable water or an antifreeze solution such as propylene glycol. Examples of wet collectors are flat-plate, evacuated-tube, and swimming pool collectors.

Operating Temperatures

A solar collector is engineered to work in either a low-temperature range or a high-temperature range. The operating temperature depends on the application the heated fluid will be used for and the type of collector.

Low-Temperature Collectors. A *low-temperature collector* is a solar collector that collects and transfers heat in the operating temperature range of 70°F to 180°F (21°C to 82°C). Typical flat-plate collectors and swimming pool collectors are considered low-temperature collectors. Common uses for low-temperature collectors are pools (70°F to 85°F), car washes (100°F to 110°F), and general home use (120°F to 140°F). Low-temperature collectors are primarily for residential or light commercial use.

High-Temperature Collectors. A *high-temperature collector* is a solar collector that collects and transfers heat in operating temperatures above 180°F (82°C). These collectors cover a wide range of designs and complexity by incorporating focusing lenses, focusing and concentrating mirrors, and tracking devices. Most high-temperature collectors are used for commercial and industrial hot water purposes and power generation.

Some evacuated-tube collectors can be classified as high-temperature collectors, but most high-temperature collectors are concentrating collectors. **See Figure 3-6.**

Collector Glazing

A solar collector may be classified as unglazed or glazed. The need for glazing and the type of glazing used depend on the design of the collector, climate in which the collector is used, and time of year the collector is used. Glazing is either tempered glass or plastic. If plastic glazing is used, it must be resistant to ultraviolet (UV) radiation so that it will not crack or yellow over time.

Unglazed Collectors. An *unglazed collector* is a solar collector that does not incorporate an enclosure or glazing to protect the absorber. Unglazed collectors are the simplest form of solar collectors. An unglazed collector is simply an absorber with an HTF flowing through it. They are less efficient than glazed or enclosed collectors due to radiant heat loss to the ambient air. Because these collectors do not protect the absorber from the elements, they are used mostly in moderate temperature locations or only during the spring or summer months.

One example of an unglazed collector is a swimming pool collector. **See Figure 3-7.** There are also some unglazed collectors used for domestic hot water (DHW), but these are only used in drainback systems. A *drainback system* is a solar water heating system that drains heat transfer fluid from the collectors into a storage tank located inside the building to protect the system from freezing temperatures.

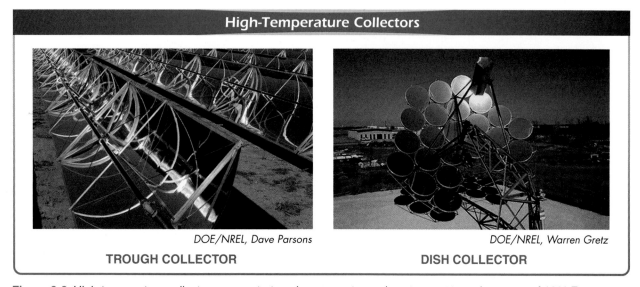

High-Temperature Collectors

DOE/NREL, Dave Parsons

TROUGH COLLECTOR

DOE/NREL, Warren Gretz

DISH COLLECTOR

Figure 3-6. High-temperature collectors concentrate solar energy to produce temperatures in excess of 180° F.

DOE/NREL, Gen-Con, Inc.

Figure 3-7. Unglazed collectors do not incorporate enclosures or glazing to protect the absorbers and are often used in swimming pool heating applications.

Glazed Collectors. A *glazed collector* is a solar collector that encloses the absorber and piping in a frame and uses a glass or plastic cover. The cover, or glazing, must allow solar radiation into the absorber and also contain most of the radiant heat released from the absorber. Therefore, both glass and plastic glazing are designed to have high light transmittance but low reflection characteristics. Glazing also provides protection from ambient temperatures, weather, and damage to the absorber. Most flat-plate collectors and evacuated-tube collectors are glazed collectors. **See Figure 3-8.**

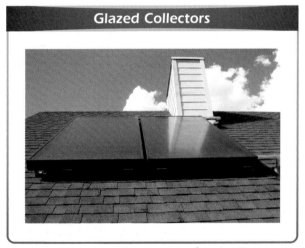

Viessmann Manufacturing Company Inc.

Figure 3-8. Glazed collectors use glass or plastic to protect the absorbers and to increase the efficiency of the collectors.

SOLAR THERMAL COLLECTOR CATEGORIES

The three classifications of solar collectors can be further broken down into specific categories based on their design and construction. Typical design categories for solar collectors include the following:

- integral collector storage (ICS) collectors
- swimming pool collectors
- flat-plate collectors
- hybrid photovoltaic thermal (PV/T) collectors
- evacuated-tube collectors
- concentrating collectors

Integral Collector Storage Collectors

An *integral collector storage (ICS) collector* is a solar collector that uses one or more storage tanks as its absorber. The ICS collector is one of the simplest solar collectors. It is an affordable solar collector system that is commonly used in North American campgrounds and summer cabins. It can also be used in a passive solar water heating system or to supplement or preheat a DHW heater. **See Figure 3-9.** Most solar collectors are used in conjunction with storage tanks that collect the heated water for future use. However, the ICS collector is the only collector designed as the primary storage system, thus eliminating the need for a separate tank. The ICS collector can be combined with a secondary storage tank, increasing the storage capacity of the system.

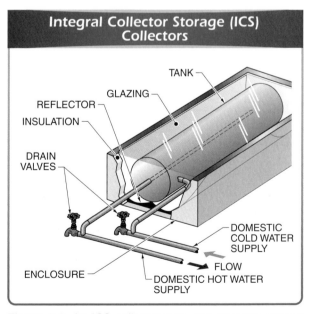

Figure 3-9. An ICS collector uses one or more storage tanks as the absorber to heat water.

The two types of ICS collectors are single-tank ICS collectors and multiple-tank ICS collectors. A *single-tank ICS collector* is a solar collector that consists of a single storage tank framed within an enclosure. A single-tank ICS collector is also known as a batch collector. A *multiple-tank ICS collector* is a solar collector that consists of a series of small tanks or tubes that are also framed within a single enclosure. **See Figure 3-10.** A typical multiple-tank ICS collector consists of rows of 4″ copper tubing. A multiple-tank ICS collector has a larger absorber area than a single-tank ICS collector, making it more efficient but also more expensive.

Multiple-Tank ICS Collectors

ABSORBER/STORAGE TANK
GLAZING
PIPING
ENCLOSURE
INSULATION
GLAZING GASKETS
OUTLET

Figure 3-10. A multiple-tank ICS collector uses several smaller storage tanks connected in series to provide a larger absorber area to increase efficiency.

The tank, or tanks, of an ICS collector is usually painted black or, in more sophisticated systems, has a selective coating applied to it to increase its absorptivity. The top of the enclosure is glazed to allow solar radiation into the enclosure. In some cases, multiple layers of glazing and insulation are used to reduce emissivity and radiant heat losses.

The HTF flowing through the ICS collector is the potable DHW. The DHW is heated and stored within the ICS collector and is ready for use during the day. Because the DHW is stored outdoors, it will normally cool down during the night. ICS collectors can be used only in areas of moderate, nonfreezing temperatures because they are exposed to the weather.

The capacity of an ICS collector can range from 30 gal. to 50 gal. (114 L to 189 L). Therefore, the entire system of the HTF, enclosure, and piping can weigh up to 500 lb (227 kg). When installing an ICS collector system, care should be taken to ensure that the mounting surface can support the weight of the system, especially if it is on a roof. If not, the surface will need to be reinforced to carry the weight safely. Another consideration when mounting the ICS collector on the roof is its profile. A batch collector can be up to 18″ tall, but a multiple-tank ICS collector may be only 8″ tall. The batch collector adds more to the weight of the load on the roof because of the force of the wind acting against its tall profile. This extra wind load should be taken into account when planning the installation of a batch collector.

Facts

Integral collector storage (ICS) collectors can weigh almost 600 lb when full. Extra care must be taken to ensure the roof can handle the concentrated load. Structural certification or roof framing reinforcement may be required. A structural engineer should be consulted for the proper method and location of framing reinforcement.

Swimming Pool Collectors

A *swimming pool collector* is an unglazed solar collector that is specifically designed to use solar radiation to heat swimming pool water. In most cases, the absorber is not enclosed in a framed enclosure. Instead, it is mounted directly to the installation surface, which is usually a building roof.

A swimming pool collector is a low-temperature collector. It is designed to heat swimming pool water to only 70°F to 85°F (21°C to 29°C). Since the absorber itself is the collector, it does not require a complicated design or expensive layered selective coatings. The absorber material can consist of polymer, plastic, or metal tubing or piping. **See Figure 3-11.** Polymers and plastics are the most used materials because of their low cost, flexibility, light weight, and noncorrosive properties.

The types of polymer and plastic tubing and piping materials frequently used for swimming pool collectors include acrylonitrile butadiene styrene (ABS), polybutylene (PB), polyethylene (PE), polyvinyl chloride (PVC), polypropylene (PP), and ethylene propylene diene monomer (EPDM). PP and EPDM are the most frequently used absorber materials. Collector manufacturers also design these materials with stabilizers and UV inhibitors to withstand exposure from the destructive wavelengths of solar radiation.

Swimming Pool Collector Absorbers

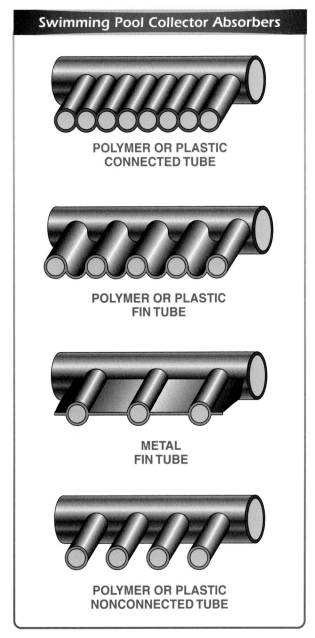

POLYMER OR PLASTIC CONNECTED TUBE

POLYMER OR PLASTIC FIN TUBE

METAL FIN TUBE

POLYMER OR PLASTIC NONCONNECTED TUBE

Figure 3-11. Swimming pool collector absorbers come in several designs and are made from either metal, polymer, or plastic tubing.

The absorber of a swimming pool collector is similar to the absorber of a flat-plate collector. It is constructed of large manifold piping connected to multiple smaller diameter tubes or passageways. The tubing may be connected by fins or connected to the other tubes. The manifold evenly distributes pool water to the absorber, allowing for a large area for solar radiation absorption. **See Figure 3-12.** The collector can be piped in series with other collectors if more absorption area is needed.

Flat-Plate Collectors

A *flat-plate collector* is a solar collector that contains a flat absorber mounted with attached flow tubes within an insulated and glazed framed enclosure. **See Figure 3-13.** Flat-plate collectors are the most commonly used collectors for heating DHW. The absorber is mounted within a framed enclosure. The top of the enclosure is glazed with one to three layers of clear plastic or glass. The sides and bottom of the enclosure are insulated to prevent radiant heat loss. The absorber flow tubes are piped to a manifold that is also contained within the enclosure. The insulated manifold piping also contains piping connections in the form of brass or copper nipples exiting from the sides of the enclosure.

The absorber of a flat-plate collector can be a simple design or it can be a complex design that utilizes superconductive metals with sophisticated selective coatings. The type of absorber design selected for the collector depends on the system application. A simple design for DHW in a single residence generally requires a less complicated and less expensive absorber and glazing. A large commercial solar thermal system may require a more efficient and thus more complex and costlier design.

A flat-plate collector is normally a low-temperature collector designed to heat the HTF to an average temperature of 140°F (60°C). However, if the HTF is not flowing, the absorber of a flat-plate collector can reach temperatures of 300°F to 400°F (148°C to 204°C). Even uninstalled and unfilled collectors can reach these temperatures, so care should be taken to cover or mask the collector during installation, while the collector is being filled, and until it is ready for use.

Tube Flow Designs

The flow of HTF through a flat-plate collector is an integral part of the solar water heating system design and is dependent on the type of absorber tube design. There are several tube flow designs, so care must be taken when selecting the design to be used to ensure that it is compatible with the type of solar water heating system being installed. Complete drainage of the collector is important in drainback systems and closed loop systems that use steamback systems. A *steamback system* is a solar water heating system that uses steam pressure created by high temperatures within the collector to remove the HTF from the exposed areas of the solar loop.

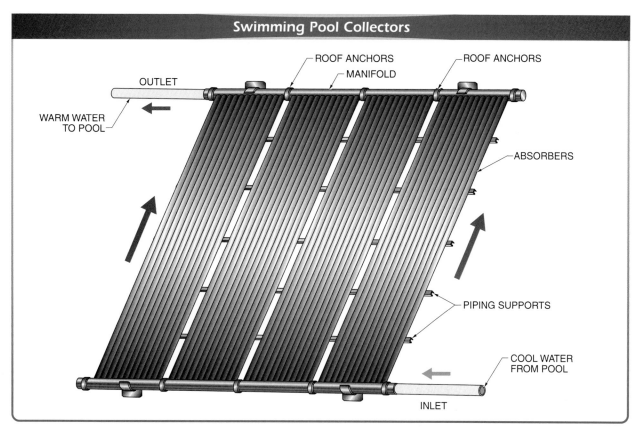

Swimming Pool Collectors

OUTLET

ROOF ANCHORS

ROOF ANCHORS

MANIFOLD

WARM WATER
TO POOL

ABSORBERS

PIPING SUPPORTS

COOL WATER
FROM POOL

INLET

Figure 3-12. A manifold evenly distributes pool water to the absorber, allowing for a larger area for solar radiation absorption.

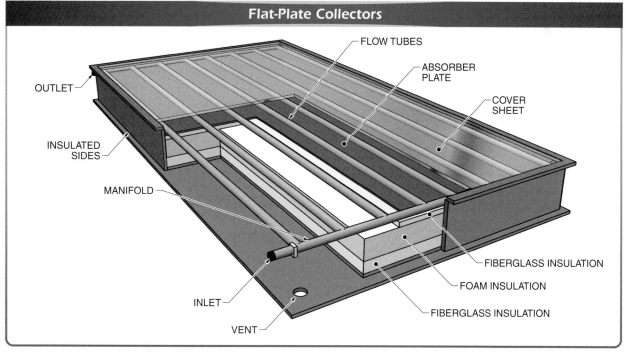

Flat-Plate Collectors

FLOW TUBES

ABSORBER
PLATE

COVER
SHEET

OUTLET

INSULATED
SIDES

MANIFOLD

FIBERGLASS INSULATION

FOAM INSULATION

INLET

FIBERGLASS INSULATION

VENT

Figure 3-13. A flat-plate collector contains a flat absorber mounted within an insulated and glazed framed enclosure and is the most commonly used collector for heating DHW.

It is critical to know which type of tube design will drain properly when needed. **See Figure 3-14.** The four common types of tube designs used for flat-plate collectors include the following:

• harp-style grid design
• blocked-header harp design
• serpentine design
• header serpentine design

Harp-Style Grid Designs. A harp-style grid design is a tube flow design that uses manifold or header tubes on the top and bottom of the collector to which multiple flow tubes are connected. The manifold ends protrude from the enclosure, allowing for piping connections. This also allows for ease in connecting two or more collectors in parallel or in series. This design ensures drainage of the HTF from either side of the collector. The HTF flows through the collector at the bottom manifold and flows up through the grid tubes to the top header. The harp-style grid design is suitable for drainback systems.

Blocked-Header Harp Designs. A blocked-header harp design is a tube flow design that uses manifold or header tubes on the top and bottom of the collector to which multiple flow tubes are connected, but the upper manifold is separated or blocked in the middle. The lower manifold does not provide connecting nipples but only connections to the flow tubes. HTF flows through one side of the upper manifold, down through the flow tubes, and then back up the other side of the collector to the upper manifold outlet. This design is not used for drainback systems, but it can be used for steamback systems. Also, it can only be connected in series with other collectors.

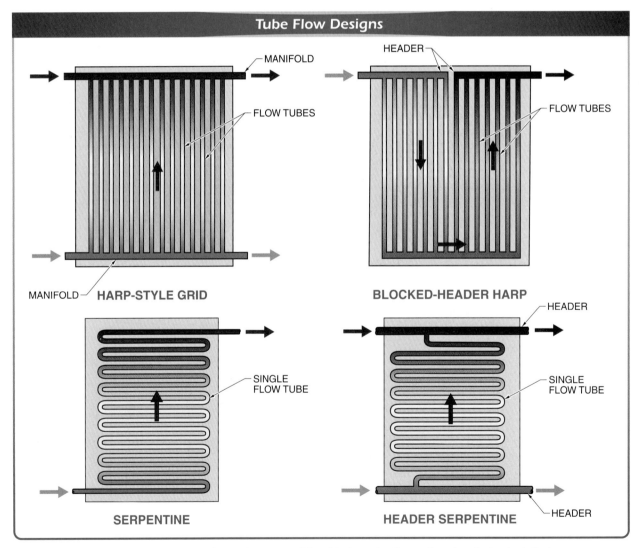

Figure 3-14. The tube flow design used must be compatible with the solar thermal system being installed.

Serpentine Designs. A *serpentine design* is a tube flow design in which one continuous tube snakes from one side to the other side for the length of the collector. There is only an inlet piping connection at the bottom of the collector and an outlet piping connection at the top of the collector. The HTF flows through the collector at the bottom, through the single tube, and out of the collector at the top. This design can only be connected in series with other collectors using an external manifold. It can be used in steamback systems but not in drainback systems because the HTF may not completely drain and could freeze.

Header Serpentine Designs. A *header serpentine design* is a tube flow design in which one continuous tube snakes from one side to the other side for the length of the collector but uses a manifold at the top and bottom of the collector instead of one inlet and outlet. The HTF flows through the collector at the bottom and out at the top. These collectors can be connected in parallel and should not be used in drainback systems.

Hybrid Photovoltaic Thermal Collectors

A *hybrid photovoltaic thermal (PV/T) collector,* also known as a hybrid flat-plate collector, is a photovoltaic flat-plate collector that converts solar radiation into both electricity and thermal heat in one panel. **See Figure 3-15.** A hybrid PV/T collector combines photovoltaic cells that convert electromagnetic solar radiation into electricity with a flat-plate absorber to capture the remaining solar radiant energy and waste heat energy from the PV module to heat domestic hot water.

Hybrid PV/T collectors can produce the same amount of electricity and in some cases more, as conventional photovoltaic collectors. Their thermal yield is in the range of flat-plate collectors without absorber selective coatings. One major advantage of a hybrid PV/T collector is that the circulation of water through the collector keeps the PV panel cool to help maintain efficiency. Hybrid PV/T collectors have been used in Europe and Asia and are now being installed in North America. The ability of the hybrid PV/T collector to provide two forms of solar energy use in one footprint makes it an attractive option.

> **Facts**
>
> *Hybrid photovoltaic thermal (PV/T) collectors can increase the amount of energy produced by the panels during peak loads by up to 30% due to the cooling effect of the water circulation in the pipes located on the back of the PV panels.*

Evacuated-Tube Collectors

An *evacuated-tube collector* is a solar collector that uses a glass tube to enclose an absorber plate or fin and attached tubing. The glass tube provides glazing for access to solar radiation as well as protecting and insulating the absorber assembly. The glass tube holds a vacuum that protects the absorber from convection and conduction heat losses. Multiple glass tubes attach to a manifold at the top of the collector and to a support rail at the bottom of the tubes. The manifold is contained in an insulated enclosure with piping connections protruding from the enclosure. **See Figure 3-16.**

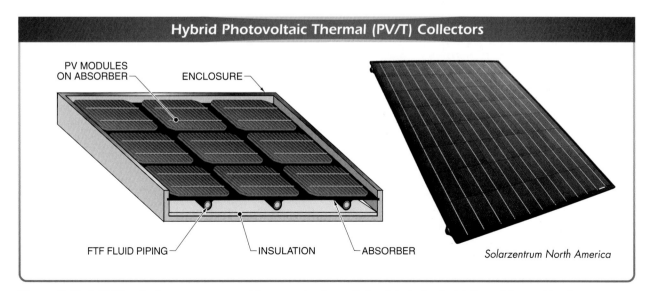

Hybrid Photovoltaic Thermal (PV/T) Collectors

PV MODULES ON ABSORBER

ENCLOSURE

FTF FLUID PIPING

INSULATION

ABSORBER

Solarzentrum North America

Figure 3-15. A PV/T collector uses one panel to convert solar radiation into both electricity and thermal heat.

Evacuated-Tube Collectors

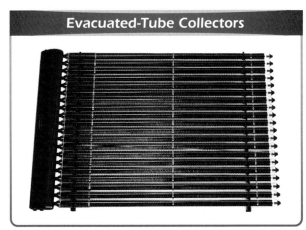

Viessmann Manufacturing Company Inc.

Figure 3-16. An evacuated-tube collector has multiple glass tubes that are mounted to a manifold contained in an insulated enclosure.

An evacuated-tube collector can reach temperatures up to 450°F (232°C) or more if the HTF is stagnant. However, because the glass tube is such an efficient insulator, the exterior of the tubes does not get hot to the touch.

One advantage of an evacuated-tube collector is that the effect of wind loading is less than that of a flat-plate collector. This is because the space between the tubes allows for the passage of wind and because some models can be installed horizontally on the roof. Installation of these collectors is somewhat easier than other types because they tend to weigh less than flat-plate collectors. Also, some models of the heat pipe collector are assembled during installation instead of the entire array being lifted to a roof.

Viessmann Manufacturing Company Inc.
Evacuated-tube collectors can be mounted either vertically or horizontally depending on the need of the building owner and available mounting area.

There are several different designs of evacuated-tube collectors. However, these designs fall into two main categories based on the configuration of the absorber assembly within the glass tubing. The two main categories are direct flow-through evacuated-tube collectors and heat pipe evacuated-tube collectors.

Direct Flow Evacuated-Tube Collectors. A *direct flow evacuated-tube collector*, also known as a flooded collector, is a solar collector that contains a flow pipe attached to an absorber in which heat transfer fluid flows through the manifold, down the pipe, and back up into the manifold within a glass enclosure. They are called "direct flow" collectors because the HTF flows within the glass tube. The flow pipe is either a coaxial arrangement (tube within a tube) or a pipe with a U-shaped bend at the bottom of the glass tube. **See Figure 3-17.**

Another variation of a direct flow evacuated-tube collector is the Sydney collector. A *Sydney collector* is a direct flow evacuated-tube collector that has a double-walled glass tube design with a vacuum created between the two glass walls. The vacuum between the two glass walls is similar to a thermos bottle. The inside of the flasklike tube has a selective coating applied to the glass, with the absorber and U-shaped pipe placed within the tube. The HTF flows down through the U-shaped pipe and back up into the manifold. The glass tube may have a reflective mirror or surface below it to increase solar radiation. **See Figure 3-18.**

A direct flow evacuated-tube collector can be mounted in a vertical or horizontal position on a roof. Because of the efficiency of the collector, it is not necessary to tilt it to the optimum position. In some models, the absorber itself can be tilted within the tube to meet the proper angle of incidence. This is also true of some models of heat pipe evacuated-tube collectors.

Heat Pipe Evacuated-Tube Collectors. A *heat pipe evacuated-tube collector* is a solar collector that transfers heat from a set of pipes that are attached to a manifold that contains the heat transfer fluid. Unlike direct flow evacuated-tube collectors, heat pipe evacuated-tube collectors do not use the HTF flowing through the glass tube. The HTF flows only through the manifold at the top of the collector. Heat is transferred to the HTF in the manifold by the conductors located at the top of the heat pipes. The glass tube that contains the heat pipe can be either single-walled glass or double-walled glass tubes of the Sydney design.

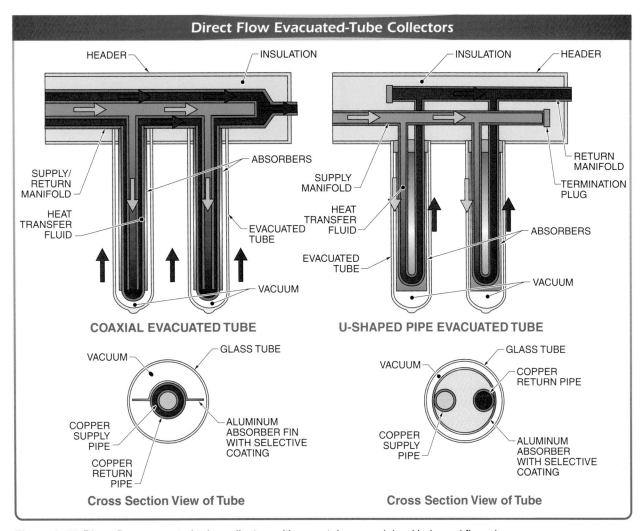

Direct Flow Evacuated-Tube Collectors

HEADER — INSULATION — INSULATION — HEADER

SUPPLY/RETURN MANIFOLD

ABSORBERS

HEAT TRANSFER FLUID

EVACUATED TUBE

VACUUM

SUPPLY MANIFOLD

RETURN MANIFOLD

TERMINATION PLUG

HEAT TRANSFER FLUID

EVACUATED TUBE

ABSORBERS

VACUUM

COAXIAL EVACUATED TUBE   **U-SHAPED PIPE EVACUATED TUBE**

VACUUM

GLASS TUBE

COPPER SUPPLY PIPE

COPPER RETURN PIPE

ALUMINUM ABSORBER FIN WITH SELECTIVE COATING

VACUUM

GLASS TUBE

COPPER RETURN PIPE

COPPER SUPPLY PIPE

ALUMINUM ABSORBER WITH SELECTIVE COATING

Cross Section View of Tube   **Cross Section View of Tube**

Figure 3-17. Direct flow evacuated-tube collectors either contain a coaxial or U-shaped flow pipe.

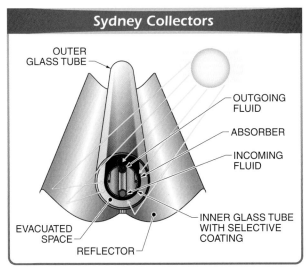

Sydney Collectors

OUTER GLASS TUBE

OUTGOING FLUID

ABSORBER

INCOMING FLUID

INNER GLASS TUBE WITH SELECTIVE COATING

EVACUATED SPACE

REFLECTOR

Figure 3-18. A Sydney collector uses a double-walled glass tube design with a vacuum between the two walls of glass.

The heat pipe is sometimes attached to an absorber plate or fin. In other models, the absorber surrounds the heat pipe or an absorber coating is applied to the inner glass wall. The heat pipe ends are plugged, screwed, or fitted into the manifold.

The heat pipe contains a low-boiling liquid that, when heated by solar radiation, evaporates at low temperatures. The gas, now at a relatively high temperature, then travels up the tube to the connection to the manifold. As the heat is transferred to the HTF in the manifold, the gas in the heat pipe condenses and flows back down to the bottom of the heat pipe. **See Figure 3-19.**

Facts

Sydney collectors were developed in the 1980s by the Solar Energy Research Group at the University of Sydney in Australia.

The heat pipe is connected to the manifold using a dry or wet connection method. In the dry connection method, the heat pipe is attached to the outside of the manifold. The HTF does not come in contact with the condensing portion of the heat pipe. One advantage of a dry connection is that an individual heat pipe or tube can be replaced without draining the system. In the wet method, the heat pipe condenser enters the manifold and comes into direct contact with the HTF. A disadvantage of the wet method is that if there is a defective tube in the array, the manifold will have to be drained before the heat pipe is removed and replaced. This could lead to increased maintenance and repair costs in the future.

Concentrating Collectors

A concentrating collector is solar thermal collector technology that is increasing in importance as energy resources continue to be diversified in North America. The high temperatures needed for commercial and industrial use can only be created using engineered, large-scale solar thermal systems that concentrate direct solar radiation on absorber systems. Concentrating collectors can create these high temperatures for commercial or industrial process heat and utility power generation by using reflecting and focusing mirror systems. The types of concentrating collector systems in use today are parabolic trough systems, solar tower plants, and dish/Stirling systems.

Parabolic Trough Systems. A *parabolic trough system* is a concentrating collector that uses a linear parabolic solar-energy reflector to concentrate energy on a receiver. The reflecting concentrator is positioned facing true south (in North America), while the reflector rotates to follow the sun focusing the solar radiation on the receiver throughout the day. **See Figure 3-20.**

The receiver is usually an absorber pipe that runs along the length of the reflector. The absorber pipe contains an HTF that is heated between 300°F and 660°F (150°C and 350°C) by the reflector. The HTF then flows to a heat exchanger where the solar-generated heat is used for process heat or power generation purposes.

The first parabolic trough system was developed in the United States in 1906. The first commercially operated solar thermal parabolic trough plant was built in Southern California in 1984. New plants are continuously being built and coming on-line. For example, Nevada Solar One outside of Boulder City, Nevada, produces 64 MW of power and came on-line in July 2007. **See Figure 3-21.**

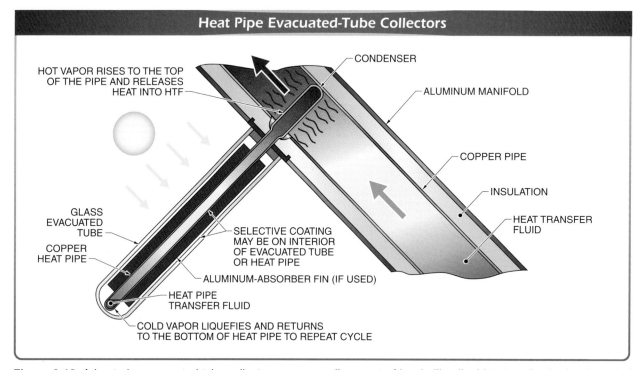

Heat Pipe Evacuated-Tube Collectors

HOT VAPOR RISES TO THE TOP OF THE PIPE AND RELEASES HEAT INTO HTF

CONDENSER

ALUMINUM MANIFOLD

COPPER PIPE

INSULATION

HEAT TRANSFER FLUID

GLASS EVACUATED TUBE

COPPER HEAT PIPE

SELECTIVE COATING MAY BE ON INTERIOR OF EVACUATED TUBE OR HEAT PIPE

ALUMINUM-ABSORBER FIN (IF USED)

HEAT PIPE TRANSFER FLUID

COLD VAPOR LIQUEFIES AND RETURNS TO THE BOTTOM OF HEAT PIPE TO REPEAT CYCLE

Figure 3-19. A heat pipe evacuated-tube collector uses a small amount of low-boiling liquid to transfer the heat created from solar energy into the HTF flowing through the manifold.

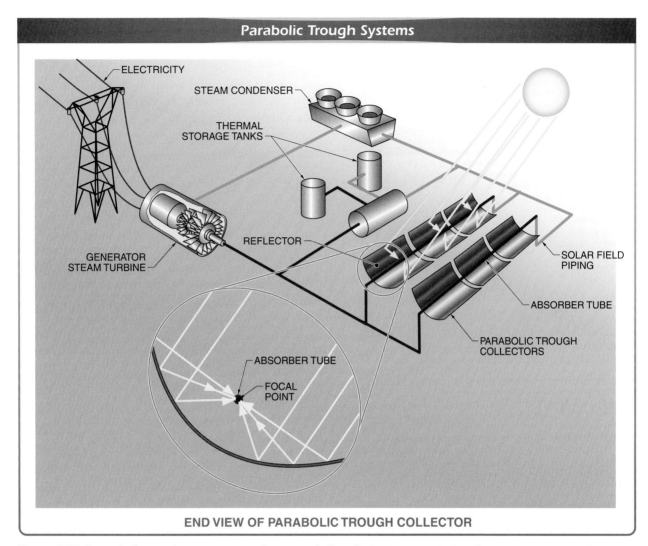

Parabolic Trough Systems

ELECTRICITY

STEAM CONDENSER

THERMAL STORAGE TANKS

GENERATOR STEAM TURBINE

REFLECTOR

SOLAR FIELD PIPING

ABSORBER TUBE

PARABOLIC TROUGH COLLECTORS

ABSORBER TUBE

FOCAL POINT

END VIEW OF PARABOLIC TROUGH COLLECTOR

Figure 3-20. A parabolic trough system uses a linear parabolic reflector to concentrate solar energy onto a receiver.

Nevada Solar One

Schott Solar

Figure 3-21. The Nevada Solar One solar energy plant uses a large-scale parabolic trough system to generate 64 MW of power.

Solar Tower Plants. A *solar tower plant* is a concentrating collector that uses hundreds or thousands of tracking reflectors or mirrors to concentrate solar radiation on a central receiver. The tracking reflectors, or heliostats, surround a tower and focus direct solar radiation at a receiver sitting atop the tower. The receiver contains an absorber that heats an HTF to upward of 1500°F (815°C). The HTF is then used to create steam for power generation. **See Figure 3-22.** Solar tower plants are a relatively new power generation system. One of the newest solar tower plants, the Sierra Sun Tower near Lancaster, California, came on-line in the summer of 2009.

Dish/Stirling Systems. A *dish/Stirling system* is a concentrating collector that uses a concave (dishlike) reflecting surface to concentrate direct solar radiation onto a receiver. **See Figure 3-23.** The receiver absorbs the heat and uses an HTF to transfer that energy to a Stirling engine. The Stirling engine uses the heat energy from the HTF to provide combustion within the engine. The system can be used for a single building or as part of a larger power generation facility such as the Maricopa Solar Sun Catcher near Peoria, Arizona.

Solar Tower Plants

DOE/NREL, Sandia National Laboratories

Figure 3-22. A solar tower plant uses hundreds of tracking reflectors or mirrors to concentrate solar radiation onto a central receiver.

Dish/Stirling Systems

DOE/NREL, Sterling Energy Systems

Figure 3-23. A dish/Stirling system uses a concave reflecting surface to concentrate direct solar radiation onto a receiver.

SOLAR COLLECTOR RATINGS

Because there are hundreds of models of collectors and dozens of manufacturers, determining the solar collector to use in a solar water heating system can be a daunting task. Some models are more efficient than others. Accordingly, there are less costly models and more expensive models. When the type of collector (unglazed/glazed and flat-plate/evacuated-tube), absorber, and other system components must be decided upon as well, the task can become almost overwhelming.

Fortunately, there are methods that can be used to help decide which collector to use, or at least to assist in comparing the efficiency versus the cost of a collector. The easiest method is to use a pre-engineered solar water heating system. Solar water heating system manufacturers design and assemble the system components as a complete package that includes the collector, storage tanks, heat exchanger, controls, pumps (if used), and sometimes even piping systems. These systems are customized to various climate and location parameters. The solar installer only has to provide the information on water use and the location of the system to be installed. The manufacturer will recommend the type and size of system and components that should be installed.

The choice of systems is made easier with the help of the Solar Rating and Certification Corporation (SRCC). The SRCC rates complete solar water heating systems based on several parameters. Each component and system listed in the SRCC database is certified as meeting baseline parameters and will perform to those parameters. Therefore, when the decision is made to use an SRCC-certified system, the installer is not required to decide which collector to use.

Solar Collector Efficiency Graphs

Another form of comparison can assist in determining the appropriate collector for the location parameters. It involves an overall comparison of collector efficiencies using a collector efficiency graph. The collector efficiency graph plots the percent of collector thermal efficiency based on the inlet fluid parameter. **See Figure 3-24.**

On the vertical axis, or y-axis, is the percentage of collector thermal efficiency in decimal form. This is the percentage of insolation (solar radiation) that is converted to heat energy by the collector. For example, if the collector line begins at the y-axis point of approximately 0.78, or 78% efficiency, this means that the specific collector is, at this point, converting 78% of solar insolation to thermal heat.

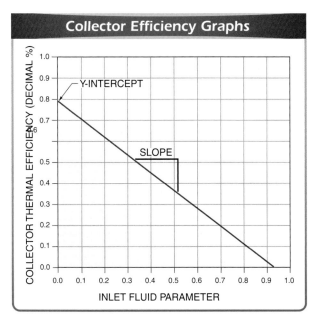

Figure 3-24. A solar collector efficiency graph demonstrates the change in thermal efficiency as the collector operates in increasingly difficult conditions.

On the horizontal axis, or x-axis, is the inlet fluid parameter (IFP). The IFP is based on three conditions that can affect the solar collector the most. These conditions include the following:

- temperature of the HTF fluid at the collector inlet
- temperature of the ambient, or outside, air at the collector
- intensity of the insolation striking the collector

To determine the inlet fluid parameter, the following formula is applied:

$$p = (T_i - T_a) \div I.$$

where

p = inlet fluid parameter

T_i = temperature of HTF (in °F)

T_a = temperature of ambient air (in °F)

I = intensity of insolation (in Btu/hr/ft²)

The inlet fluid parameter is representative of the conditions under which the collector is operating. At point 0.0 on the x-axis, the collector is working under ideal conditions. At point 0.0 for the inlet fluid parameter, termed the y-intercept, the collector inlet temperature is the same value as the outside air temperature. For example, 70°F – 70°F = 0. The collector will not lose heat to the outside air and will thus be able to operate at its peak efficiency. That is, 78% for a y-intercept of 0.78.

As the inlet fluid parameter increases as it moves to the right, the collector will be operating under increasingly more difficult conditions. As this happens, collector efficiency decreases. The sample graph shows the collector efficiency drop from the point of most efficiency (the y-intercept) to the point of least efficiency as the inlet fluid parameter increases. The speed at which this happens can be represented as the slope of the collector. The larger the negative slope number, the less efficient the collector is over a range of conditions.

Fortunately, the SRCC has computed these collector performance values by using the test results from the SRCC collector certification and rating process. The SRCC has also plotted the average efficiencies for the three types of solar collectors that installers will encounter the most, which are unglazed flat-plate collectors, glazed flat-plate collectors, and evacuated-tube collectors. **See Figure 3-25.**

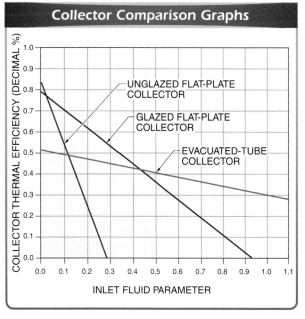

Figure 3-25. A collector comparison graph demonstrates that an evacuated-tube collector performs more efficiently than an unglazed or glazed flat-plate collector when conditions deteriorate.

As can be seen on the collector comparison graph, an unglazed flat-plate collector will work well under ideal conditions, but as conditions change, such as the day becoming cloudy, its efficiency drops dramatically. A glazed flat-plate collector has a relatively high y-intercept and a less steep slope. It will perform very well in moderate conditions. An evacuated-tube collector has a flatter slope but is slightly less efficient in good conditions. However, as conditions deteriorate, an evacuated-tube collector performs much better.

This comparison can help an installer determine the type of solar collector to use. For example, in cooler, less sunny conditions, an evacuated-tube collector might be an ideal choice. In moderate conditions with less temperature change and more sun, a flat-plate collector could be a better choice. Of course, an unglazed flat-plate collector, such as a swimming pool collector, works well in ideal summer conditions but would not be of much use in cold conditions.

SRCC COLLECTOR RATING AND CERTIFICATION PROCESSES

Comparing an individual collector to other collectors is also made easier by the SRCC collector certification process. The following information about the SRCC process is from the SRCC website, www.solar-rating. org, on the Solar Facts page. It is reprinted with permission, and the copyright is owned by the Solar Rating and Certification Association.

About SRCC, Rating, and Certification

The SRCC currently operates two major solar programs: collector certification (OG-100) and heating system certification (OG-300). The OG-100 collector certification program applies to that part of a solar energy system that is exposed to the sun and collects the sun's heat. The collectors can be used to heat water, air, or other heat transfer media. The OG-300 rating and certification program for solar hot water systems integrates results of collector tests with a performance model for entire systems and determines whether systems meet minimum standards for system durability, reliability, safety, and operation. Factors affecting total system design, installation, maintenance, and service are also evaluated.

A direct comparison of an SRCC-rated collector to an SRCC-rated solar water heating system is not possible. The reason for this is two-fold. First, the collector rating shows the performance of one component in the solar package while the system rating shows the performance of an entire solar package. Second, each rating, whether a collector rating or a system rating, is developed using a separate set of assumed conditions.

Viessmann Manufacturing Company Inc.
The SRCC does not provide ratings for large commercial water heating systems. However, they do provide guidelines for such systems on their website.

The OG-100 directory contains information about solar collectors that have been certified and rated by the SRCC. **See Figure 3-26.** The information in the directory will provide you with reliable and comparable data for solar water heating collectors you may be considering buying. The rating information is a helpful tool for comparing the efficiency of the various solar collectors on the market. While you can, and should, compare collector ratings, you cannot compare collector ratings with system ratings. All collectors which have been certified by SRCC will bear the SRCC label, which is your assurance that an independent party has verified the performance and basic durability of the solar product you are considering. Copies of SRCC labels are shown in the directory.

The directory contains descriptive information about the solar collectors and also "performance" information about them. "Performance" data relates to the energy output of the collector. The SRCC performance information contained in this directory provides a way to compare the relative performance of different solar water heating collectors, not the actual performance you can expect from a given collector. This is because the collectors and systems are tested under standard laboratory conditions, which are certain to be different from those in your home. Think of the SRCC ratings as you do the mpg ratings for cars—a benchmark but not necessarily the same performance you will experience.

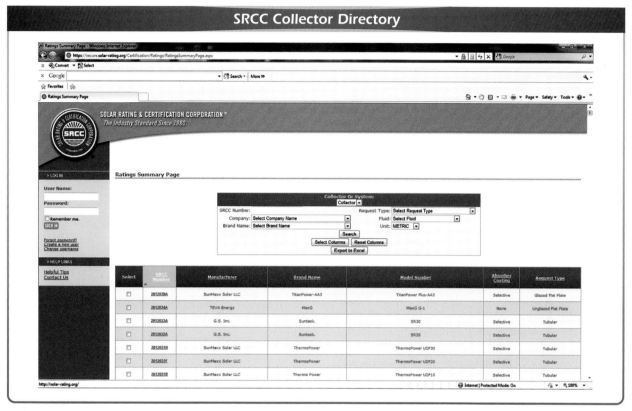

Figure 3-26. The OG-100 directory contains information about solar collectors that have been certified and rated by the SRCC.

Remember, too, that performance (or energy output) is only one criteria in choosing a solar energy collector. Quality of installation, cost, availability of service and parts, and the expected life of the equipment are also important points to consider. Equipment which is well-designed and well-built, but poorly installed, cannot perform according to the manufacturer's specifications.

How Collectors Are Rated

Each time SRCC allows a solar manufacturer to attach the SRCC label to its product, very specific steps have been followed to assure consumers that the product meets SRCC's approval and that the performance information provided to you is correct. First, SRCC selects a solar collector at random from the manufacturer's facility. The collector is then sent for testing to an independent laboratory accredited by SRCC. When the collector is received by the lab, it is inspected to document the materials used. (You will see much of this information in the directory pages.) Then, the collector is subjected to a variety of durability tests to reveal any leaks, to check the integrity of construction, and to assess the collector's resistance to sudden expansion and contraction and changes in water temperature.

Following the durability tests, the energy output of the collector is measured to determine the performance of the collector under the standard laboratory conditions. These measurements result in the performance figures found in the box at the top of each collector's rating page in the directory. Finally, when the testing is complete, the lab partially disassembles the collector and inspects it for any hidden problems.

Facts

Under the SRCC OG-100 certification program, collector certification expires 12 years after the test report date. The collector manufacturer must reapply for new certification under the certification procedures in effect at the time of application renewal. While the new application is in process, the current certification may be extended an extra year.

When the last inspection is completed, the lab sends the test report to the SRCC for review and calculation of the figures, which appear in the rating directory. The SRCC also checks the collector design for reliability and durability. When the collector is certified, the manufacturer is notified and required to begin affixing the SRCC label to the solar collector. Also, the manufacturer must provide a copy of the Certification Award with each certified collector.

Types of Solar Collectors

As you shop for a solar collector, you may see several different types. They are:

1. Unglazed liquid-type collectors are those in which a liquid is heated by the sun in a stationary collector which does not have glass or other transparent covering. These collectors are commonly used for swimming pool heating systems, but are also used in domestic water heating systems.

2 Glazed liquid-type solar collectors are those in which a liquid is heated by the sun in a stationary collector which has a cover of glass or other transparent material. They are the most common type of collectors, and are often used for domestic water heating and space heating systems.

3. Air-type collectors are those in which the sun heats air rather than water in the collector. They are most commonly used for space heating applications.

All three types of collectors work well and can be compared with others of the same type, using the data in this directory.

Alternate Energy Technology
A solar collector may have a label that gives important manufacturer and certification information.

Performance Data

The performance data about a given collector appears in the box at the top of each rating page. **See Figure 3-27.** The data on the left (*kilowatt-hours per panel per day*) is in metric (or SI units) and the data on the right (*thousands of Btu per panel per day*) is in English (or inch-pound units). The data, whether you read it in metric or English units, provides the total energy produced by that collector in a standard "rating day," that is, under the test conditions used to define a day.

Across the top of the chart are three categories which represent various weather conditions and seasons of the year. **See Appendix** for a listing of average daily total solar radiation in several U.S. cities. The amount of sunlight striking the collector (or "irradiance") is an important factor in how much energy the collector can produce. Also important is how much the energy output of the collector declines as the sunlight declines. Irradiance is measured in megajoules per square meter per day (or in Btu per square foot per day). Generally, a clear sky would be characterized by the 23 MJ/(m²/d) [2,000 Btu/(ft²/day)] column, while a cloudy sky would be characterized by the 11 MJ/(m²/d) [1,000 Btu/(ft²/day)] column. The 17 MJ/(m²/d) [1500 Btu/(ft²/day)] column characterizes mildly cloudy conditions. *Note:* To find the correct insolation available in a given area, use a national insolation chart or map. **See Figure 3-28.**

Once you have determined the correct weather column, you will need to choose the correct category. The categories are listed down the left side of the box, using letters A through E. The accompanying numbers are the difference between the temperature of the water or air entering the collector and the temperature of the air around the collector. These temperature differences are important factors in the ability of the solar collector to produce energy. To use the rating chart, it is easier to refer to the Collector Thermal Performance Ratings Categories table for the correct category. **See Figure 3-29.**

The collector with the higher number in the box which reflects your climate and category produces more energy than those with lower numbers. While such a comparison should not be the only basis for your choice of a solar energy system, you may find it helpful. Remember, too, that the energy output of these collectors in the directory has been measured under test conditions, which are almost certainly not the same as the collector will be subjected to in your home. The remainder of the system and the quality of the installation are also critically important factors in how well your solar system works, and how much energy and money you save.

Collector Certification Records

CERTIFIED SOLAR COLLECTOR

SUPPLIER:
Rheem Water Heaters
101 Bell Rd
Montgomery, AL 36117 USA
www.rheem.com

BRAND:	Rheem
MODEL:	RS32-BP
COLLECTOR TYPE:	Glazed Flat Plate
CERTIFICATION #:	2009057B
Original Certification:	July 15, 2009
Expiration Date:	January 05, 2019

The solar collector listed below has been evaluated by the Solar Rating & Certification Corporation™ (SRCC™) in accordance with SRCC OG-100, Operating Guidelines and Minimum Standards for Certifying Solar Collectors, and has been certified by the SRCC. This award of certification is subject to all terms and conditions of the Program Agreement and the documents incorporated therein by reference.

COLLECTOR THERMAL PERFORMANCE RATING

Kilowatt-hours (thermal) Per Panel Per Day				Thousands of BTU Per Panel Per Day			
Climate ->	High Radiation (6.3 kWh/m².day)	Medium Radiation (4.7 kWh/m².day)	Low Radiation (3.1 kWh/m².day)	Climate ->	High Radiation (2000 Btu/ft².day)	Medium Radiation (1500 Btu/ft².day)	Low Radiation (1000 Btu/ft².day)
Category (Ti-Ta)				Category (Ti-Ta)			
A (-5 °C)	12.5	9.3	6.3	A (-9 °F)	42.6	31.9	21.3
B (5 °C)	11.7	8.6	5.5	B (9 °F)	39.9	29.3	18.7
C (20 °C)	10.0	7.0	3.9	C (36 °F)	34.2	23.8	13.5
D (50 °C)	5.8	3.0	0.7	D (90 °F)	19.8	10.4	2.2
E (80 °C)	1.4	0.0	0.0	E (144 °F)	4.8	0.0	0.0

A- Pool Heating (Warm Climate) **B-** Pool Heating (Cool Climate) **C-** Water Heating (Warm Climate)
D- Space & Water Heating (Cool Climate) **E-** Commercial Hot Water & Cooling

COLLECTOR SPECIFICATIONS

Gross Area:	3.051 m²	32.84 ft²	Dry Weight:	48 kg	106 lb
Net Aperture Area:	2.769 m²	29.81 ft²	Fluid Capacity:	3.8 liter	1.0 gal
Absorber Area:	0.000 m²	0.00 ft²	Test Pressure:	1103 kPa	160 psi

TECHNICAL INFORMATION

Tested in accordance with:

ISO Efficiency Equation [NOTE: Based on gross area and (P)=Ti-Ta]

SI UNITS:	η= ( )	( )*(P)/G	( )*(P)²/G	Y Intercept:	0.744	Slope:	-5.162 W/m².°C
IP UNITS:	η= ( )	( )*(P)/G	( )*(P)²/G	Y Intercept:	0.744	Slope:	-0.909 Btu/hr.ft².°F

Incident Angle Modifier								Test Fluid:		Water	
θ	10	20	30	40	50	60	70	Test Mass Flow Rate:		76.000 kg/(s m²)	55920.800 lb/(hr/ft²)
Kτα	-1.29	-1.31	-1.33	-1.38	-1.45	-1.58	-1.85	Impact Safety Rating:			

REMARKS:

Jim Huggins
Technical Director

SRCC
OG-100 CERTIFIED

Figure 3-27. An SRCC collector certification record contains performance information pertaining to a specific solar collector.

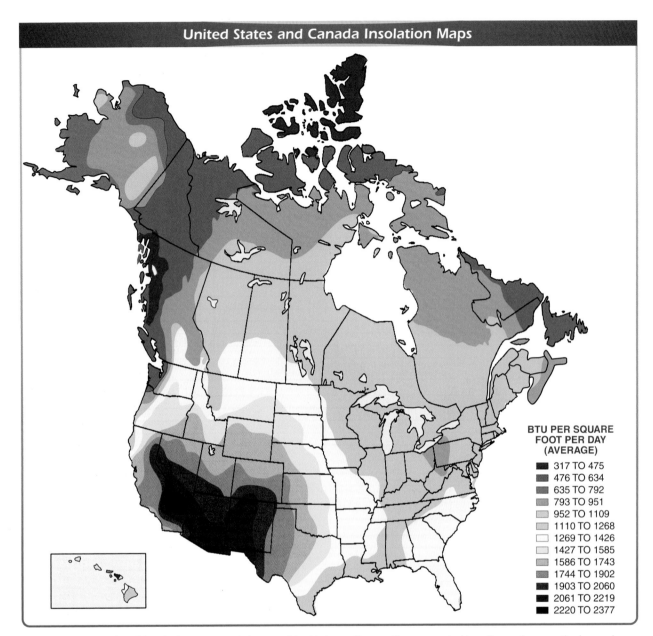

Figure 3-28. A national insolation map contains graphical information on the amount of irradiance in a particular region.

Collector Thermal Performance Rating Categories			
Category	Temperature Difference		Application
A—Pool heating (warm climate)	−9°F	−5°C	Certain types of solar-assisted heat pumps; swimming pool heating
B—Pool heating (cool climate)	9°F	5°C	Liquid collectors with certain types of solar-assisted heat pumps; swimming pool heating; space-heating air-systems
C—Water heating (warm climate)	36°F	20°C	Service hot water systems; space-heating air systems
D—Water heating (cool climate)	90°F	50°C	Service hot water systems; space-heating liquid systems; air conditioning
E—Air conditioning	144°F	80°C	Space-heating liquid systems; air conditioning; industrial process heat

Figure 3-29. The collector thermal performance rating categories are identified by the temperature difference between the water or air entering the collector and the temperature of the air surrounding the collector.

Descriptive Information

Included in the descriptive information (*Collector Specifications*) is the size of the collector. The gross area is the size of the top face of the collector; the net aperture is the size of the glass or other glazing material that sunlight can enter. The size of the collector may be relevant when comparing energy output and price.

Also, the "dry weight" of the collector combined with the "fluid capacity" (for liquid systems; a gallon of water weighs 8.3 pounds) will give you a rough idea of how much weight the solar system will be adding to your roof, if that is where the system is to be installed. Remember to multiply the dry weight plus the fluid weight by the number of collectors in the system.

Note: Included on the rating page is the Collector Materials section, which describes the materials the collector is made from, including the absorber and insulation. Also shown is the type of absorber coating used. There is also a Technical Information area, which states the y-intercept and the slope of the collector. Below this area is the type of HTF used and the flow rate the collector was tested with.

Comparing Collector Efficiency and Cost

With the ratings discussed above, it is easy to compare the energy output of one collector to another. It can be difficult however, to take into account the price of the different collectors.

One method is to compare the energy output for each dollar spent on different collectors. Or, in other words, how many Btu (or MJ) does a dollar buy if spent on Collector #1 versus Collector #2? This question can be answered by dividing the energy output by the cost of the collector. For example, you are considering a solar water heating application. Collector #1 has a rating in Category C (for water heating) under the correct climate column of 29 MJ (per collector per day) or 21,000 Btu (per collector panel per day). Collector #1 sells for $387. Collector #2 is rated at 35 MJ or 33,000 Btu; it sells for $675. Thus:

Collector #1

$$\frac{29\,MJ}{\$387} = \mathbf{0.07\,MJ\,/\,\$} \text{ or } \frac{27,500\,Btu}{\$387} = \mathbf{71\,Btu\,/\,\$}$$

Collector #2

$$\frac{35\,MJ}{\$675} = \mathbf{0.05\,MJ\,/\,\$} \text{ or } \frac{33,000\,Btu}{\$675} = \mathbf{49\,Btu\,/\,\$}$$

Collector #1 is the better buy, based on performance under the test conditions alone. The higher the number of MJs or Btu per dollar, the more cost-effective the collector is...all other things being equal. Remember, though, that the design and quality of the rest of the system and the installation are also critical to a good solar energy system.

Technical Explanation of the Collector Testing and Rating Program

The SRCC solar collector thermal performance test is based on the American Society of Heating, Refrigerating, and Air Conditioning Engineers (ASHRAE) Standard 96-1980, *Methods of Testing to Determine the Thermal Performance of Unglazed Flat-Plate Liquid-Type Solar Collectors,* for unglazed liquid collectors and on ASHRAE Standard 93-1986, *Methods of Testing to Determine the Thermal Performance of Solar Collectors,* for glazed flat-plate liquid collectors, air collectors, linear tracking concentrators, and other collector devices which fall within the scope of the test standard.

Based on the thermal performance data derived from the ASHRAE 96-1980 or ASHRAE 93-1986 test methods, SRCC then calculates the collector ratings according to SRCC Document RM-1, *Methodology for Determining the Thermal Performance Rating for Solar Collectors.* This rating methodology accounts for diffuse irradiance, which is assumed to be distributed isotropically throughout the view of the collector. The methodology is applicable to all non-tracking collector panels.

Before a collector model is issued certification and ratings, SRCC requires that an individual collector be selected at random from the manufacturer's inventory. That unit is then sent to an independent laboratory accredited by SRCC for testing according to SRCC Standard 100-81, *Test Methods and Minimum Standards for Certifying Solar Collectors.* The SRCC test sequence for collectors is a combination of durability and performance tests.

The Required Tests and the Purpose of Each Are Described Below:

- **Receiving Inspection.** To inspect and document the condition of the collector prior to formal testing.
- **Static Pressure Test.** To determine if a loss of pressure occurs or evidence of fluid leakage or fluid path deterioration.
- **30-Day Exposure Test.** To verify integrity of construction after at least 30 days exposure to adverse conditions.
- **Thermal Shock/Water Spray Test.** To verify that the collector structure and performance will not be degraded due to sudden thermal expansion or contraction.

- **Thermal Shock/Cold Fill Test.** To determine the reaction of a hot collector after the introduction of cold water.
- **Post Exposure Static Pressure Test.** To determine if a loss of pressure occurs or evidence of fluid leakage or fluid path deterioration after a collector has been stagnated under worst case conditions.
- **Time Constant Determination Test.** To determine the transient behavior of the collector or the time required to respond to abrupt changes in either insolation or inlet temperature.
- **Thermal Performance Test.** To determine the instantaneous efficiency of the collector over a wide range of operating temperatures. ("Efficiency" is defined as the ratio of collected energy to the available energy falling on the entire collector area.)
- **Incident Angle Modifier Test.** The incident angle modifier needs to be determined in order to predict collector performance over a wide range of conditions. The modifier algorithm is used to modify the efficiency curve to account for changes in performance as a function of the sun's incidence angle.
- **Disassembly and Final Inspection.** To visually inspect the major components and subassemblies and to report their conditions after testing has been completed.

Once the collector test unit has completed the above sequence of tests, the results are sent to SRCC for evaluation and computation of the thermal performance ratings.

Durability Requirements

A collector is judged by SRCC to have successfully completed the durability-type tests if none of the following conditions occurred during the testing:
- Severe deformation of the absorber.
- Severe deformation of the fluid flow passages.
- Loss of bonding between fluid flow passages and absorber plates.
- Leakage from fluid flow passages or connections.
- Loss of mounting integrity.
- Severe corrosion or other deterioration caused by chemical action.
- Crazing, cracking, blistering or flaking of the absorber coating or delamination of reflective surface.
- Retention of water in the insulation.
- Excessive retention of water anywhere in the collector.

- Swelling, severe outgassing or other detrimental changes in collector insulation, which adversely affect the collector performance.
- Leakage or damage to hoses inside the collector enclosure or leakage from mechanical connections.
- Cracking, crazing, permanent warping, or buckling of the cover plate.
- Cracking or warping of the collector enclosure material.

In addition, in order to qualify for collector certification and ratings, manufacturers must document to SRCC that their collectors meet the SRCC requirements for durability in design and construction. For example, all collectors must be designed to prevent condensation build-up and all glass cover plates must be of a non-shattering or tempered type.

A Word about Flow Rates

The SRCC solar collector thermal performance ratings are valid only for the fluid and flow rate used to generate the ASHRAE test data. Since performance of a collector may vary with changes in flow rate, in order to allow for an even more direct comparison of the thermal performance of various collector models, SRCC adopted the requirement beginning in April of 1983 that all thermal performance testing of solar collectors be conducted at the ASHRAE standard recommended flow rates except as noted below.

For unglazed flat-plate liquid-type solar collectors, the ASHRAE standard flow rate per unit area (transparent frontal or aperture) is 0.07 kg/(s m^2) [51.5 lb/(hr ft^2)]. For glazed flat-plate liquid-type solar collectors, the ASHRAE standard flow rate per unit area (transparent frontal or aperture) is 0.02 kg/(s m^2) [14.7 lb/(hr ft^2)]. When air is the transfer fluid, the ASHRAE standard flow rate is 0.01 m^3/(s m$_2$) [2 cfm/ft$_2$] or 0.03 m^3/(s m^2) [6 cfm/ft^2], inclusive.

For those collectors which have been designed for a specific flow rate other than the ASHRAE standard recommended flow rate, the manufacturer may petition to have the collector rated at its design flow rate. The flow rate at which each solar collector model was tested is provided on each directory listing.

SRCC Certification Labels

All solar products certified by SRCC are required to be labeled with an approved SRCC certification label within sixty (60) days of receipt of certification. The label shown below should be on each collector certified under SRCC's OG 100 protocol. **See Figure 3-30.**

American Solar Living LLC

Figure 3-30. Collectors that are certified by the SRCC must be labeled with a SRCC certification label that gives the manufacturer's information, model number, gross area, and the Clear Day Rating.

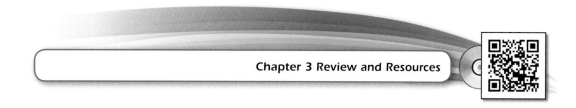

Chapter 3 Review and Resources

Solar Water Heating System Components

The components of a solar water heating system include a water heater, a storage tank, pumps, valves, and piping. These components are needed to harness the solar energy received from the collectors. All components of a solar water heating system must be rated for the high temperatures that can occur in these systems.

SOLAR WATER HEATING STORAGE

The storage tank is an important component of the solar water heating system. The purpose of a storage tank is to hold the water heated directly or indirectly by the solar collectors. The two basic types of storage tanks are pressurized and atmospheric tanks. Typical pressurized storage tanks include conventional storage water heaters, solar storage tanks, and drainback tanks. Some pressurized tanks may contain heat exchangers or back-up heating elements. Atmospheric tanks are usually used for larger residential or commercial applications since they are designed to hold several hundred to several thousand gallons. Both pressurized and atmospheric tanks may contain internal heat exchangers that are used to transfer the solar energy to the domestic hot water (DHW).

PRESSURIZED STORAGE TANKS

A *pressurized storage tank* is a storage tank built to withstand domestic cold water (DCW) pressure without rupturing. Pressurized storage tanks are usually either dedicated solar storage tanks or gas or electric storage water heaters. The advantage of using an existing or new storage water heater is the cost. They are typically less expensive than dedicated solar storage tanks.

However, they usually do not have as much insulation as dedicated solar storage tanks and do not come with internal heat exchangers. Therefore, they are limited to direct solar water heating systems unless an external heat exchanger is used.

The size required by a solar water heating system depends on the expected hot water demand. Factors such as the number of people in a household and whether the heated water will be used for additional purposes such as space heating will affect the size of the tank. Pressurized storage tanks for residential solar applications range in size from 40 gal. to 120 gal. (150 L to 450 L). **See Figure 4-1.** Commercial pressurized tanks can be several hundred to several thousand gallons in size.

Pressurized tanks over 120 gal. (450 L) are almost never used in residential solar applications because such tanks must meet National Sanitation Foundation (NSF) and American Society of Mechanical Engineers (ASME) standards. They are also much more expensive than tanks of 120 gal. or less. If the heated water storage needs are in excess of 120 gal., multiple tanks can be used. In most cases, it is more cost-effective than using one large tank. However, the cost of a multiple-tank system must be weighed against the extra material and labor cost of piping multiple tanks. In addition, there may be a loss of efficiency due to the extra piping and tank surface area.

Pressurized Storage Tanks

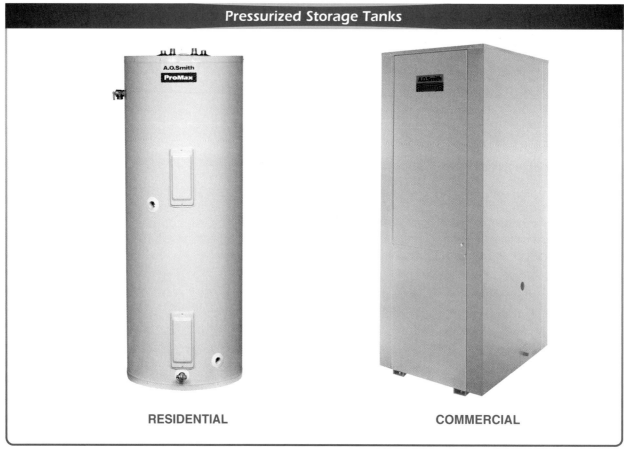

RESIDENTIAL　　　　　**COMMERCIAL**

A. O. Smith

Figure 4-1. Pressurized storage tanks are available for residential and commercial applications.

Storage Water Heaters

The storage tank of a storage water heater is welded from sheet metal to resist pressure and temperature. Most tanks are made from steel, but stainless steel, Monel®, aluminum, and copper alloys can be used also. **See Figure 4-2.** The inside of a storage tank is lined with a material that resists the temperature and corrosive effects of the heated water. A ceramic lining, often referred to as glass, is the most common lining applied to the interior of steel water heater tanks. Some manufacturers also glass-coat the exterior of gas-fired water heater tanks to protect them from the corrosive gases of combustion.

Galvanization was once the most popular tank lining method. However, its use has been virtually discontinued in favor of methods using other materials such as cross-linked polyethylene (PEX) and copper plating, which provide longer life under all water conditions. Cementlike coatings, known as stone coatings, are available but may be limited in use due to their high weight and relative fragility.

Storage Water Heaters

CORROSION-RESISTANT LINING

TANK

INSULATION

SHEET-METAL JACKET

WATER

Figure 4-2. A water heater storage tank consists of several layers, the thickest of which is the insulation that minimizes heat loss from the stored hot water.

The tank is surrounded by insulation in order to minimize heat loss and wrapped in a sheet metal jacket. However, the major drawback of older storage water heaters is that standby heat loss tends to be higher than that of dedicated solar storage tanks due to thinner tank insulation. However, modern water heaters can provide almost the same efficiency as dedicated storage tanks.

Water Heater Connections/Ports. A water heater storage tank has several threaded connections or ports. These connections are common to all storage water heaters regardless of the type of heat source. Normally the hot and cold water connections are located on the top. **See Figure 4-3.** The hot water outlet is connected to the supply pipe that conveys hot water to fixtures in the building. As water is heated within the tank, it rises and ensures that the water reaching the hot water outlet is at the proper temperature.

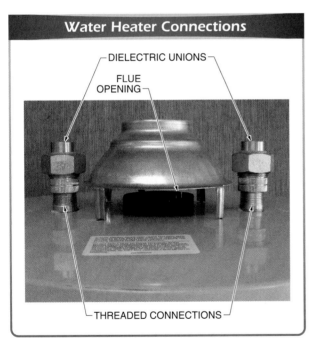

Figure 4-3. Most water heater connections are at the top of a cylindrical tank.

Electric Water Heaters

An *electric water heater* is a storage water heater that uses heat produced by the flow of electricity through a heating element to heat water. **See Figure 4-4.** There are several advantages to electric water heaters. Electric water heaters are generally considered 100% efficient because virtually all of the heat generated by the heating element is transferred into the water. These devices

do not require a supply of combustion air or venting of combustion gases to the outside. They are inherently clean systems, do not generate soot or grime, and may be installed in areas of a building where a heater fired by fossil fuel may be impractical. An electric water heater is often smaller than a gas-fired water heater of similar capacity and has no risk of fuel leakage.

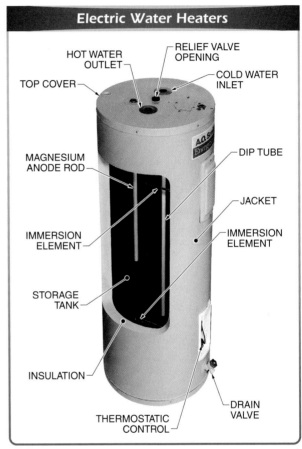

A. O. Smith

Figure 4-4. Electric water heaters use electric heating elements as heat sources.

Heating Elements. A *heating element* is an electrical device within a water heater or solar storage tank with an intentionally high resistance that produces heat when connected to a power supply. Heating elements are available in different voltage, wattage, and watt density ratings.

Watt density is the amount of heat generated per surface area unit of a heating element. A high watt-density element is approximately 180 W/in². Low watt-density elements are 75 W/in² to 100 W/in². Generally, a lower watt-density heating element has a longer expected life.

Heating elements also vary in physical style. Two types of heating elements used for water heating are immersion and wraparound heating elements. **See Figure 4-5.** The most common type of heating element is an immersion heating element.

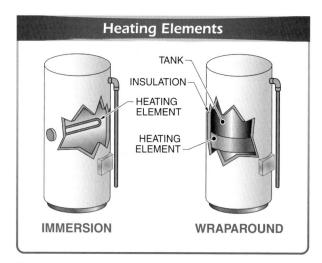

Figure 4-5. The two common types of electric heating elements are immersion and wraparound heating elements.

An *immersion heating element* is a heating device within a water heater or solar storage tank that makes direct contact with the substance to be heated. At least one immersion heating element is inserted into a water heater storage tank and makes direct contact with the water. Each immersion element has a separate automatic thermostat. Electric water heaters must be filled with water whenever the immersion elements are operated. Dry firing (operating the elements without water) will cause the elements to quickly overheat and burn out because there is no water to carry the heat away from the element.

Immersion heating elements require additional penetrations through the side of a storage tank for each of the immersion elements. At least one flanged or threaded opening is provided for the element mounting.

A less common type of heating element is the wraparound heating element. A *wraparound heating element* is a heating element that surrounds the substance to be heated. Wraparound heating elements offer the advantage of element replacement without draining the water. In addition, openings in the tank are not required, so there is less risk of water leaks. However, the heat from the heating elements must first be transferred through the tank wall to be absorbed by the water, which is a less efficient process.

Wiring. Heating elements are available in a variety of standard voltages and wattages to meet the specific requirements of the installation. Residential water heaters typically operate at 120 V or 240 V. The higher voltage is better for electric water heaters, although 120 V elements are available. Commercial and industrial units typically operate between 208 V and 440 V. The wire must be sized correctly to safely carry the necessary current.

Gas-Fired Water Heaters

A *gas-fired water heater* is a storage water heater that uses heat produced by the combustion of natural or liquefied petroleum (LP) gas to heat water. **See Figure 4-6.** Natural or LP gas is ignited in the main burner at the bottom of the storage tank, and the combustion gases flow through the flue inside the tank to heat the water. The flue is usually fitted with a baffle to slow the flow of hot gases, which improves heat transfer. The gases exit the water heater through a draft hood and are vented to the outside of the building.

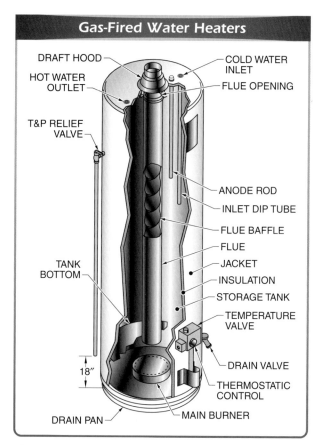

Figure 4-6. A gas-fired water heater uses the products of combustion, which flow through the water heater, as a heat source.

Gas-fired water heaters are particularly susceptible to efficiency losses and malfunctions (puffs or explosions) if the source of combustion air is inadequate. Gas-fired water heaters require properly sized and installed gas vents or chimneys. A supply of air for combustion is required for gas-fired water heaters. The supply must be adequate if the efficiency of the heater is to be maintained. The amounts specified in state, local, or provincial code requirements should be considered as the absolute minimum values.

Gas Controls. A thermostatic control circuit keeps the water near a set temperature by controlling the firing and extinguishing of the main burner. **See Figure 4-7.** The element senses the temperature decrease, and the controls open the main gas supply to the main burner. A pilot flame ignites the gas at the main burner. The thermostatic control circuit shuts off gas flow when the water reaches the desired temperature.

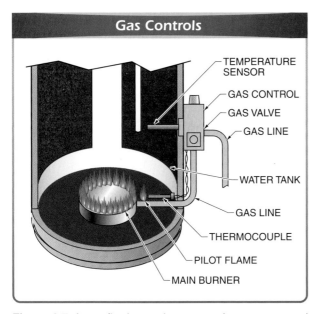

Gas Controls

TEMPERATURE SENSOR

GAS CONTROL

GAS VALVE

GAS LINE

WATER TANK

GAS LINE

THERMOCOUPLE

PILOT FLAME

MAIN BURNER

Figure 4-7. A gas-fired water heater requires sensors and controls to ensure the safe and efficient operation of the burner assembly.

If the pilot flame is extinguished or becomes too small to ignite the main burner, the gas supply to the water heater must be automatically shut off to avoid creating a dangerous situation. Therefore, a thermocouple-controlled valve is used to control the gas supply. A thermocouple is a junction of two different metals that generates a small voltage in proportion to temperature and used to open or close the gas water heater pilot valves. When the pilot flame is sufficient, the nearby thermocouple is heated enough to generate the voltage necessary to hold a gas valve open against the force of a spring. If the heat is not sufficient, such as when the pilot flame is extinguished, the thermocouple does not produce enough voltage and the spring closes the gas valve.

Vent Systems. A vent system is an arrangement of combustion equipment and pipes that manage the input of fresh air and the exhaust of the products of combustion. Gas-fired water heaters may use natural, direct, or sealed box vent systems.

A *natural vent system,* also known as a gravity vent, is a vent system that uses the natural buoyancy of the products of combustion to carry exhaust to the outside. Fresh air from the room is brought into the combustion chamber by the low pressure caused by rising exhaust.

A *direct vent system* is a vent system that uses separate piping for combustion air and the products of combustion. This dual piping system offers higher efficiencies than natural vent systems. There are two piping methods used in these dual systems. With the concentric method of direct venting, a coaxial pipe, which is a pipe within a pipe, is used to bring in combustion air from the outside through the outer chamber and exhaust the products of combustion to the outside through the inner chamber. The other method of direct venting involves the use of a collinear pipe, which consists of two separate pipes. One pipe brings in the outside air for combustion, while the second pipe carries away the products of combustion to the outside. The objective of the direct vent system is to have no openings within the building, although the appliances do not necessarily have sealed combustion chambers.

A *sealed box system* is a combustion system that uses the direct vent method and includes a sealed combustion chamber. With this system, all openings to the indoor space are eliminated.

Solar Storage Tanks

A *solar storage tank* is a pressurized storage tank specifically designed for use in a solar water heating system. These tanks typically have larger capacities and better insulation than conventional storage water heaters. **See Figure 4-8.** They usually have multiple sets of connections or ports for piping connections for heat exchangers, back-up heating systems, or hydronic heating. There may also be multiple locations for the insertion of sensors to monitor the water temperature at various levels in the tanks. A solar storage tank may also include heating elements to heat the water in the tank when the solar system cannot meet the heating demand or at night when it is not running.

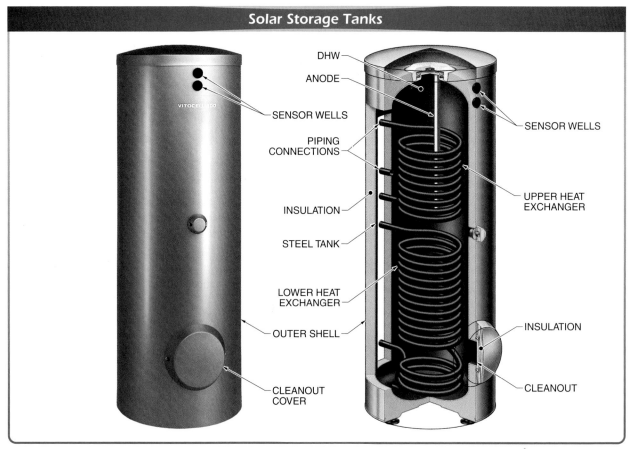

Figure 4-8. Solar storage tanks have larger capacity and better insulation than conventional storage water heaters and may also contain one or more heat exchangers.

Typical sizes of solar storage tanks for residential systems are 60 gal. (225 L), 80 gal. (300 L), or 120 gal. (450 L). The dimensions of the tanks vary by manufacturer, but most are able to fit through standard doorway openings.

Like conventional water heaters, solar storage tanks are constructed of welded steel or stainless steel and may be glass-lined or stone-lined. However, new polymer tanks wrapped with fiberglass have been introduced that are less prone to corrosion. Insulation of 2″ to 3″ is used in between the interior tank and the exterior shell. Some manufacturers ship their storage tanks with the insulation and shell unattached. The installer attaches these after the tanks are set into place.

Like a conventional water heater, a pressurized solar storage tank is required to have a T&P relief valve for overheating and overpressure protection. Most tanks also have an anode for corrosion protection. Storage tanks for direct solar water heating systems do not usually have any internal heat exchangers. However, storage tanks for indirect systems have one, two, or three internal heat exchangers depending on the requirements of the system.

Drainback Tanks

A *drainback tank* is a small storage tank that is used to hold HTF when a drainback system is not in operation. **See Figure 4-9.** The drainback tank allows the HTF from collectors to be stored, which prevents HTF stagnation and freezing in the collectors. Common drainback tanks hold 5 gal. to 15 gal. (19 L to 57 L) of fluid. However, drainback tanks of up to 30 gal. (114 L) are available. In addition, some solar storage tanks up to 120 gal. (455 L) have been used as drainback tanks. Depending on the construction of the drainback tank, the tank may contain heat exchangers, anode rods, or a sight glass. Drainback tanks must be sized to hold all of the HTF that is contained in the collectors and any vulnerable piping. In addition, the drainback tank must allow for additional reserve HTF capacity for correct operation when the system is active.

Drainback Tanks

DRAINBACK TANK

SIGHT GLASS

SOLAR
STORAGE TANK

Figure 4-9. A drainback tank is a small storage tank that is used to hold HTF when a drainback system is not in operation.

Anode Rods

Galvanic corrosion is a significant potential problem in water heaters and certain solar storage tanks because it is accelerated by elevated temperatures. Galvanic corrosion is an electrochemical process that causes an electrical current to flow between two dissimilar metals in an electrically conductive environment, which eventually corrodes one of the materials (the anode).

The inside of a water heater storage tank is coated with a nonmetallic lining that prevents electrical conduction between the tank metal and other metal parts. However, the lining can degrade over time or become damaged by sediment, leaving the metal structure vulnerable to corrosion. Because tank integrity is critical, a second system consisting of a sacrificial anode is used to control any galvanic corrosion in a way that minimizes its damaging effects. A *sacrificial anode* is a piece of metal that is more susceptible to galvanic corrosion than the metal structure to which it is attached.

Anodes for water heaters are normally rods of magnesium installed in a port in the top of the storage tank. If bare metal becomes exposed inside the tank, the anode corrodes first. A film of magnesium is transferred by electroplating from the rod to the inside of the tank wall. This corrosion protection lasts only as long as the anode. In some areas, because of the composition of water, the anode may dissipate too rapidly for long-term protection. In this case, a secondary sacrificial anode may be supplied on some models. Some anodes are segmented to allow insertion into the tank in confined spaces with limited room above the tank. **See Figure 4-10.**

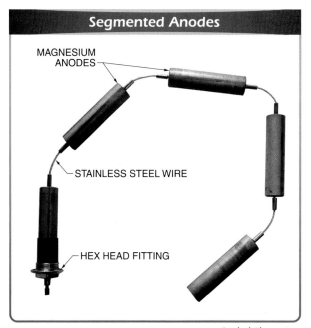

Segmented Anodes

MAGNESIUM
ANODES

STAINLESS STEEL WIRE

HEX HEAD FITTING

Stiebel Eltron, Inc.

Figure 4-10. Segmented anodes allow insertion into the tank in confined spaces with limited room above the tank.

Temperature and Pressure Relief Valves

A water heater or solar storage tank can develop dangerously high temperatures and pressures if the heating controls fail to shut off the heat source. When water is heated inside a closed vessel, its boiling point increases in direct proportion to the pressure within the vessel. Because of this, water may be heated above its normal boiling point of 212°F (100°C) at atmospheric pressure without becoming steam. When this occurs, the water is said to be superheated.

Superheated water is water under pressure that is heated above 212°F (100°C) without becoming steam. When the pressure within a container of superheated water is suddenly released, such as by a pipe rupture, the superheated water may immediately convert to steam, which expands by about 1700 times. This rapid expansion can result in an explosion of the water heater. Excessive pressure can also develop within a water heater storage tank from high water service pressure, water hammer, or thermal expansion.

In order to prevent damage and injury from excessive water heater temperatures, a temperature and pressure relief valve is installed. A *temperature and pressure (T&P) relief valve* is an automatic self-closing safety valve installed in the opening of a water heater tank that releases heated water and relieves pressure in a controlled manner. **See Figure 4-11.** Under normal conditions, the T&P relief valve remains closed. However, if the temperature or pressure reaches certain levels, the valve automatically opens to relieve pressure and to release heated water. The release of heated water allows cooler water to enter the tank, reducing the temperature within the tank.

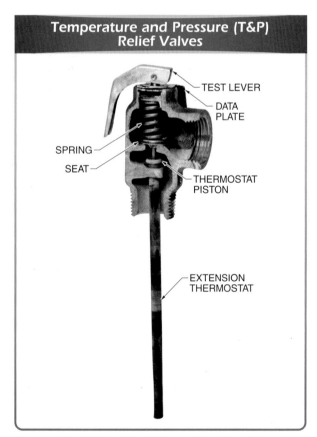

Figure 4-11. A T&P relief valve is a safety device that prevents damage and injury due to excessive water heater temperature or pressure by releasing hot water in a controlled manner.

The temperature relief portion of a T&P relief valve must open before the water temperature reaches the atmospheric boiling point. T&P relief valves for domestic water heaters are set to open at 210°F. The temperature sensing portion of the valve is a probe mounted at the bottom of the valve and placed into the storage tank or the hot water outlet. The pressure relief portion of the valve must open before the pressure within the system exceeds the working pressure of the system component with the lowest pressure rating. T&P relief valves are set to open at 150 psi. The data plate or label on a T&P relief valve must contain information about the temperature and pressure setting of the valve, relief capacity of the valve, and recognized standards authority such as ASME International.

The minimum inlet and outlet size of a T&P relief valve is ¾″. A full-size drain line extends from a T&P relief valve to a safe place of disposal. The end of the drain cannot have threads, which would facilitate the installation of any device that may prevent free flow through the drain.

A test lever on a T&P relief valve is used to verify the proper operation of the valve. T&P relief valves should be tested once a year by manually opening the valve. Water should flow from the valve if the T&P valve is operating properly. If water does not flow from the valve, it must be replaced immediately.

Expansion Tanks

An *expansion tank* is a tank with an air-filled cushion that provides additional volume for water or HTF under thermal expansion. Because air is compressible, the air cushion compresses as the increased volume of expanded water enters the tank. **See Figure 4-12.** As water enters the tank, it displaces some of the space that was occupied by the air cushion.

A diaphragm expansion tank, commonly used in domestic water heater installations, uses a diaphragm barrier or bladder to separate the air cushion from the system water. This prevents the loss of air through the absorption of water and eliminates the possibility of a water-logging condition in the tank and subsequent loss of pressure control.

Expansion tanks are commonly used in closed piping systems in order to accommodate the expansion of potable water or other HTF when heated. Potable water system expansion tanks include a separate rigid polypropylene liner on the water side of the diaphragm. The liner allows potable water to be used without the risk of corrosion or contamination.

Expansion Tanks

EXPANSION TANK
SUPPORT BRACKET

EXPANSION
TANK

Figure 4-12. An expansion tank is a metal tank that holds both water/HTF and air, which are separated with a flexible diaphragm. Expansion tanks are connected to piping systems and used to provide space for hot water/HTF to expand.

In order to operate properly, an expansion tank must be properly sized for its intended application. The expansion tank manufacturer's sizing and installation instructions should be used to properly size the tank for a specific installation.

ATMOSPHERIC STORAGE TANKS

An *atmospheric storage tank,* also known as an open system tank, is an unpressurized storage tank that is designed to hold large quantities of solar heated water. It is also called an open system tank because the tank is open to the atmosphere. Atmospheric storage tanks are designed to hold large quantities of heated water in large residential or commercial applications. They typically hold from 100 gal. to 500 gal. (380 L to 18,930 L) of unpressurized water. **See Figure 4-13.** One advantage that an atmospheric tank has over a pressurized storage tank is that the price per gallon of storage is significantly less for an atmospheric tank.

Atmospheric Storage Tanks

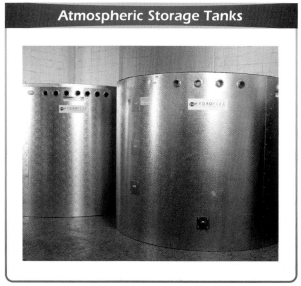

Hydroflex Systems, Inc.

Figure 4-13. Atmospheric storage tanks are designed to hold large quantities of heated water in large residential or commercial applications.

Atmospheric storage tanks are manufactured with materials that can withstand temperatures that reach above 200°F (94°C). However, some tanks may be only warranted by their manufacturer for lower temperatures. Atmospheric tanks are comprised of the tank walls, insulation, and in some cases a liner. The tank walls are often made of stainless steel, aluminum, polypropylene, or fiberglass. These materials are strong and corrosion-resistant.

The tank walls are usually surrounded by several inches of expanded polystyrene, polyurethane, or polyisocyanurate insulation. The insulation may be bonded to the tank during the manufacturing process or installed during the installation process. Insulation may be added to the tank as required.

Some atmospheric storage tanks use an inner liner that is supported by the tank walls. The most popular liner material is ethylene propylene diene monomer (EPDM). EPDM is a synthetic polymer that is waterproof and has good heat and chemical resistance. PVC and fiberglass are also used as liner materials but not to the extent of EPDM.

Within the tank is at least one immersed heat exchanger. These heat exchangers are typically made of corrosion-resistant materials such as stainless steel or copper tubing. The heat exchangers are usually installed after the tank has been set in place. **See Figure 4-14.**

Atmospheric Storage Tank Heat Exchangers

Hydroflex Systems, Inc.

Figure 4-14. Atmospheric storage tank heat exchangers are typically made from corrosion-resistant materials such as stainless steel or copper tubing and are usually installed after the tank has been set in place.

BACK-UP WATER HEATING SOURCES

Along with conventional water heaters and the integral heaters in solar storage tanks, there are two other sources of back-up water heating. These two sources are tankless water heaters and boilers. The amount of water needed, intended use, and cost will determine which of these two options are right for a particular solar water heating system. Small systems may only need a tankless water heater, but a large system, like those heating DHW and water for hydronic heat, may require a small boiler.

Tankless Water Heaters

A tankless water heater is a good option as a back-up water heating source for a solar water heating system. A *tankless water heater,* also known as a demand-type or instantaneous water heater, is a direct water heater that heats water flowing through it on demand and does not include a hot water storage tank. The water heater heats only the water that is needed and continues to heat the water until the demand is satisfied. When sized correctly, a tankless water heater can produce an endless supply of hot water when the water heated by the solar collectors cannot meet the demand.

Tankless water heaters have some advantages over storage water heaters. When there is low or no hot water demand, storage water heaters lose heat through the insulating jacket of the tank and then use energy to maintain the desired water temperature. In contrast, tankless water heaters have no tanks to store water and therefore do not have the standby heat losses of storage water heaters. Heating water on demand generally requires considerably less energy in tankless water heaters.

The heat source of a tankless water heater can be either gas or electric. **See Figure 4-15.** The types of connections, components, and control systems used in tankless water heaters are similar to those of the storage water heaters that use the same type of heat source. For example, like other gas-fired water heaters, gas-fired tankless water heaters require proper venting of the products of combustion.

Boilers

Another method of supplying back-up heated water for a solar water heating system involves the use of a boiler. Small boilers for residential systems are able to produce more heated water than tankless water heaters. Boilers are often used in hydronic heating applications when there is an increased need for heated water or HTF. Like conventional and tankless water heaters, boilers can use either gas or electricity as a heat source. Some manufacturers have even integrated a small boiler into the housing of their solar storage tanks. **See Figure 4-16.**

HEAT EXCHANGERS

A *heat exchanger* is a component of a solar water heating system that is used to transfer solar-generated heat from one liquid to another. Heat exchangers are made of steel, copper, bronze, stainless steel, aluminum, or cast iron. The heat exchangers used in solar water heating systems are usually made of copper because copper is a good thermal conductor and resists corrosion better than other metals. Heat exchangers for solar water heating are available in several types, designs, and sizes.

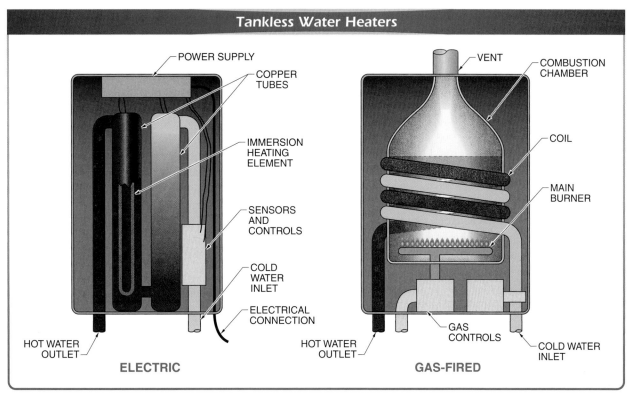

Figure 4-15. An electric or gas-fired tankless water heater can produce an endless supply of hot water when the water heated by the solar collectors cannot meet the demand.

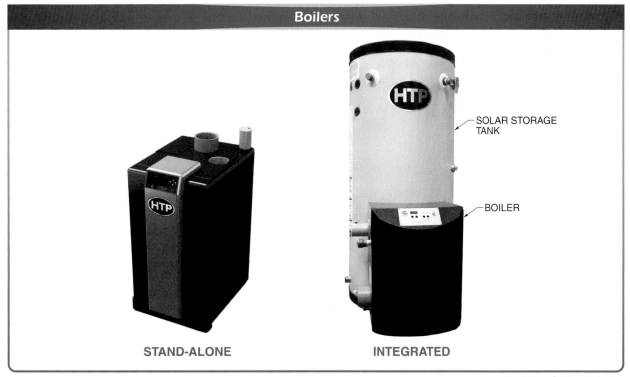

Heat Transfer Products, Inc.

Figure 4-16. Stand-alone or integrated boilers can be used to supply back-up heated water for solar water heating applications.

Types of Heat Exchangers

Heat exchangers are classified as internal and external heat exchangers and have either single- or double-wall construction. The type of heat exchanger used depends on the type of HTF used, system design, and intended use of the HTF.

Single-Wall Heat Exchangers. A *single-wall heat exchanger* is a heat exchanger that has only one wall separating the DHW and the HTF. **See Figure 4-17.** Single-wall heat exchangers are used when the HTF is a nontoxic fluid such as distilled water or propylene glycol. Single-wall heat exchangers are more efficient than double-wall heat exchangers of equal size. For this reason, the size of a larger double-wall heat exchanger must be increased to match the heat transfer capability of a single-wall heat exchanger.

The use of single-wall heat exchangers may be limited by local plumbing and mechanical codes. Generally, if the system uses nontoxic HTF and the normal operating pressure of the solar water heating system is less than the domestic potable water pressure, single-wall heat exchangers are permissible by most jurisdictions.

Double-Wall Heat Exchangers. A *double-wall heat exchanger* is a heat exchanger that has two walls separating the DHW and the HTF. Double-wall heat exchangers are used when the HTF is considered a toxic fluid. Double-wall heat exchangers provide greater protection from leaks than single-wall heat exchangers. While double-wall heat exchangers increase safety, they are less efficient because heat must transfer through two surfaces rather than one. To transfer the same amount of heat, a double-wall heat exchanger must be larger than a single-wall exchanger.

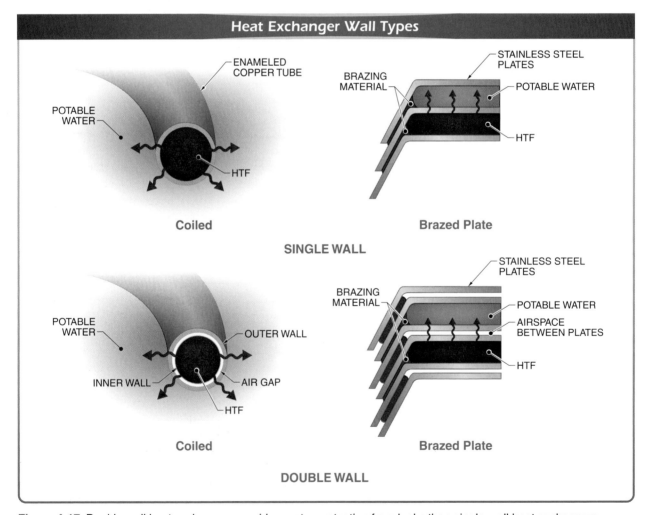

Heat Exchanger Wall Types

Figure 4-17. Double-wall heat exchangers provide greater protection from leaks than single-wall heat exchangers.

Internal Heat Exchangers

An *internal heat exchanger*, also known as an in-tank or immersed heat exchanger, is a heat exchanger that resides in a solar storage tank and allows the transfer of solar-generated heat from the HTF to the potable water without permitting the two liquids to mix. The most common type of internal heat exchanger is the coiled heat exchanger. **See Figure 4-18.** A solar storage tank may include one, two, or even three heat exchangers.

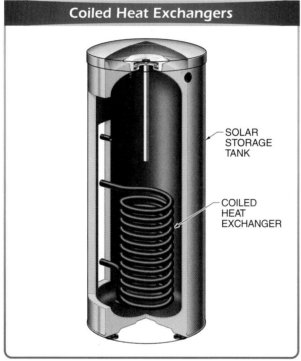

Coiled Heat Exchangers

SOLAR
STORAGE
TANK

COILED
HEAT
EXCHANGER

Viessmann Manufacturing Company Inc.

Figure 4-18. Coiled heat exchangers are the most common type of internal heat exchangers.

External Heat Exchangers

An *external heat exchanger* is a heat exchanger that resides outside of a storage tank. External heat exchangers are used when installing systems that use existing hot water heaters or for larger systems that require more heat transfer. There are many types of external heat exchangers, but the ones most commonly used in solar water heating systems are the brazed plate, tube-in-tube, and shell-and-tube heat exchangers.

Brazed Plate Heat Exchangers. A *brazed plate heat exchanger* is an external heat exchanger that consists of stacked steel plates that are brazed together. **See Figure 4-19.** A stacked plate design creates a series of parallel channels through which the HTF and the

DHW can flow without ever coming into contact with each other. Brazed plate heat exchangers may have ten or more plates. In most cases, they are more efficient at transferring heat than coiled internal heat exchangers because the fluids flow in opposite directions. This action is called counterflow. However, the fluids are subject to fouling, which may decrease their flow capacity.

Tube-in-Tube Heat Exchangers. A *tube-in-tube heat exchanger* is an external heat exchanger that consists of one tube with HTF flowing in one direction surrounded by another tube containing DHW flowing in the other direction. Tube-in-tube heat exchangers are not used very often and are limited to smaller systems.

Shell-and-Tube Heat Exchangers. Shell-and-tube heat exchangers are generally used for swimming pool or large commercial solar water heating systems. A *shell-and-tube heat exchanger* is an external heat exchanger that consists of a large pressure vessel with multiple tubes that run through it. Shell-and-tube heat exchangers provide a large ratio of heat transfer area to volume and are capable of high flow rates. One advantage of the shell-and-tube heat exchanger is that it can be cleaned easily.

SOLAR WATER HEATING SYSTEM PUMPS

An active solar water heating system requires a pump to circulate the HTF through the collectors, piping, storage tanks, and heat exchangers. The pump is controlled by an operational controller. The controller is set to start the pump whenever the HTF drops to a predetermined minimum temperature. The pump then operates until the HTF reaches its predetermined maximum temperature. The pump and the rest of the components that comprise a solar water heating system must be able to handle the high temperatures of a solar water heating system.

The most commonly used pump in solar water heating systems is the centrifugal pump. A *centrifugal pump* is a pump in which the pressure is developed principally by the action of centrifugal force. *Centrifugal force* is the force that tends to impel a thing or parts of a thing outward from a center of rotation. The inlet is an opening in the pump housing that is at the center of the impeller. The outlet is an opening in the pump housing that is located on the outer perimeter of the pump housing. **See Figure 4-20.** The rotating impeller moves water to the outside edge of the rotating impeller and throws it against the pump housing. The pump housing directs the water to piping connected to the system. Pressure produced by the centrifugal pump must overcome the system resistance to maintain an adequate flow in all operating conditions.

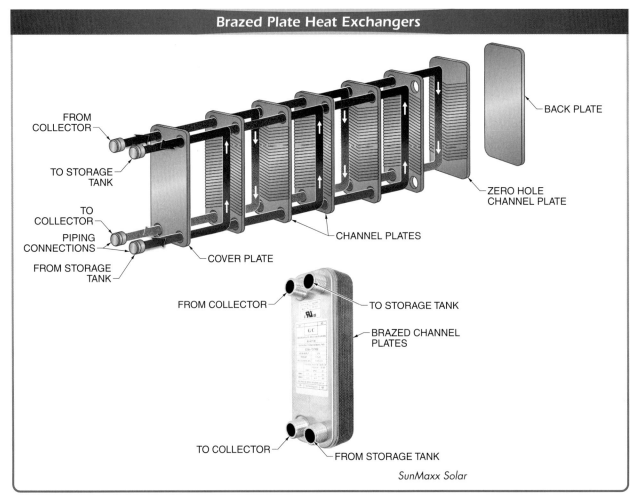

Brazed Plate Heat Exchangers

FROM COLLECTOR

TO STORAGE TANK

TO COLLECTOR

PIPING CONNECTIONS

FROM STORAGE TANK

COVER PLATE

CHANNEL PLATES

BACK PLATE

ZERO HOLE CHANNEL PLATE

FROM COLLECTOR

TO STORAGE TANK

BRAZED CHANNEL PLATES

TO COLLECTOR

FROM STORAGE TANK

SunMaxx Solar

Figure 4-19. The stacked plate design of a brazed plate heat exchanger creates a series of parallel channels through which HTF and DHW can flow without ever coming into contact with each other.

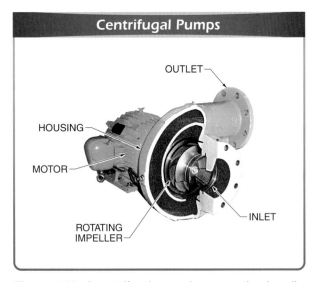

Centrifugal Pumps

OUTLET

HOUSING

MOTOR

ROTATING IMPELLER

INLET

Figure 4-20. A centrifugal pump has a rotating impeller inside a housing.

Pumps are classified by the material of their construction, their flow rate, and the amount of static, or maximum, head. *Pump flow rate* is the amount of liquid a pump can move without any restrictions and is measured in gallons per hour (gal./hr) or liters per hour (L/hr). *Static head* is the maximum vertical height a pump can push a liquid and is measured in feet or meters.

Pump Construction

Pump bodies are made of cast iron, stainless steel, or bronze. Cast iron pumps should only be used with glycol because the inhibitors prevent the corrosion that would occur if the pump were used with only water. Stainless steel and bronze are more resistant to corrosion, so they are typically used in potable water applications. However, stainless steel and bronze pumps tend to be more expensive than cast iron pumps.

Pump Curves

For a given pump, as the head height increases, the flow rate decreases. This relationship is typically nonlinear and is represented by a pump, or performance, curve. A *pump curve* is the relationship between flow rate and head. **See Figure 4-21.** The curve varies by pump size, power, and manufacturer.

It is important that the pump is correctly sized for the system and is able to maintain the proper flow rate for the head. For most residential systems, the size of the pump is not an issue. This is because most types of solar water heating systems are completely filled with fluid. The pump needs to only circulate the fluid and overcome the friction of moving the fluid through the piping and components. However, a drainback system is an exception.

A drainback system is not completely filled when the system is not operating, and the pump must push the fluid up into the empty piping and collector. Therefore, a more powerful pump may be needed for drainback systems. A multispeed pump may also be used. The multispeed pump in a drainback system operates at a higher speed when it is first energized to force the fluid up into the empty piping and collectors. It then slows down once the fluid is fully circulating through the system.

Pump Applications

Powerful centrifugal pumps are used in large residential and commercial applications. Complex solar water heating systems may require the installation of multiple pumps. **See Figure 4-22.** The pumps used in the majority of residential solar water heating systems are small centrifugal pumps called circulators.

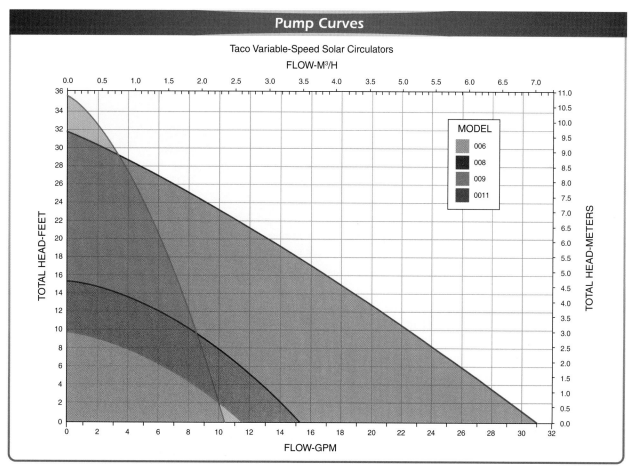

Taco, Inc.

Figure 4-21. For a given pump, as the head height increases, the flow rate decreases. This relationship is typically nonlinear and is represented by a pump, or performance, curve.

Multiple Pumps

Alternate Energy Technology

Figure 4-22. Complex solar water heating systems may require the installation of multiple pumps.

A *circulator* is a line-mounted centrifugal pump used to move HTF or DHW through system piping and solar collectors. **See Figure 4-23.** Circulators may be integrated into pump stations or may be stand-alone components. Stand-alone circulators are usually installed between isolation flanges. An *isolation flange* is a combination isolation ball valve and companion flange for circulators. Isolation flanges allow pumps to be easily serviced or replaced without the need to drain the system.

Circulators typically run on 120 VAC power, but circulators that run on DC power are also available. AC circulators may be hardwired into the existing electrical power system or wired with a plug and plugged into an electrical outlet. A licensed electrician should be consulted if there are any questions related to powering a circulator.

Facts

In order for a circulator to operate correctly, the circulator manufacturer designates the proper mounting positions for installation. Failure to follow the manufacturer's instructions can lead to loss of system efficiency or circulator failure.

Circulators

Taco, Inc.

Figure 4-23. Circulators that move HTF or DHW through system piping and solar collectors may be integrated into pump stations or may be stand-alone components.

DC Circulators. A *DC circulator* is a small magnetic drive pump that can be powered by a photovoltaic module in a closed-loop solar water heating system. **See Figure 4-24.** DC circulators run on either 12 V or 24 V. These pumps can be used in solar water heating systems, individual space heating zones, and individual radiant floor loops. However, they often lack the power of AC circulators and must be properly sized for the system. Also, a battery or house-current back-up system may be needed when the photovoltaic panel cannot supply enough power to run the pump effectively. This may cause a loss of overall system efficiency and reliability.

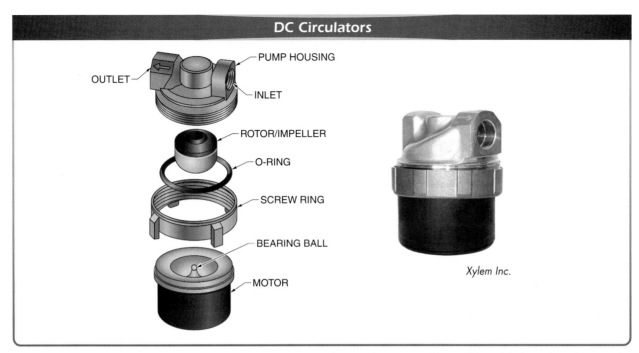

DC Circulators

OUTLET

PUMP HOUSING

INLET

ROTOR/IMPELLER

O-RING

SCREW RING

BEARING BALL

MOTOR

Xylem Inc.

Figure 4-24. A DC circulator is a small magnetic drive pump that can be powered by a PV module in a closed-loop solar water heating system.

OPERATIONAL CONTROLLERS

The operational controller is the brain of a solar water heating system. An *operational controller* is an electronic device used to monitor and control the HTF in a solar water heating system. **See Figure 4-25.** Operational controllers are primarily used for active systems, but some passive systems may use them as well. The operational controller monitors the temperatures at various locations within the system such as the storage tank and the collector outlet. Using this data and the parameters programmed into the controller, the controller energizes and deenergizes the circulators or pumps. An operational controller can be a stand-alone component or integrated into a pump station that includes the controller, pump, and other components and fittings in a single module.

The three basic types of operational controllers used in the solar water heating industry include timer controllers, photovoltaic controllers, and differential controllers. A *timer controller* is a basic operational controller that switches an electrically powered device on and off based on the time of day. A *photovoltaic (PV) controller* is an operational controller that uses energy generated in the form of direct current electricity via a PV module or panel to energize the pump of a solar water heating system. A *differential controller,* also known as a differential temperature controller, is an operational controller used to energize an electrically

powered component based on a setpoint difference between two temperature values. The type of control device that should be used for a specific system depends on the type and sophistication of the solar water heating system and building owner needs.

Operational Controllers

Figure 4-25. An operational controller is an electronic device used to monitor and control the HTF in a solar water heating system.

VALVES

There are many types of valves used in solar water heating systems. A *valve* is a device used to regulate fluid or gas flow within a system. Valves are used to turn the fluid (liquid or gas) flow on and off or to regulate the direction, pressure, and/or temperature of a fluid within the system. Valve bodies are available as cast bronze or cast iron. In general, valve bodies for 2″ and smaller valves are manufactured from cast bronze with bronze internal components. Valve bodies for 2″ and larger valves are typically manufactured from cast iron and have bronze internal components. All valves used in a solar water heating system must be designed for use with the high temperatures within the solar water heating system.

Shutoff Valves

Valves turn fluid flow on and off or regulate the direction, pressure, and/or temperature of a fluid. A *full-way valve,* also known as a shutoff valve, is a valve designed to be used in its fully open or fully closed position. The disk and seat of full-way valves, such as gate valves, may be damaged if the valve is used to throttle fluid flow. In general, full-way valves installed on water supply piping are the same size as the pipe on which they are installed.

Ball Valves. A *ball valve* is a valve in which fluid flow is controlled by a ball that fits tightly against a resilient (pliable) seat in the valve body. As the ball is turned, water flows through a port in the ball. **See Figure 4-26.** A ball valve requires only a 90° rotation of the handle to open or close. Ball valves are full-way valves that can be used for throttling fluid flow and are common in solar water heating system piping and water supply piping (instead of gate valves or globe valves).

RESOL

Actuator-controlled two- and three-way ball valves are used to automatically control the flow of fluid in solar water heating systems.

Three-Way Ball Valves. A *three-way ball valve* is a ball valve used in diverter applications suitable for fluid transfer in certain solar water heating systems or as a mixing valve in hydronic systems. Three-way ball valves may be manually or electronically controlled. The rotation of the ball determines the direction of fluid flow. **See Figure 4-27.**

Globe Valves. A *globe valve* is a valve used to control fluid flow by means of a pliable disk that is compressed against a valve seat surrounding the opening through which water flows. **See Figure 4-28.** Due to the internal water passage configuration, fluid flowing through the valve changes direction several times. This results in turbulence, resistance to fluid flow, and pressure drops in the system. Globe valves are recommended on installations requiring frequent operation, throttling, and/or a positive shutoff when closed, including plumbing fixture supply pipes. Globe valves must be installed with the flow direction arrow pointing downstream.

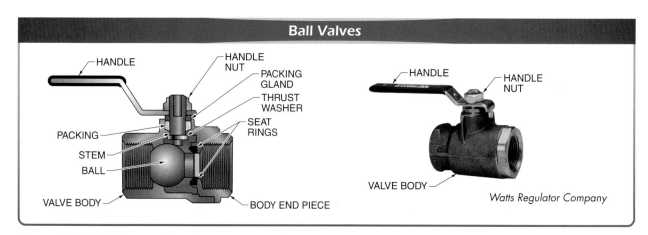

Figure 4-26. Ball valves require only a 90° rotation of the handle to open or close.

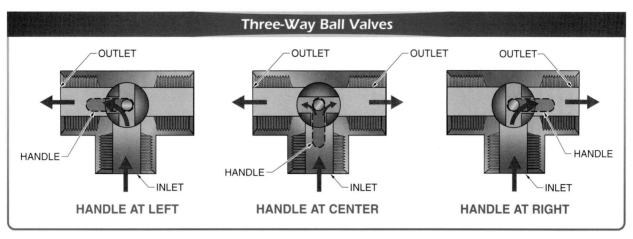

Figure 4-27. Three-way valves can direct flow to the left, right, or both depending on the position of the lever.

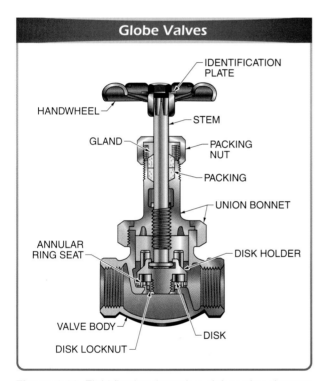

Figure 4-28. Fluid flowing through a globe valve changes direction several times, which results in turbulence, resistance to fluid flow, and pressure drops.

Gate Valves. A *gate valve* is a full-way valve used to regulate fluid flow in which a threaded stem raises and lowers a wedge-shaped disk that fits against a smooth machined surface, or valve seat, within the valve body. **See Figure 4-29.** Gate valves are typically installed in piping systems where they remain completely open or completely closed most of the time such as on each side of a water meter. When the wedge-shaped disk is retracted from the seat, the gate valve permits a straight and unrestricted fluid flow. Gate valves may use a split-wedge disk, solid-wedge disk, flexible-wedge disk, or double disk.

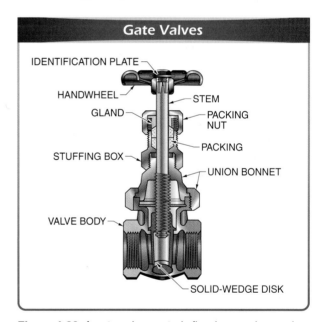

Figure 4-29. A gate valve controls flow by opening or closing a wedge-shaped disk.

Balancing Valves

A *balancing valve* is a valve designed to regulate the rate of fluid flow through a solar water heating system to meet system requirements. The flow rate is controlled by turning the flow rate limiter, which is a ball valve in most cases. **See Figure 4-30.** Most balancing valves are combined with a flow meter to form a single component. These components are installed on many pump stations.

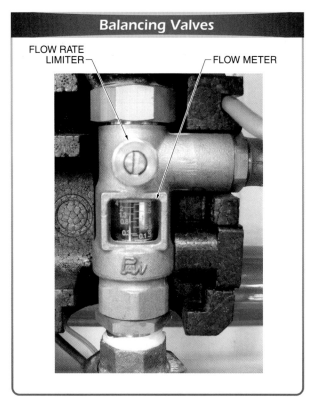

Figure 4-30. A balancing valve is a valve designed to regulate the rate of fluid flow through a solar water heating system to meet system requirements.

Pressure-Reducing Valves

A *pressure-reducing valve,* also known as a pressure-regulating valve, is an automatic device used to convert high and/or fluctuating inlet water pressure to a lower or constant outlet pressure. Pressure-reducing valves are installed near the water meter on a building water service to reduce excessive water main pressure. A pressure-reducing valve has an adjustment screw to adjust the outlet pressure. A strainer must be installed with a pressure-reducing valve to prevent dirt and debris from entering the valve mechanism, although many pressure-reducing valves contain an integral strainer. **See Figure 4-31.** The strainer should be checked and cleaned whenever the pressure-reducing valve or water meter is serviced. Pressure-reducing valves should be serviced annually by checking for leaks.

Thermostatic Mixing Valves

The temperature of the water heated by solar collectors can easily exceed the maximum safe temperature for residential use. Many building codes limit the temperature of the hot water supplied for residential use

and require that thermostatic mixing valves be used. A *thermostatic mixing valve (TMV),* also known as a tempering valve, is a valve that blends hot water with cold water to produce a safe and constant-temperature hot water flow at the point of use. **See Figure 4-32.**

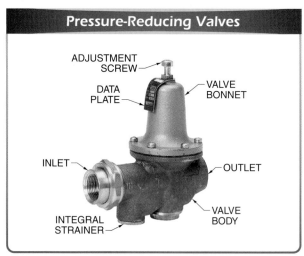

Watts Water Technologies, Inc.

Figure 4-31. Pressure-reducing valves are automatic devices used to convert high and/or fluctuating inlet fluid pressure to low or constant fluid pressures.

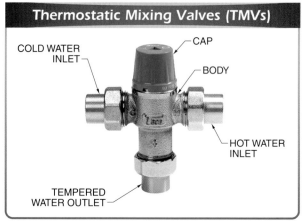

Taco, Inc.

Figure 4-32. TMVs blend hot water with cold water, which results in a safe and constant-temperature hot water flow at the point of use.

Check Valves

A *check valve* is a valve that permits fluid flow in only one direction and closes automatically to prevent backflow. Check valves react automatically to changes in the pressure of the fluid flowing through the valve and close when pressure changes occur.

Check valves are available as spring check, swing check, and lift check valves. **See Figure 4-33.** The most commonly used valve in a solar water heating system is the spring check valve. This is due to its simple design, ability to prevent thermosiphoning, and ability to be installed in any orientation since it does not rely on gravity for closure.

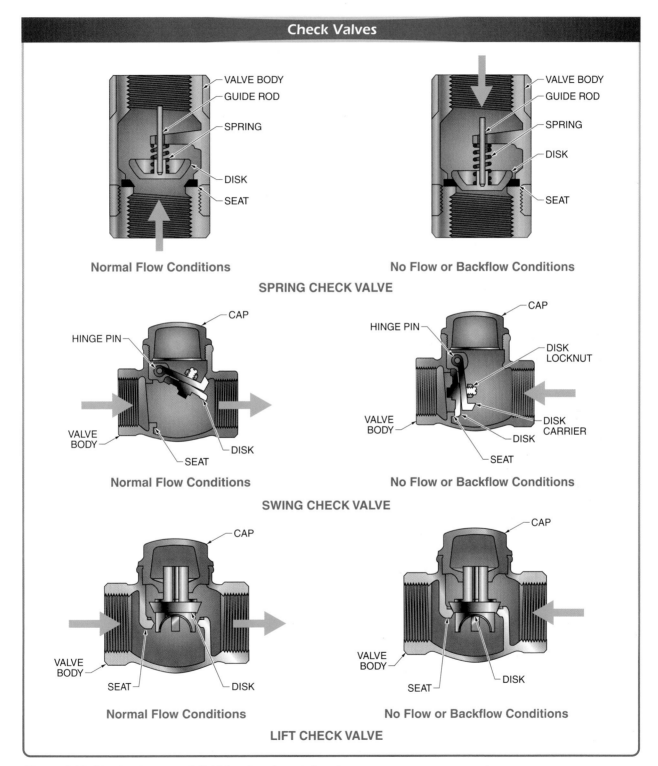

Figure 4-33. Check valves permit fluid flow in only one direction.

Spring Check Valves. A *spring check valve,* also known as a silent check valve, is a check valve that prevents backflow through the use of a conical or cylindrical brass disk held in place by a spring. Under normal operating conditions, the normal fluid flow pushes against the disk and compresses the spring. The valve remains open as long as there is sufficient flow to compress the spring. When the pressure drops or the direction of flow reverses, the spring decompresses and holds the disk against the valve seat. Spring check valves should not be used in low flow rate applications such as PV circulator systems.

Swing Check Valves. A *swing check valve* is a check valve that prevents backflow through the use of a hinged disk within the valve body. Under normal operating conditions, fluid flows straight through the valve and holds open the hinged disk. When backflow occurs, the hinged disk swings down into position. Swing check valves provide little resistance to fluid flow. They are commonly used with gate valves in installations in which fluids are moving at a low velocity and there is seldom a change in fluid flow direction.

Lift Check Valves. A *lift check valve* is a check valve that prevents backflow through the use of a disk that moves vertically within the valve body. Under normal operating conditions, fluid pressure forces the disk from its seat, which allows fluid to flow. When backflow occurs, the disk drops onto its seat, and stops the backflow. Lift check valves have a high resistance to fluid flow due to the fluid passageway within the valve body. Lift check valves are commonly used with globe and angle valves when frequent changes in flow direction can occur. Lift check valves are used in water supply systems but may also be used in drainage systems.

Drain Valves

A *drain valve* is a valve that contains a globe valve and hose threads and is used to drain or flush tank or system piping. Drain valves are installed on water heaters, on storage tanks, and in system piping. **See Figure 4-34.** A drain valve can be a separate component such as a boiler drain, integrated by the manufacturer into a water heater or storage tank, or combined with other components in the form of a multivalve.

Automatic Drain Valves. An *automatic drain valve* is a specialized drain valve placed at the bottom of a solar panel array that allows fluid in the array to drain when it is not in use. Draining the solar panel array prevents the fluid from overheating or freezing. An automatic drain valve is located within a climate-protected area. Automatic drain valves are rarely used today due to their unreliability. However, older existing systems may be equipped with them, so installers must be aware of these valves in case such systems require service.

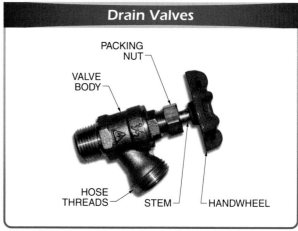

Watts Water Technologies, Inc.

Figure 4-34. Drain valves are installed on water heaters, on storage tanks, and in system piping to allow draining or flushing of the solar water heating system.

Pressure Relief Valves

A *pressure relief valve (PRV)* is a safety device that is used to relieve excessive pressure in a solar water heating system. Excessive system pressure can damage collectors and components. They can also cause property damage or injuries. Relief pressure on most PRVs is factory set by the manufacturer and ranges from 25 psi to 150 psi. **See Figure 4-35.** However, field-adjustable PRVs are available.

Atmospheric Vacuum Breakers

An *atmospheric vacuum breaker (AVB),* also known as a vacuum breaker, is a device that allows air to enter a piping system to facilitate the draining of HTF out of the piping when necessary. It consists of a body, a float-type check valve member that operates on atmospheric pressure, and an air vent that closes when the device is pressurized and opens when the inlet pressure is atmospheric. **See Figure 4-36.**

Pressure Relief Valves (PRVs)

PRV

Figure 4-35. A PRV is a safety device that is used to relieve excessive pressure in a solar water heating system.

Facts

An atmospheric vacuum breaker is a backflow prevention device designed to prevent back-siphonage. Backflow prevention devices are installed to protect a potable water supply from contamination due to a cross connection. The proper operation of backflow prevention devices is critical for health and safety reasons. Many state and local jurisdictions require that the devices be tested annually. A vacuum relief valve is not designed or approved as a backflow prevention device and should not be used as such.

Vacuum Relief Valves

A *vacuum relief valve* is a plumbing valve that is used to automatically allow air into the piping system should a vacuum occur. **See Figure 4-37.** There are several different styles of vacuum relief valves, but they all function the same. Positive system pressure and a spring hold a float tight against a seat. When negative pressure occurs, such as when a drain valve is opened somewhere in the piping loop, the negative pressure pulls on the float and compresses the spring. This unseats the float and allows air to enter the piping system. Air entering the system allows fluid to leave the piping more quickly.

Atmospheric Vacuum Breakers (AVBs)

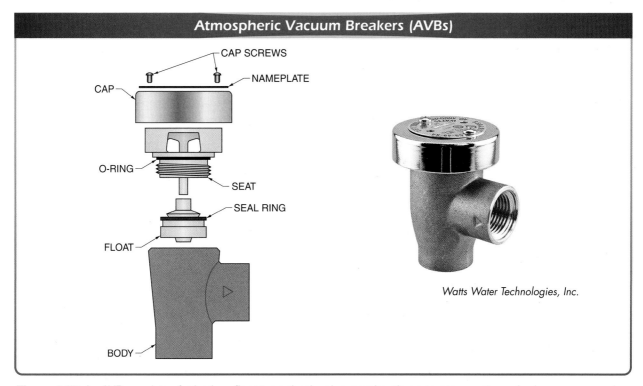

Watts Water Technologies, Inc.

Figure 4-36. An AVB consists of a body, a float-type check valve member that operates on atmospheric pressure, and an air vent that closes when the device is pressurized and opens when the inlet pressure is atmospheric.

Air Vents

Manually or automatically operated air vents allow trapped air to bleed from solar water heating piping and collectors. Manual air vents allow the air to be purged by the installer during the HTF filling process. Air vents should be installed vertically at the highest point in the system. This process requires an installer on the roof to manually operate and observe the vent while another installer fills the piping system below. An automatic air vent allows the piping to be filled without manually opening the vent. An *automatic air vent* is a valve that is controlled by a float that opens to allow air to escape from a closed piping system. **See Figure 4-38.** Either type of air vent can be used with direct systems. However, only manual air vents should be used with indirect systems.

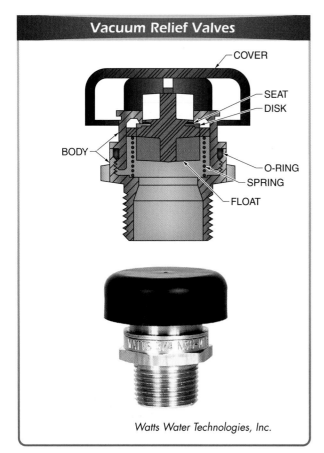

Figure 4-37. A vacuum relief valve is a plumbing valve that is used to automatically vent a system should a vacuum occur.

Vacuum relief valves are used to aid in the manual or automatic draindown of solar water heating systems, drain pool collectors, and prevent water heaters from collapsing when water is siphoned out of the tank. Vacuum relief valves are not designed or approved for use as backsiphon or backflow preventers.

AIR REMOVAL COMPONENTS

Trapped air in a solar water heating system can prevent fluid from circulating properly through the solar collectors. This results in reduced heating capacity. Air in the piping system can also cause excessive noise and a pump condition known as air lock. *Air lock* is a condition that occurs when trapped air surrounds a pump impeller and does not allow the pump to push any liquid through a system. Air removal components allow the manual or automatic removal of air from the solar water heating system.

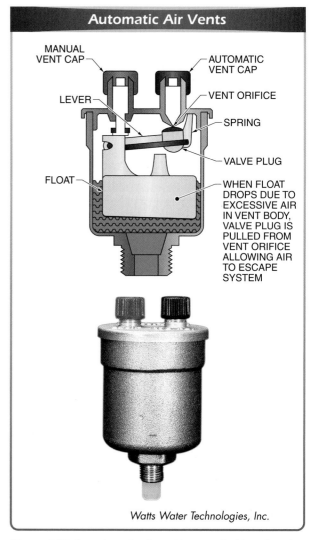

Figure 4-38. An automatic air vent is controlled by a float that opens to allow air to escape from a closed piping system.

Air Separators

Entrained air in solar water heating systems can cause noise, corrosion, and pump malfunctions. Removing the air is accomplished by using an air separator. An *air separator* is a solar water heating component that uses mechanical means, such as baffles or screens, to separate entrained air from HTF. They may also function as air vents. **See Figure 4-39.** Air separators must be installed vertically and should be located before the pump on the collector return piping to eliminate air bubbles that could cause the pump to malfunction. There are several air separator types that can be used on vertical or horizontal piping.

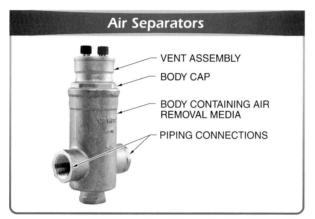

Air Separators

- VENT ASSEMBLY
- BODY CAP
- BODY CONTAINING AIR REMOVAL MEDIA
- PIPING CONNECTIONS

Watts Water Technologies, Inc.

Figure 4-39. An air separator is a solar water heating component that uses mechanical means, such as baffles or screens, to separate entrained air from HTF.

METERS AND GAUGES

Solar water heating systems must be monitored in order to maintain efficiency, and it must be verified that the systems are functioning properly. Meters and gauges allow installers to verify the rate of the fluid flow in the piping system, level of fluid in the piping system and drainback tanks, and temperature and pressure in the piping system and solar storage tanks. Meters and gauges should be installed at locations that allow easy viewing by the installer, service technician, and system owner.

Flow Meters

A *flow meter* is a solar water heating component that is used to measure the flow rate of fluid in piping. The types of flow meters commonly used in solar water heating systems include variable-area, inline, and electronic flow meters. Regardless of the type of meter used, meters must be installed in the correct direction of flow. **See Figure 4-40.**

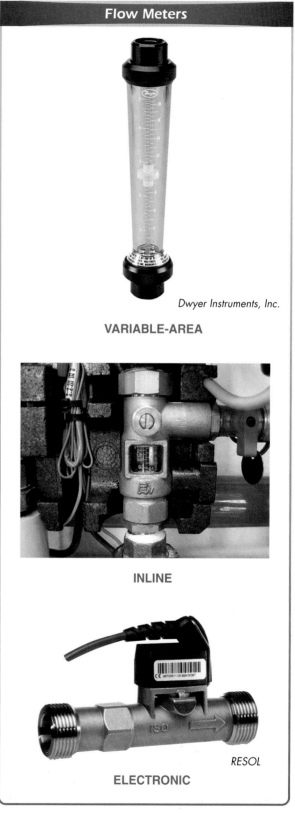

Flow Meters

Dwyer Instruments, Inc.

VARIABLE-AREA

INLINE

RESOL

ELECTRONIC

Figure 4-40. Flow meters are used to measure the flow rate of fluid in solar water heating systems.

Variable-Area Flow Meters. A *variable-area flow meter* is a flow meter that maintains a constant differential pressure and allows the flow area to change with the flow rate. A variable-area flow meter measures flow using the amount of resistance created by a float as it changes the area (size) of the flow path. As fluid flows through the tube, the float rises and changes the flow area. The movement of the float varies depending on the resistance against the flow. The most common type of variable-area flow meter is the rotameter.

A *rotameter* is a variable-area flow meter that consists of a clear tapered tube and a float with a fixed diameter. Because they are clear, they are sometimes used in place of sight glasses to determine the level of fluid in drainback tanks. The top of the rotameter is installed at the same level as the top of the tank. When the pump is deenergized, the fluid in the collectors and piping above the tank flow back into the drainback tank and the solar loop piping. The fluid level indicated in the rotameter will correspond to the fluid level in the tank.

Inline Flow Meters. An *inline flow meter* is a flow meter installed in a piping system that measures the displacement of a tapered piston or float. In some inline flow meters, the piston is held into place with a calibrated spring. During system operation, the greater the fluid flow rate, the higher on the scale the piston or float moves. Inline flow meters are not as large as rotameters, so they cannot be used as sight glasses. Some inline flow meters have an integrated flow rate limiter that can be used to adjust the flow rate in the system. Inline flow meters are commonly used in solar pump stations.

Electronic Flow Meters. An *electronic flow meter* is a flow meter that uses electronic sensors to measure fluid flow in a system. Electronic flow meters are connected to operational controllers and often can relay fluid temperature as well as fluid flow. They have several advantages over other types of flow meters. For example, electronic flow meters contain no moving parts, so there is no cause for mechanical wear on the sensors, and there is little chance that the meters will become blocked due to debris in the piping systems. Also, they do not need to be placed where they must be visible since they send electronic signals directly to the controllers via long cables. This allows installers increased flexibility when piping solar water heating systems. One disadvantage of electronic flow meters is that they typically cannot be repaired and therefore must be replaced.

Sight Glasses

A *sight glass* is a plumbing component that is composed of a transparent tube through which the level of liquid in a drainback tank or piping can be visually checked. **See Figure 4-41.** A sight glass is installed at the level of the drainback tank so that the fluid level can be determined. A site glass can be integrated into the drainback tank by the manufacturer or added during piping installation by the installer. The sight glass must be made of materials that are able to withstand the high temperatures that occur in the solar loop.

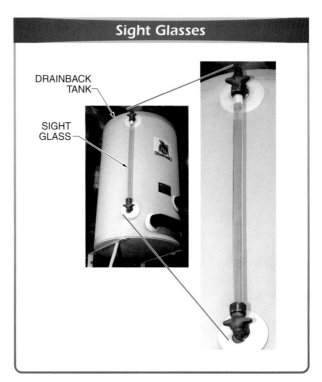

Figure 4-41. A sight glass is a plumbing component that is composed of a transparent tube through which the level of liquid in a drainback tank or piping can be visually checked.

Temperature and Pressure Gauges

Temperature and pressure gauges are installed in solar water heating systems to easily monitor the operating condition of the systems. These gauges may be separate components installed at specific locations by the installer or they may be integrated into a pump station. **See Figure 4-42.** Temperature and pressure gauges should be installed for easy viewing. Some systems may use a combination temperature-pressure gauge that has two scales. The top scale measures the pressure of the system, while the bottom scale indicates the temperature of the fluid.

Temperature and Pressure Gauges

TEMPERATURE

PRESSURE

COMBINATION

Watts Water Technologies, Inc.

Figure 4-42. A temperature-pressure gauge indicates the temperature and pressure of a fluid.

PIPING

There are several types of piping suitable for conveying water and other liquids in solar water heating systems from one location to another. Metal piping, such as copper tube and corrugated stainless steel tubing (CSST), is suitable for use in almost every part of the solar water heating system due to its ability to withstand high temperatures and resist corrosion. Plastic pipe, such as PEX, PEX-AL-PEX, PVC, and CPVC, does not have corrosion issues but is not capable of handling the high heat generated in the solar loop. Plastic pipe is more suitable for lower temperature applications, such as conveying DCW and DHW, hydronic heating, and swimming pool heating, since the maximum temperatures of the systems are within the acceptable range for plastic piping.

Local plumbing codes specify the type of piping material that can be used for each plumbing system. Local plumbing codes are based on local conditions such as soil types, ground conditions, local rainfall, and frost or freezing conditions. The local plumbing code should be consulted to ensure that only code-approved material is being used for plumbing systems.

Copper Tubing

Copper tubing is the most commonly used piping material in solar water heating systems. It is lightweight and easy to handle. It is manufactured in hard tubing lengths of 20′ and in soft, relatively flexible rolls in lengths up to 100′. Hard tubing should be used when grading the collector piping is critical, such as in drainback systems, and in open visible areas, such as mechanical rooms. It is easy to install and support hard copper tubing at a specific grade, plus the installation looks straight, neat, and professional. Rolled copper tubing can be used when grading is not necessary and where it is not visible such as in attic spaces. It is also ideal for use underground.

Copper tubing suitable for solar use is manufactured in three wall thicknesses: Type M, L, and K. During the manufacturing process, drawn and annealed copper tubes are permanently stamped every 18″ with the tube type, manufacturer name or trademark, and the country of origin. In addition, drawn copper tubing is identified with a colored stripe and lettering. Type M, identified by red markings, is the thinnest tube and thus the least expensive. Because of its thinness, Type M copper tubing is not manufactured in soft rolls. Type L, identified by blue markings, is the next larger size in wall thickness and is typically used for potable water piping. Type K, identified by green markings, is the thickest copper tube and thus the most expensive. It is typically used for high-pressure systems and underground burial.

All three tube types are manufactured to the same nominal or standard diameters in sizes ⅜″, ½″, ¾″, 1″, 1¼″, 1½″, 2″, 2½″, 3″, 4″, and up to 12″. For any given diameter of copper tubing, the outside diameter (OD) of copper tube types M, L, and K is ⅛″ larger than the nominal or standard size. The inside diameter of copper tube is determined by the wall thickness. For example, the outside diameter of ½″ type K and L copper tube is ⅝″ (0.625″). The inside diameters of ½″ type K and L copper tubes are 0.527″ and 0.545″, respectively. **See Figure 4-43.**

When copper tube is joined to another metal pipe of a different material, a dielectric union must be used. A *dielectric union* is a pipe fitting used in between two dissimilar metal piping materials. Failure to use a dielectric union could result in a galvanic reaction which can cause corrosion. The dielectric union has an internal gasket that keeps the dissimilar metals from touching and causing premature corrosion. **See Figure 4-44.**

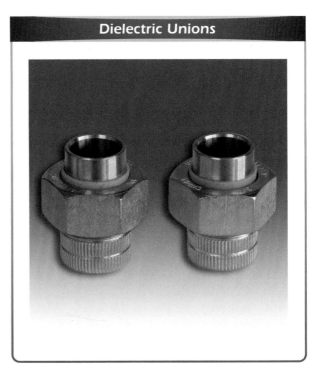

Dielectric Unions

Figure 4-44. A dielectric union separates two dissimilar metal pipes to prevent premature corrosion.

Sizes and Weights of Copper Tubing									
Nominal Sizes*	Outside Diameter*	Inside Diameter*			Wall Thickness*			Weight†		
	Types K-L-M	Type K	Type L	Type M	Type K	Type L	Type M	Type K	Type L	Type M
¼	0.375	0.305	0.315	0.325	0.035	0.030	0.025	0.145	0.126	0.106
⅜	0.500	0.402	0.430	0.450	0.049	0.035	0.025	0.269	0.198	0.145
½	0.625	0.527	0.545	0.569	0.049	0.040	0.028	0.344	0.285	0.204
⅝	0.750	0.652	0.666	0.690	0.049	0.042	0.030	0.418	0.362	0.263
¾	0.875	0.745	0.785	0.811	0.065	0.045	0.032	0.641	0.455	0.328
1	1.125	0.995	1.025	1.055	0.065	0.050	0.035	0.839	0.655	0.465
1¼	1.375	1.245	1.265	1.291	0.065	0.055	0.042	1.04	0.884	0.682
1½	1.625	1.481	1.505	1.527	0.072	0.060	0.049	1.36	1.14	0.940
2	2.125	1.959	1.985	2.009	0.083	0.070	0.058	2.06	1.75	1.46
2½	2.625	2.435	2.465	2.495	0.095	0.080	0.065	2.93	2.48	2.03
3	3.125	2.907	2.945	2.981	0.109	0.090	0.072	4.00	3.33	2.68
3½	3.625	3.385	3.425	3.459	0.120	0.100	0.083	5.12	4.29	3.58
4	4.125	3.857	3.905	3.935	0.134	0.110	0.095	6.51	5.38	4.66
5	5.125	4.805	4.875	4.907	0.160	0.125	0.109	9.67	7.61	6.66
6	6.125	5.741	5.845	5.881	0.192	0.140	0.122	13.9	10.2	8.92
8	8.125	7.583	7.725	7.785	0.271	0.200	0.170	25.9	19.3	16.5
10	10.125	9.449	9.625	9.701	0.338	0.250	0.212	40.3	30.1	25.5
12	12.125	11.315	11.565	11.617	0.405	0.280	0.254	57.8	40.4	36.7

* in in.
† in lb/ft

Figure 4-43. Copper tubing is available in many standard sizes.

Corrugated Stainless Steel Tubing

Corrugated stainless steel tubing (CSST) was originally designed for use as natural gas piping. However, it has been adapted for use as piping for solar water heating systems. CSST is flexible tubing that allows easy installation in difficult areas such as attics and ceilings. **See Figure 4-45.** One drawback of CSST is that, due to its flexibility, it is difficult to eliminate sags in the material unless it is fully supported or used in vertical lines. Therefore, care should be taken when using it for drainback systems.

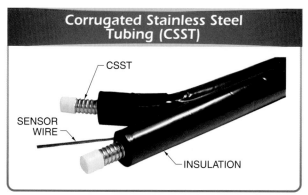

Bosch Thermotechnology Corp.

Figure 4-45. CSST is flexible and easy to install.

CSST is manufactured in various lengths up to 200′ rolls and in ½″, ⅝″, ¾″, and 1″ diameters. CSST for solar water heating use is often preinsulated and may also contain sensor wiring. CSST is typically rated for temperatures up to 300°F (150°C), but it can withstand temperatures up to about 350°F (177°C) for short periods of time. The manufacturer should be consulted regarding the specifications of a particular product.

Underground CSST Installations. Some manufacturers place the CSST, sensor wire, and additional insulation into a 3″ to 6″ high-density polyethylene (HDPE) corrugated pipe that is used for underground applications. The additional insulation and the HDPE pipe ensure a moisture-free barrier once the line set is buried. This type of line set is simple and results in watertight underground installations.

Cross-Linked Polyethylene Pipe and Fittings

Cross-linked polyethylene (PEX) is a thermosetting plastic made from medium- or high-density cross-linkable polyethylene that is used for water service piping and cold and hot water distribution piping. PEX tubing and fittings are available in sizes ranging from ¼″ to 2″. **See Figure 4-46.** Tubing is available in straight lengths of 20′ and coils of 100′, 300′, 400′, 500′, and 1000′. PEX tubing is often used in domestic cold and hot water distribution and hydronic floor heating applications. PEX tubing is manufactured in a variety of colors, but red PEX is typically used for DHW and blue PEX for DCW.

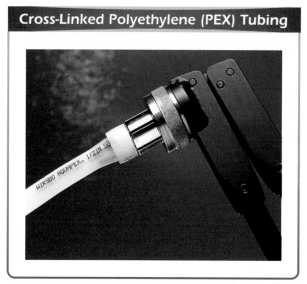

Uponor Wirsbo

Figure 4-46. PEX tubing is manufactured in long coils that are easy to cut and connect.

PEX-AL-PEX Tubing

PEX-AL-PEX tubing is PEX tubing that has a layer of aluminum sandwiched between an inner and outer layer of cross-linked polyethylene. The aluminum layer of PEX-AL-PEX tubing allows for greater operating pressure than conventional PEX tubing. PEX-AL-PEX holds its shape when bent and is subject to less thermal expansion than conventional PEX.

PEX-AL-PEX is often used for hydronic heat piping because the aluminum layer also acts as an oxygen barrier. The oxygen barrier is beneficial because most plastic piping is permeable to oxygen. Oxygen can cause corrosion in the hydronic heating system components that contain iron or steel. The PEX-AL-PEX used for hydronic applications is often colored orange.

Polyvinyl Chloride Pipe and Fittings

Polyvinyl chloride pipe and fittings are commonly used for swimming pool solar water heating applications. *Polyvinyl chloride (PVC)* is a plastic piping material used for swimming pool solar water heating, sanitary drainage and vent piping, aboveground and underground stormwater drainage, water mains, and water service lines. PVC pipe and fittings arc joined by solvent cement.

PVC pipe is available as schedule 40 or 80 pipe. Schedule 40 PVC pipe is standard-weight pipe, and the walls of schedule 80 pipes are approximately one-third thicker than schedule 40 pipes. Schedule 40 PVC pipe and fittings are white. Schedule 80 PVC pipe and fittings, commonly used for industrial pressure applications, are dark gray in color.

PVC pipe is typically manufactured using the extrusion process in a variety of sizes ranging from 1¼″ to 6″ in diameter and in 10′ and 20′ lengths. PVC pipe up to 16″ in diameter is available for use in underground drainage piping. PVC pipe and fittings have outstanding physical properties including excellent corrosion and chemical resistance. PVC pipe and fittings must not be used to store and/or convey compressed air or other compressed gases.

The installation of PVC piping for swimming pool solar water heating systems involves installing the solar loop only. Schedule 40 PVC pipe and fittings are primarily used for the solar loop of a swimming pool system. Temperatures involved in this system are about 90°F (32°C) at the very high end. Schedule 40 PVC pipe is suitable for temperatures up to 140°F (60°C). Schedule 80 PVC pipe is thicker pipe and is used primarily in higher pressure commercial installations. **See Figure 4-47.**

Chlorinated Polyvinyl Chloride Pipe and Fittings

Chlorinated polyvinyl chloride pipe and fittings are commonly used for hot and cold water distribution systems. *Chlorinated polyvinyl chloride (CPVC)* is a cream-colored thermoplastic material specially formulated to withstand higher temperatures than other plastic materials and used for potable water distribution, corrosive industrial fluid handling, and fire suppression systems. **See Figure 4-48.** CPVC hot- and cold-water distribution systems are typically rated for 180°F at 100 psi of pressure. CPVC pipe is joined by solvent cement.

Polyvinyl Chloride (PVC) and Chlorinated Polyvinyl Chloride (CPVC) Pipes

Schedule 40

Nominal Pipe Size*	Outside Diameter*	Minimum Wall Thickness*	Nominal Inside Diameter*	Weight† PVC	Weight† CPVC
½	0.840	0.109	0.622	0.16	0.17
¾	1.050	0.113	0.824	0.21	0.23
1	1.315	0.133	1.049	0.32	0.34
1¼	1.660	0.140	1.380	0.43	0.46
1½	1.900	0.145	1.610	0.51	0.55
2	2.375	0.154	2.067	0.68	0.74
2½	2.875	0.203	2.469	1.07	1.18
3	3.500	0.216	3.068	1.41	1.54
4	4.500	0.237	4.026	2.01	2.20
5	5.563	0.258	5.047	2.73	
6	6.625	0.280	6.065	3.53	3.86

Schedule 80

Nominal Pipe Size*	Outside Diameter*	Minimum Wall Thickness*	Nominal Inside Diameter*	Weight† PVC	Weight† CPVC
½	0.840	0.147	0.546	0.20	0.22
¾	1.050	0.154	0.742	0.27	0.30
1	1.315	0.179	0.957	0.41	0.44
1¼	1.660	0.191	1.278	0.52	0.61
1½	1.900	0.200	1.500	0.67	0.74
2	2.375	0.218	1.939	0.95	1.02
2½	2.875	0.276	2.323	1.45	1.56
3	3.500	0.300	2.900	1.94	2.09
4	4.500	0.337	3.826	2.75	3.05
5	5.563	0.375	4.813	3.87	
6	6.625	0.432	5.761	5.42	5.82

* in in.
† in lb/ft

Figure 4-47. PVC piping is commonly used for swimming pool solar water heating applications.

CPVC pipe is available in sizes ranging from ¼″ to 12″ and 10′ in length. CPVC pipe for plumbing systems is manufactured using the extrusion process in sizes ranging from ½″ to 2″ copper tube size (CTS). Industrial CPVC pipe is manufactured using the extrusion process in sizes ranging from ¼″ to 12″ in schedule 40 and schedule 80 wall thicknesses.

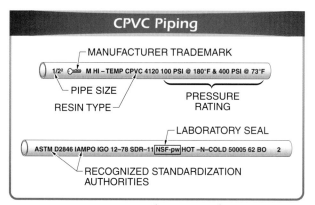

CPVC Piping

MANUFACTURER TRADEMARK

1/2² ⊂⊐ M HI – TEMP CPVC 4120 100 PSI @ 180°F & 400 PSI @ 73°F

PIPE SIZE

RESIN TYPE

PRESSURE RATING

LABORATORY SEAL

ASTM D2846 IAMPO IGO 12–78 SDR–11 [NSF-pw] HOT –N–COLD 50005 62 BO　2

RECOGNIZED STANDARDIZATION AUTHORITIES

Plastic Pipe and Fitting Association

Figure 4-48. CPVC pipe and fittings are commonly used in plumbing because they are easy to install.

PIPING AND FITTING INSULATION

The piping and fittings for solar water heating systems must be insulated to prevent heat loss. The insulation material used for solar water heating installations must be rated for high-temperature applications. Pipe insulation that is typically used for DHW applications may not be suitable for solar water heating piping. The insulation manufacturer should be consulted if there are any questions regarding the temperature ratings. Two common insulation materials used in solar water heating systems include fiberglass and closed-cell elastomeric foam. **See Figure 4-49.**

Some insulation materials are rated for interior and exterior use, while others are rated for interior use only. Fiberglass insulation is only for use on interior piping and fittings due to its tendency to absorb water. When insulation absorbs water, it loses its ability to properly insulate as designed. Rock wool can also be used indoors only, but it is far less common than fiberglass insulation.

Closed-cell elastomeric foam is piping and fitting insulation that is acceptable for interior and exterior applications. Since it is closed-cell foam, it does not absorb water. Closed-cell elastomeric foam typically has a higher insulation value per inch than fiberglass insulation but is more expensive.

Any type of insulation material will break down when exposed to ultraviolet (UV) radiation for long periods of time. For this reason, any insulation used on the exterior of a building must be covered with a suitable UV-resistant covering such as certain paints, PVC covers, or metal cladding. **See Figure 4-50.** All piping, regardless of the location, should also have labels and flow arrows.

Fittings must also be properly insulated to avoid excessive heat loss. The insulation should cover any voids or spaces of the piping and fittings. However, any gauges should be left exposed for easy viewing, and the pump should never be insulated.

FLASHING

Flashing is a piece of metal or thermoplastic placed at roof penetrations that are vulnerable to leakage. Flashing must be properly installed to protect the roof and the areas below the roof from water damage. Water that penetrates a roof can damage the roof framing and sheathing and possibly lead to mold or rot. The type of flashing used depends on the size of the penetrations, the type of roofing material, and the slope of the roof.

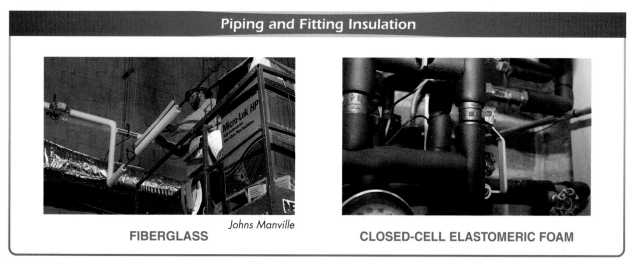

Piping and Fitting Insulation

Johns Manville

FIBERGLASS　　**CLOSED-CELL ELASTOMERIC FOAM**

Figure 4-49. Insulating piping increases the efficiency of heat transfer from the solar collectors to the storage tank.

UV Protection

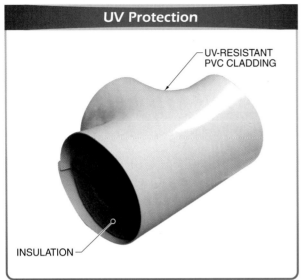

UV-RESISTANT PVC CLADDING

INSULATION

Stiebel Eltron, Inc.

Figure 4-50. Any insulation used on the exterior of a building must be covered with a suitable UV-resistant covering such as a PVC cover.

Pipe Flashing

Regardless of the type of system installed, there will be piping penetrations into the roof or exterior walls. In order to properly maintain the integrity of a roof or wall, pipe flashing should be used. Pipe flashing is available as a prefabricated component, but it may be fabricated in the field. The most commonly used type of prefabricated pipe flashing is made of metal or a thermoplastic plate with a rubber boot on top. **See Figure 4-51.** The metal used for pipe flashing is galvanized steel, aluminum, or copper. Also, some flashing may be colored to match the existing roofing material for aesthetic purposes. One type of pipe flashing used for copper piping is coolie cap flashing.

Coolie Cap Flashing. *Coolie cap flashing* is flashing typically made of copper and provides a watertight seal for solar piping and wiring. **See Figure 4-52.** The flashing material must be compatible with the solar piping. Coolie cap flashing may be purchased preassembled, or it may be fabricated onsite to conform to the slope of the roof. Coolie caps are comprised of a flashing plate with at least one small pipe section soldered in place. The solar piping is designed to run through the small section of pipe. A small copper cap is placed over the solar piping. Once the piping connections are complete, the cap is then soldered to the solar piping to seal the opening.

Pipe Flashing

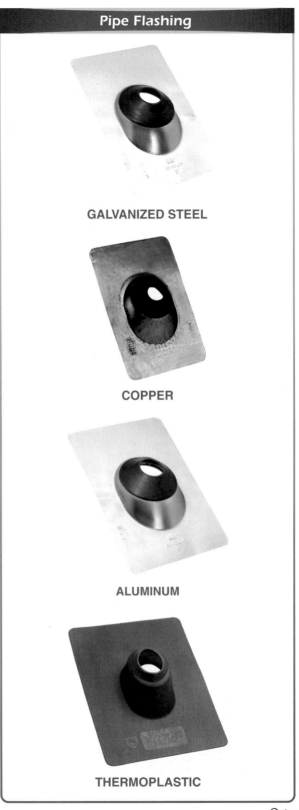

GALVANIZED STEEL

COPPER

ALUMINUM

THERMOPLASTIC

Oatey

Figure 4-51. Popular prefabricated pipe flashing is made of metal or a thermoplastic plate with a rubber boot on top.

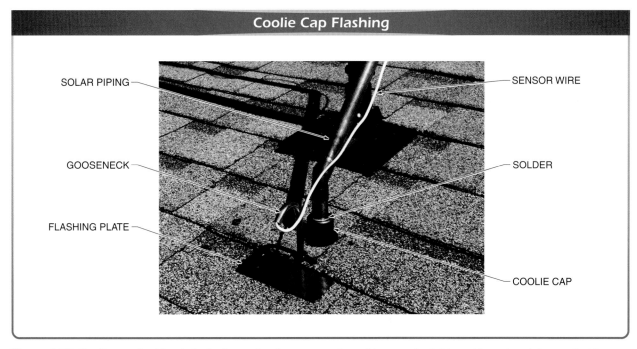

Coolie Cap Flashing

SOLAR PIPING

SENSOR WIRE

GOOSENECK

SOLDER

FLASHING PLATE

COOLIE CAP

Figure 4-52. Coolie cap flashing is typically made of copper and provides a watertight seal for solar piping and wiring.

A thinner section of pipe formed into a gooseneck shape is used for routing the sensor wire through the roof. The gooseneck prevents any kinking or snagging of the wire. Once the wire has been pulled through the piping and the appropriate connections are made, a compatible sealant is used to seal the end of the piping.

Collector Support Flashing

Solar collectors are most commonly mounted on south-facing roofs. In order to securely attach a collector and its support structure, penetrations must be made directly into the roof surface. Depending on the attachment method, either generic or specialized collector support flashing is used. **See Figure 4-53.** The flashing designs used for rooftop PV systems can also be used for solar water heating systems. When selecting the type of flashing to install, a qualified roofing contractor should be consulted to ensure that there are no water infiltration issues.

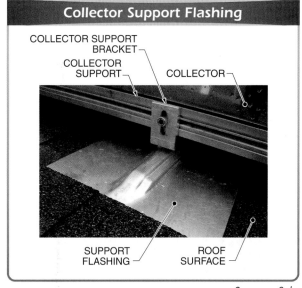

Collector Support Flashing

COLLECTOR SUPPORT
BRACKET
COLLECTOR
SUPPORT

COLLECTOR

SUPPORT
FLASHING

ROOF
SURFACE

Sunmaxx Solar

Figure 4-53. Collector support flashing provides protection for collector supports that are attached directly to a roof surface.

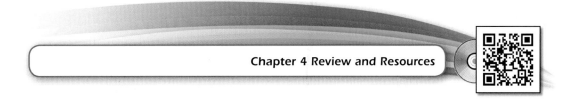

Chapter 4 Review and Resources

5

Solar Water Heating System Operation Fundamentals

Like other plumbing and mechanical systems, there are many different ways to install solar water heating systems. Each installer will also have a unique method and style of installation. However, these methods and styles must meet the basic requirements of the particular type of solar water heating system being installed as well as the local plumbing and mechanical codes. It is imperative that the installer know these requirements and the fundamental operation of these systems to ensure safe and proper installations.

SOLAR WATER HEATING SYSTEM TYPES

A solar water heating system can range from a small, simple system with few components to a large, complex system with multiple collectors and large storage capacities. Regardless of the size, a basic understanding of solar water heating system operation is required to properly configure a system. The system configuration for a particular installation is selected based on several factors such as the system size, location characteristics, heat transfer fluid (HTF) use, storage needs, and system cost. Solar water heating systems are classified by the methods used to create HTF flow, transfer collected heat to the domestic hot water (DHW), and protect the HTF from extreme temperatures.

Creating HTF Flow

A solar water heating system uses either nonmechanical or mechanical means to create HTF flow through the solar collector and storage vessel. Each method corresponds to a particular type of system. The use of nonmechanical means is indicative of a passive system, while the use of mechanical means is indicative of an active system.

Passive and active solar water heating systems are also categorized by how the systems transfer heat to the DHW. A direct solar water heating system, whether passive or active, uses the DHW itself as the HTF. In most cases, potable cold water is heated in the system. An indirect solar water heating system uses a heat exchanger to heat the DHW. The heat exchanger is either an internal or external heat exchanger and uses a fluid other than the potable cold water for heat transfer to the DHW. **See Figure 5-1.**

Passive Systems

A *passive system* is a solar water heating system that circulates HTF through a solar collector circuit or loop without the use of a pump or other mechanical means. A passive system relies on gravity or natural convection of the HTF to create circulation through the circuit or loop, which is a process called thermosiphoning. Passive systems contain no moving parts, pumps, or electronic controls, and thus use no electricity.

The most recognizable passive system configuration is a solar collector that also serves as the storage system. With this system, there is no piping other than the domestic cold water (DCW) inlet and the DHW outlet from the collector.

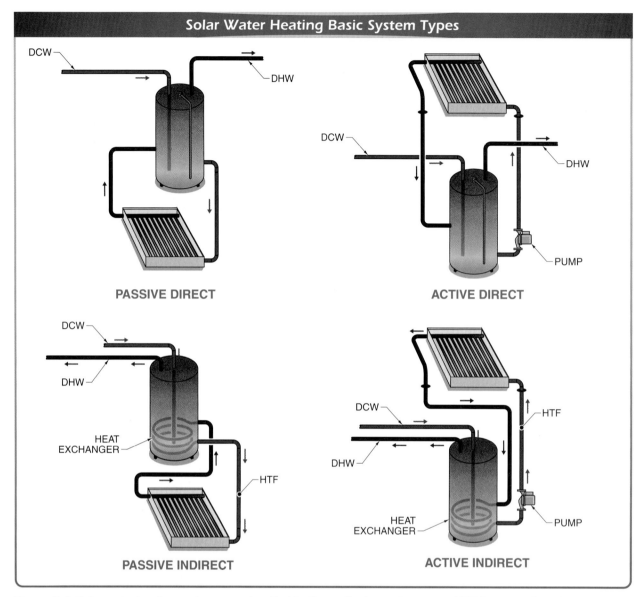

Figure 5-1. Solar water heating systems are classified by the methods used to create HTF flow, transfer collected heat to the DHW, and protect the HTF from extreme temperatures.

However, there can be other configurations of passive solar water heating systems. In some cases, the storage system is located above the solar collector. In either configuration as the solar-heated HTF in the storage system cools the heavier, cooler HTF naturally flows downward to the inlet of the collector. Then, as that HTF is heated within the solar collector, it becomes less dense and rises to the storage system inlet. This circulation of cool HTF being heated in the collector and then rising to the storage tank is the thermosiphon principle in action. Thermosiphoning within the solar loop continues until the temperature at the bottom of the storage tank is as hot as the temperature at the solar collector outlet. A *solar loop* is the piping that carries HTF into and out of a solar collector and to and from the storage tank or heat exchanger.

With another passive system configuration, the collector can be installed above the storage system as long as there is a method of moving the heated HTF to the storage system. In most cases, this method is normal DCW line pressure, which is fed into the solar collector inlet. The DCW rises as it is heated in the collector and is eventually circulated to the storage system located below the collector when DHW is used at fixtures.

Passive solar water heating systems were originally referred to as batch or thermosiphon systems. An example is the Climax batch system. As passive systems became more varied and sophisticated, the solar water heating industry created separate categories for the different configurations of passive systems.

Passive solar water heating systems are placed in three general categories within the SRCC OG-300 directory. These categories are integral collector storage (ICS) systems, thermosiphon systems, and self-pumping systems.

Integral Collector Storage Systems. An *integral collector storage (ICS) system* is a passive solar water heating system that consists of either a single storage tank or a series of multiple large tubes or tanks mounted within a glazed enclosure.

Single-Tank ICS Systems. A *single-tank ICS system* is a passive solar water heating system in which the collector is a single storage tank mounted within a glazed enclosure. The single-tank ICS system is still sometimes referred to as a batch system. The single-tank, or batch, system is the simplest type of solar water heating system. **See Figure 5-2.**

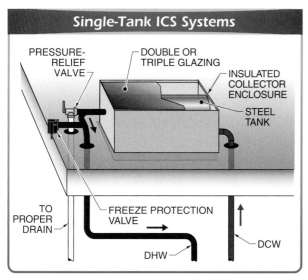

Figure 5-2. The single-tank ICS system is the simplest type of solar water heating system.

Multiple-Tank ICS Systems. A *multiple-tank ICS system* is a passive solar water heating system in which the collector is a series of multiple large pipes or small tanks mounted within a glazed enclosure. **See Figure 5-3.**

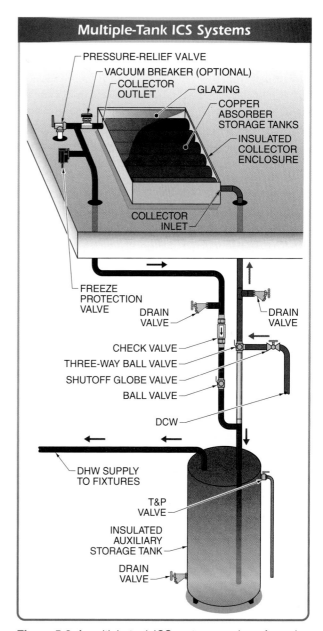

Figure 5-3. A multiple-tank ICS system consists of a series of large tubes or tanks mounted within a glazed enclosure. The collector serves as the primary storage system for the HTF, but the system may use an auxiliary storage tank to increase the hot water capacity.

In both types of ICS systems, the HTF, normally the DHW, is heated and stored within the tank located in the solar collector. The collector outlet feeds the DHW fixtures. Thermosiphoning occurs within the collector, while normal water pressure provides the DHW flow to fixtures. In some cases, gravity is used to convey the DHW to fixtures. Pumps and controls are not needed.

The collector may serve as the only storage system. However, an auxiliary storage tank may be used also to increase DHW capacity.

Thermosiphon Systems. A *thermosiphon system* is a passive solar water heating system that consists of a storage tank located above a collector. A thermosiphon system may be comprised of individual components that are connected together onsite or be a prepackaged system comprised of nonseparable components. The prepackaged system is often called a nonseparable thermosiphon system. **See Figure 5-4.** A *nonseparable thermosiphon system* is a passive solar water heating system that consists of a storage tank located above the collector and connected to the collector as one piece.

SunMaxx Solar

Figure 5-4. The storage tank of a nonseparable thermosiphon system is located above the collector to facilitate the thermosiphon process.

The HTF is heated within the collector and flows by natural convection up to the storage tank. Within the storage tank, the HTF flows downward into the collector and continues the thermosiphoning process as the HTF is used or cooled.

Note: There may be some confusion related to the terminology for passive solar water heating systems other than the self-pumping system. In some cases, the terms "batch," "ICS," and "thermosiphon" are used interchangeably to describe these types of passive systems. However, the terms used in this book should be used to describe these solar water heating categories.

Self-Pumping Systems. A *self-pumping system* is a passive solar water heating system that uses a phase change (liquid to vapor) or other passive means to cause the fluid in a collector to circulate and transfer heat from the collector to the storage system. This system is relatively new and circulates the HTF through the use of "geyser pumping."

With the self-pumping system, HTF is heated to the boiling point within the collector. As the HTF changes to steam, it pushes the HTF above it within the collector up and out of the collector into a small reservoir connected to the storage tank inlet. The reservoir is located above the inlet to the collector. HTF is circulated from the collector, to the storage tank, and back to the collector by the principle of fluid gravity balance. **See Figure 5-5.**

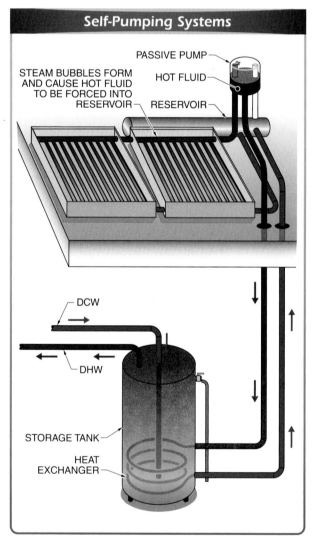

Figure 5-5. A self-pumping system uses a phase change or other passive means to cause fluid in the collector to circulate and transport heat to the storage system.

Passive System Disadvantages. One disadvantage of most passive systems is that they can be less efficient than other types of systems. This is due to the heat loss that may occur at night as cooler ambient temperatures cool the HTF stored in the rooftop storage systems. As a result of the heat loss, the HTF must be totally reheated the next day. Therefore, these systems work best if the DHW is needed in the later part of the day, because morning temperatures in the systems are cooler than afternoon temperatures. However, the self-pumping system is the one passive system that avoids the ambient heat loss because of its design.

Active Systems

An *active system* is a solar water heating system that relies on pumps or circulators to create HTF flow through its solar collector and storage system. An active solar water heating system does not rely on gravity or thermosiphoning to force the HTF through the system. **See Figure 5-6.** Unless the pump is continually running, controls and sensors must be used to activate and stop flow when necessary. Electricity is needed to operate the pumps and controls of an active system, but a photovoltaic panel can be used to reduce the cost of using electricity.

The advantage of using an active solar water heating system is that there is much more flexibility for component location than with the passive system. For example, collectors can be mounted on building roofs a considerable distance away and piped to a storage tank in a basement. Active systems also allow for more efficiency by allowing better management of heat transfer and collection within the solar loop. The disadvantage of an active system is the need for electricity and the use of moving parts that will eventually wear out or break down over time.

Heat Transfer Methods

A solar water heating system is also classified by its heat transfer method. Harvested heat is either directly transferred to the DHW circulating through the solar collector or indirectly transferred to the DHW by using a heat exchanger to extract the heat from HTF running in a closed loop through the collector. Therefore, solar water heating systems can be classified as either direct systems or indirect systems.

Direct Systems

A *direct system,* also referred to as an open system or open-loop system, is a solar water heating system that

uses potable DHW as HTF. In these systems, there is only one piping circuit or loop in which the HTF travels from the solar collector to the building or residence fixture supplies. There is no heat exchange system used other than the solar collector itself. The direct solar water heating system can be used in passive and active systems.

Active Systems

Solar Service Inc.

Figure 5-6. An active system uses a pump or circulator to force HTF through the solar collector and storage system.

The terms "open system" and "open-loop system" are used to indicate that the solar loop flows through the storage tank or water heater and supplies hot water to the DHW fixtures. This system is also "open" to the water supply to the storage tank or water heater. Normally this supply is the potable cold water.

Passive Direct Systems. A *passive direct system* is a solar water heating system that uses domestic potable water as an HTF but does not use pumps or other mechanical means to circulate the fluid through the collector. **See Figure 5-7.** In a passive direct system that does not use a separate storage tank, the single solar loop consists of the cold water supply piped to the inlet of the solar collector and then piped from the outlet of the collector and then to the DHW-supplied fixtures. A passive system that uses a separate storage tank consists of a single solar loop that is piped from the solar collector to the storage tank with a return line piped from the storage tank back to the collector. The storage tank outlet is then piped to the fixture supplies. In either case, heat exchangers are not used.

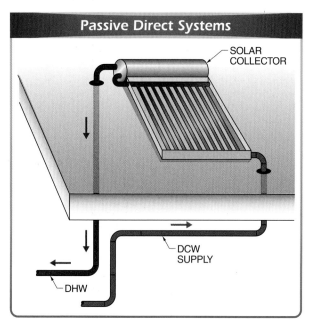

Figure 5-7. A passive direct system uses domestic hot water as the HTF, but does not use pumps or other mechanical means to circulate the fluid through the collector.

Active Direct Systems. An *active direct system* is a solar water heating system that uses domestic potable water as an HTF and employs a pump or circulator to circulate the fluid through the solar collector. Because the active direct system does not depend on the thermosiphon principle to move the HTF, the solar collector is usually installed above the storage tank and fixtures. **See Figure 5-8.** The active direct system also uses only one piping circuit to move the DHW from the collector to the storage tank and then to the fixture supplies. Heat exchange systems are not used in active direct systems.

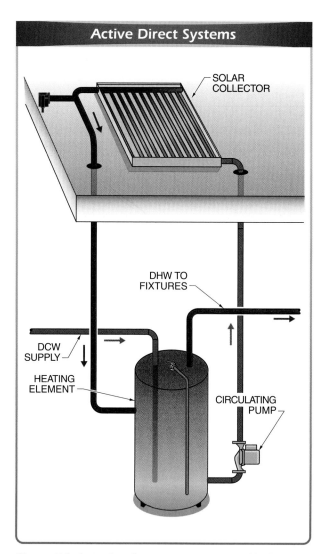

Figure 5-8. An active direct system uses potable DHW as the HTF and employs a pump or circulator to circulate the fluid throughout the solar collector.

Indirect Systems

An *indirect system*, also referred to as a closed system or closed-loop system, is a solar water heating system that uses a heat exchange system to transfer the harvested heat from the collector to the DHW. In these systems, the potable DHW is not used as the HTF in the solar collector. Usually, the HTF is a propylene glycol and water solution. Indirect systems use two loops or circuits of piping. One loop or circuit is used to transfer the heat from the HTF to the DHW and another to move the DHW to the points of use at the fixtures. Like direct systems, indirect systems can be passive or active.

The primary loop or circuit of the indirect solar water heating system consists of piping from the outlet of the solar collector to the inlet of the heat exchanger and from the return outlet of the heat exchanger to the inlet of the solar collector. The secondary loop or circuit is the piping that supplies the potable DCW to the inlet of the storage tank and the piping from the outlet of the storage tank to the DHW fixtures. The heat exchanger may be located inside or outside the storage tank.

The terms "closed system" and "closed-loop system" are used to indicate that the solar loop is closed to the storage tank and the potable cold water supply to the tank.

Passive Indirect Systems. A *passive indirect system* is a solar water heating system that uses an HTF and heat exchanger to tranfer heat to the DHW without the use of a pump or other mechanical means to circulate the HTF through the collector. In most cases, the stor-

age tank is mounted above the solar collector, which is piped to the storage tank heat exchange system. This is the primary loop or circuit. **See Figure 5-9.** The heat exchange system can be either an external or internal heat exchanger. The secondary loop or circuit consists of the potable cold water inlet, the storage tank, and the outlet piping to the DHW fixtures. Self-pumping systems are passive indirect systems. The collector is mounted above the storage system and is piped to a heat exchanger system.

Active Indirect Systems. An *active indirect system* is a solar water heating system that uses a pump or circulator to force an HTF that is not the DHW through the solar loop. **See Figure 5-10.** The heat transfer takes place in a heat exchanger. The solar loop or circuit is the primary loop consisting of the solar collector and piping connected to the inlet and outlet of the heat exchanger. The heat exchanger can be either an external or internal heat exchanger. The secondary loop, or circuit, consists of the potable cold water inlet, the storage tank, and the outlet piping to the DHW fixtures.

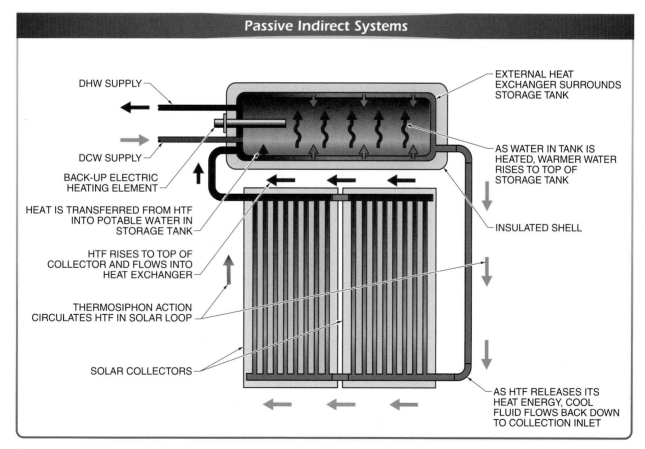

Figure 5-9. A passive indirect system is a solar water heating system that uses thermosiphoning to circulate an HTF that is not the DHW through the solar loop.

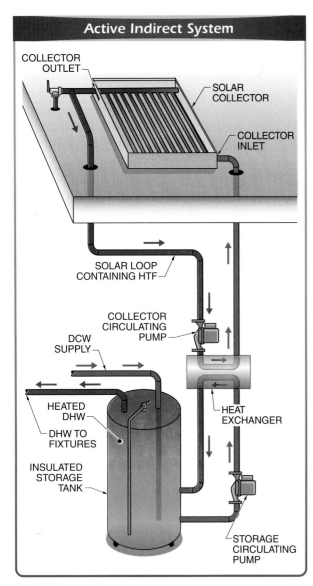

Active Indirect System

COLLECTOR OUTLET

SOLAR COLLECTOR

COLLECTOR INLET

SOLAR LOOP CONTAINING HTF

COLLECTOR CIRCULATING PUMP

DCW SUPPLY

HEAT EXCHANGER

HEATED DHW

DHW TO FIXTURES

INSULATED STORAGE TANK

STORAGE CIRCULATING PUMP

Figure 5-10. An active indirect system uses a pump or circulator to force an HTF that is not the DHW through the solar loop.

HEAT TRANSFER FLUID PROTECTION

A solar collector is installed where the aperture of the collector will receive the most solar radiation throughout the day. This requires that the collector be mounted on a roof or other area exposed to sunlight. Therefore, the collector will be exposed to ambient, or outside, air temperatures. In locations where there is no possibility for the ambient temperature to reach 32°F (0°C), there is no need to take any precautions to protect the HTF when it is not circulating in the system. However, if there is even a remote possibility for ambient air temperatures to reach the freezing point, the

HTF must be protected from freezing. Failure to adequately protect the system could result in damage to the collector, piping, and other system components. Damage to interior spaces caused by leaking or burst piping is also a possibility.

The protection of the HTF contained in a solar water heating system from freezing conditions is accomplished through the use of one or more of five fundamental methods. HTF freeze protection can be accomplished through the use of the following:

• thermal mass
• electrical heating elements
• antifreeze
• creating HTF flow
• draining down the system

Thermal Mass

Passive direct systems, such as ICS systems and batch systems, have large amounts of DHW in their storage tanks. The water in the tanks has a high thermal mass. *Thermal mass* is the capacity of a material to store thermal energy for extended periods. The thermal mass of the water prevents the storage tanks from freezing because it would take a long duration of freezing temperatures to freeze all of the DHW in the tanks. The use of thermal mass as freeze protection is only effective in milder climates where temperatures will not remain freezing for long periods. When using thermal mass for freeze protection, it is still important to protect the piping running to and from the collector.

Electrical Heating Elements

The solar water heating HTF, and in some cases the solar collector, can be protected from freezing by using electrical heating elements called heat tracing. *Heat tracing,* also known as heat taping, is an electrical heating element embedded in flexible cable that runs along the length of vulnerable piping and activates when the ambient temperature drops below a predetermined setpoint in order to keep the pipe from freezing. **See Figure 5-11.** Heat tracing alone may not generate enough heat to protect the collector. Therefore, other heating systems may be used to protect the collector.

The disadvantage of using heat tracing is a prolonged loss of power that would nullify the protection and place the system in danger of freezing. Heat tracing also consumes additional energy, which may reduce any energy efficiency gains from using a solar water heating system. Therefore, this form of HTF protection is rarely used today.

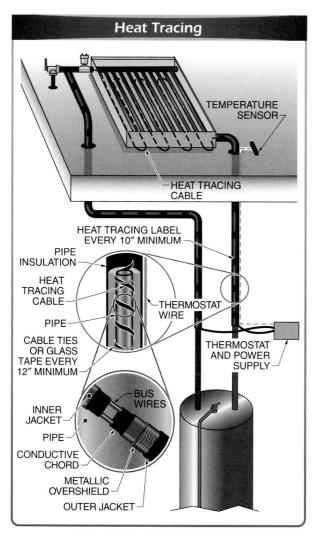

Figure 5-11. Heat tracing is an electrical heating element embedded in flexible cable that runs along the length of vulnerable piping. Heat tracing is activated when the ambient temperature drops below a predetermined setpoint.

Antifreeze

Indirect solar water heating systems usually use HTFs with freeze protection qualities to protect the systems. The most widely used HTF is a solution containing water and propylene glycol. Care must be taken to use the proper concentration for the lowest temperatures that may be encountered. The proper concentration allows the system to be protected even when the system is not circulating.

There are disadvantages to using antifreeze solutions. According to most plumbing codes, if a toxic HTF, such as ethylene glycol, is used, the heat exchange system must use a double-wall heat exchanger. This protects the potable DHW from possible contamination if the exchanger develops a leak. Another disadvantage is

that these antifreeze solutions become more viscous as temperatures drop. This changes the pump flow rates of the HTF. The change in flow rate due to temperature changes must be taken into consideration when sizing the pump in an active system to ensure that the HTF will attain the proper elevation to circulate through the system.

Creating HTF Flow

Regardless of whether the system is direct or indirect, passive or active, if the flow of HTF in the exposed solar loop and collector can be maintained, the HTF will not freeze. To create flow or to circulate the HTF and prevent it from freezing, a freeze protection valve or system pump can be used.

Freeze protection valves are used in direct systems to create a small flow of DHW through the solar loop. **See Figure 5-12.** The valve is mounted on the return side of the solar collector just before the penetration of the roof. The valve is set to open before freezing temperatures are reached, around 38°F to 42°F (3°C to 6°C), using either a spring-loaded thermostat or a bimetallic switch. The DHW pressure pushes a small amount of water, which drains onto the roof or ground, through the valve. A check valve is mounted below the freeze protection valve in the direction of flow so that water is drawn from the collector instead of the storage tank. The small flow of water through the valve maintains a slight flow of DHW through the solar loop and keeps the HTF from freezing.

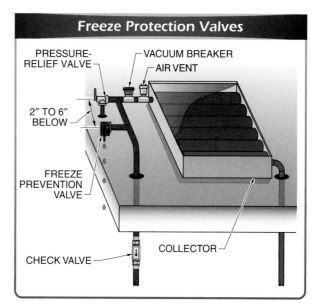

Figure 5-12. A freeze protection valve creates a small flow of DHW through the solar loop when the temperature drops near freezing.

Active systems, whether direct or indirect, can be protected from freezing simply by maintaining the flow of HTF through the solar loop with the use of the system pump or circulator. The system controls must include an outside temperature sensor that will activate the pump when freezing temperatures are sensed. If the HTF flow is maintained, the HTF will not freeze. The drawback of this method is that the loss of power for a prolonged period of time may place the system in jeopardy.

Draining Down the System

A solar water heating system can be protected from freezing by simply partially draining down the system. The complete removal of HTF from the collector and vulnerable piping in the solar loop is necessary for this to be successful. This can be accomplished by manually or automatically draining down the system or by using a drainback system.

Manual Draindown Systems. A *manual draindown system* is a solar water heating system that allows HTF to be drained out of the collector and vulnerable solar loop piping by an operator using isolation and drain valves when the ambient air temperature approaches freezing temperatures. **See Figure 5-13.**

A manual draindown system can be created by installing isolation valves on the supply and return lines from the solar collector and its piping. Drain valves must also be installed to facilitate HTF removal. The valves must be installed in an accessible area to make it as easy as possible to drain the system. The HTF can be captured to be reused in the system or, in the case of direct systems, the DHW could simply be drained away.

The major drawback of manual draindown systems is the human factor. Relying on the system owner to remember to drain the system when necessary is risky. Even a diligent owner may be out of the building or away from home when the system needs to be drained. This problem can be solved by installing an automatic draindown or drainback system.

Automatic Draindown Systems. An *automatic draindown system* is a solar water heating system that allows HTF to be drained out of the collector and vulnerable solar loop piping using a sensor-controlled automatic draindown valve. In this system, the HTF is drained away and not captured, requiring the system to be a direct system in which the potable DHW is the HTF. Once drained, the system must be refilled before it is used again. This is accomplished by using the normal potable cold water to fill the solar collector. A pump or circulator provides the water flow in the solar loop, and thus it is an active system.

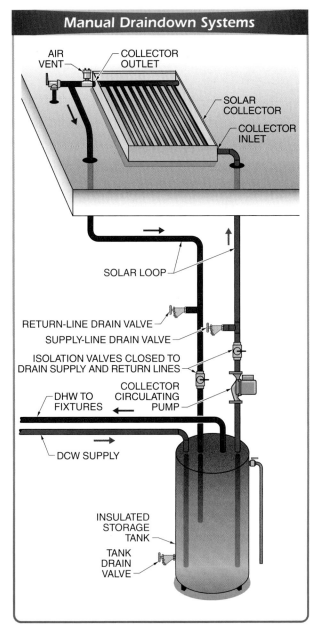

Figure 5-13. A manual draindown system requires the system owner to drain the HTF out of the system when the ambient air approaches freezing temperatures.

The operation of an automatic draindown system is relatively simple. A valve specially designed to automatically close the flow of the supply piping to the solar collector and to open a valve to drain the collector is installed in a nonfreezing area of the system. **See Figure 5-14.** The valve contains a temperature sensor that activates the valve ports when temperatures drop toward the freeze point.

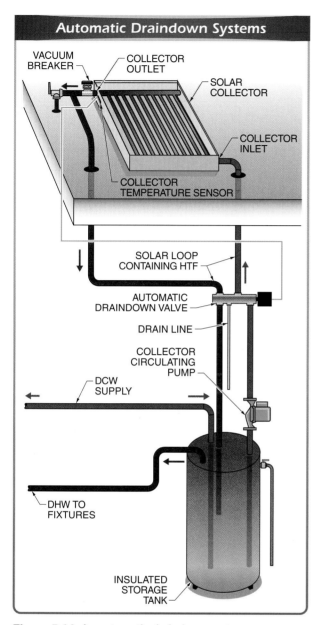

Automatic Draindown Systems

VACUUM BREAKER
COLLECTOR OUTLET
SOLAR COLLECTOR
COLLECTOR INLET
COLLECTOR TEMPERATURE SENSOR
SOLAR LOOP CONTAINING HTF
AUTOMATIC DRAINDOWN VALVE
DRAIN LINE
COLLECTOR CIRCULATING PUMP
DCW SUPPLY
DHW TO FIXTURES
INSULATED STORAGE TANK

Figure 5-14. An automatic draindown system uses a specially designed valve to automatically close the flow of the supply piping to the solar collector and to open a valve to drain the collector.

When the draindown valve is activated, the HTF within the collector is drained from the system. A vacuum breaker mounted above the collector allows air to enter the collector and facilitate the draining of the HTF. An air vent should also be mounted alongside the vacuum breaker to allow air within the collector to escape once the ambient temperature rises and the draindown valve activates the fill valve and closes the drain valve.

Although this is a simple system used to prevent the HTF from freezing, there are a few drawbacks to the system design. The DHW that flows through the collector is wasted each time the system is activated. That water may also freeze if it settles on the ground, causing a slipping hazard or other problems. The collector and its supply and return piping must be graded to drain completely or damage may occur. If the draindown valve or vacuum breaker malfunctions, the collector may not drain completely. If the valve is electronically controlled, it will not activate during a power outage. Both of these situations could result in damage to the system and/or the building.

The draindown system is rarely installed nowadays. It was a system designed and used primarily in the 1970s. However, an installer may be called upon to service one of these systems and should be familiar with its operation.

Heat Transfer Products, Inc.
Direct solar water heating systems require the proper HTF protection when they are installed where freezing temperatures may occur due to damage that may be caused by the possible freezing of DCW or DHW piping.

Drainback Systems. A *drainback system* is a solar water heating system that drains heat transfer fluid from the collectors into a storage tank located inside a building to protect the system from freezing temperatures. Drainback systems are designed to function in freezing conditions. However, when the system pump is deenergized by the system controller, the HTF is allowed to drain out of the solar loop. The drainback system is different from the draindown system in that the HTF is collected in a reservoir or tank and reused the next time the system is activated. Because the HTF must be elevated to fill the solar collector, the drainback system must use a pump. **See Figure 5-15.** Therefore, a drainback system is an active system.

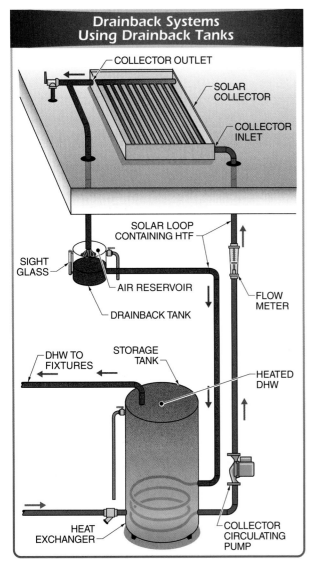

Figure 5-15. A separate drainback tank is used to contain the HTF when the controller shuts off power to the collector circulating pump.

The drainback system works by entraining an air reservoir within a tank in the solar loop. The tank can be either a dedicated drainback tank or the solar storage tank itself. **See Figure 5-16.**

When the HTF pump is activated, the HTF circulates from the tank through the collector and back into the tank. When the HTF pump is deactivated, HTF flow stops and the air within the tank begins to work its way up into the collector until the HTF in the collector and adjacent piping is drained back into the tank. **See Figure 5-17.**

For a drainback system to work correctly, the system must be properly designed and installed. The collector, or collectors, must be of the type capable of draining all

HTF inside the collector(s) such as a harp-style collector. The collector and HTF piping must be sloped, typically ¼″ per foot, to facilitate drainage and air fill of the collector. The storage tank can be pressurized or nonpressurized. However, there must be enough air space or air pad within the tank to accept the volume of the HTF from the collector plus 1 gal. in excess per collector for additional protection. This will enable the free flow of air and HTF when the pump is deactivated.

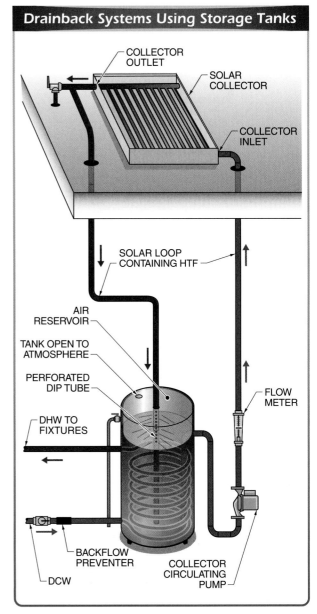

Figure 5-16. A solar storage tank can be used as a drainback reservoir, provided there is enough air space to hold the volume of HTF from the collectors plus 1 gal. in excess per collector for additional protection.

Drainback System Operation

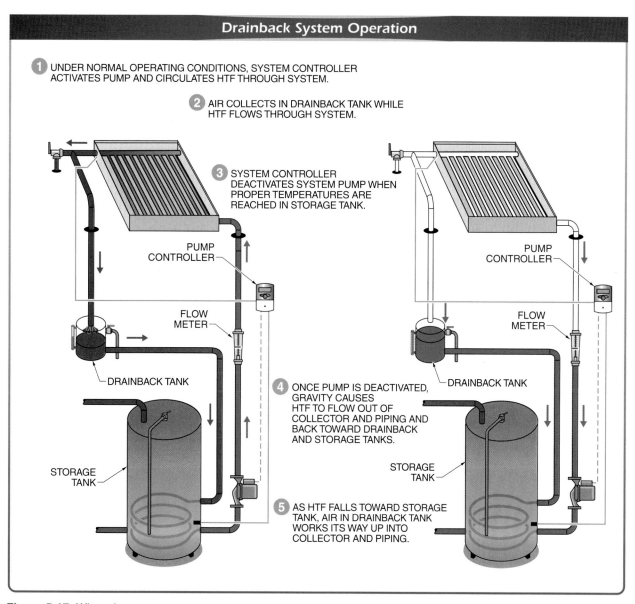

1 UNDER NORMAL OPERATING CONDITIONS, SYSTEM CONTROLLER ACTIVATES PUMP AND CIRCULATES HTF THROUGH SYSTEM.

2 AIR COLLECTS IN DRAINBACK TANK WHILE HTF FLOWS THROUGH SYSTEM.

3 SYSTEM CONTROLLER DEACTIVATES SYSTEM PUMP WHEN PROPER TEMPERATURES ARE REACHED IN STORAGE TANK.

PUMP CONTROLLER

FLOW METER

DRAINBACK TANK

4 ONCE PUMP IS DEACTIVATED, GRAVITY CAUSES HTF TO FLOW OUT OF COLLECTOR AND PIPING AND BACK TOWARD DRAINBACK AND STORAGE TANKS.

STORAGE TANK

5 AS HTF FALLS TOWARD STORAGE TANK, AIR IN DRAINBACK TANK WORKS ITS WAY UP INTO COLLECTOR AND PIPING.

Figure 5-17. When the proper storage temperature is reached in the storage tank, the pump controller deactivates the pump and causes the HTF to flow out of the collector and back toward the drainback and storage tanks.

There can be no check valves within the solar loop or in the HTF pump itself. Check valves will not allow the flow of air up into the collector and the HTF flow into the tank. The HTF pump must be sized correctly to ensure the HTF reaches the elevation of the collector. It must also be sized to provide the proper HTF flow through the collectors, especially from nonpressurized tanks. This flow rate is normally 2 fps.

The drainback system is typically an active indirect system that uses either an internal or external heat exchanger for HTF heat transfer to the DHW. In many cases, distilled water is used as the HTF because of its

very good heat transfer properties. The need for an HTF with antifreeze properties is not necessary because the HTF will drain into a tank installed within the heated areas of the building. However, a propylene glycol solution can be used for increased freeze protection.

The main advantage of the drainback system is that even if the system pump or power fails, the HTF will drain back into the reservoir, and the system will be protected. The disadvantage is that there are still mechanical components of the system, such as pumps and controllers, that must be maintained and eventually replaced.

HTF Protection from Overtemperature Conditions

In addition to protecting HTF from freezing conditions, the HTF must be protected from overtemperature conditions. Temperatures of the HTF can reach upward of 300°F (144°C) especially on high-insolation days when there is very little use of the DHW and the system becomes stagnant. *Stagnation* is a condition in which HTF no longer circulates through a solar collector. Stagnation is the result of the system controller deactivating the HTF pump because of high temperatures in the solar storage tank or because of pump failure. Failure to provide precautions for overtemperature conditions in solar water heating systems can cause component and building damage.

The protection of the HTF contained in a solar water heating system from overtemperature conditions can be accomplished through the use of one or more of five devices and systems. The five devices and systems used for HTF overtemperature protection are as follows:
- temperature relief valves
- drainback systems
- continual operation of HTF pumps
- heat dump systems
- steamback systems

Temperature Relief Valves. The simplest form of HTF overtemperature protection is the release of DHW from the solar storage tank through a temperature relief valve. This release of hot water allows cooler water into the system. Heat from the solar collector is transferred to the cooler water in the storage tank, controlling the temperature within the collector. This method is only viable in passive direct systems, such as the ICS systems, and has the disadvantage of wasting DHW to a drain or the ground.

Drainback Systems. The drainback system inherently protects the HTF from overtemperature conditions by draining the HTF back into the reservoir when the system pump is deactivated. The system pump will be deactivated when the upper-limit temperature setpoint is reached. When the system pump is deactivated, the HTF will drain from the solar collector into the reservoir, thus preventing the HTF from having an overtemperature condition.

Continual Operation of HTF Pumps. In active systems, the continual operation of the HTF pump may be enough to control the temperatures within the solar loop.

With active indirect systems, the pump is coupled with an HTF that has a high resistance to thermal degradation. This is often enough to prevent the HTF from having an overtemperature condition.

Heat Dump Systems. In cases where there is too much excess heat created by the solar water heating system to be controlled by other methods, a heat dump may be used to remove the excess heat. A *heat dump* is a solar water heating component that provides a method for heat to be transferred away from the solar collector. A heat dump consists of some form of additional heat exchange system that transfers heat to another source or to the atmosphere.

A heat dump can come in many forms. It is normally preceded by a three-way valve from the top of the solar collector. When overtemperature conditions occur, the valve opens to allow the HTF to flow to the heat dump. The heat dump could be as simple as a piping loop connected to a separate hydronic radiator coil. **See Figure 5-18.**

A heat dump can also be a proprietary component added to the solar water heating system such as a finned radiator capsule. The finned radiator capsule consists of a radiator section, pressure-relief valve, vacuum breaker valve, and HTF reservoir. **See Figure 5-19.** The finned radiator capsule can regulate the pressure in the solar water heating system by venting a small amount of the HTF into the reservoir section of the capsule. Any air bubbles that have risen to the top of the collector are driven out with the HTF and rise to the top of the fluid in the reservoir and out the vent.

The radiator section also helps lower the temperature of the HTF by transferring the heat to the surrounding air. As the system cools, the system temporarily develops negative pressure, which opens the vacuum breaker valve and causes the HTF to be drawn from the reservoir back into the system through a siphon tube. A fluid level switch located in the reservoir is used to notify the system owner if the level becomes low and requires refilling.

During stagnation, the antifreeze solution may begin to boil. Steam exiting the collector enters the radiator section of the capsule where the heat from the steam is transferred to the surrounding air. The transfer of heat to the surrounding air reduces the temperature of the steam to below the boiling point and causes the steam to condense. The condensed fluid then drains back to the collector. This process continues as long as the collector is stagnated, protecting the system from boiling dry or overheating.

Heat Dumps

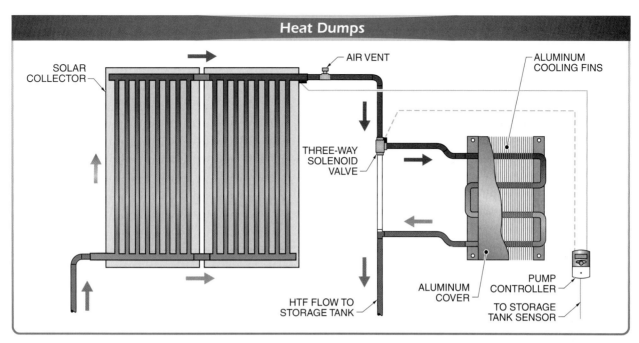

Figure 5-18. When the temperature in a storage tank reaches and exceeds a predetermined limit, the pump controller activates the three-way solenoid valve to redirect the HTF flow through the heat dump until the storage tank temperature drops below the set limit.

Finned Radiator Capsules

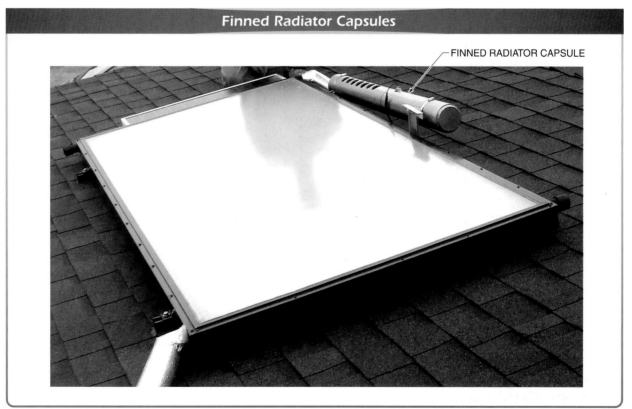

Butler Sun Solutions Inc.

Figure 5-19. A finned radiator capsule regulates the temperature of a solar water heating system by venting a small amount of HTF into the reservoir section of the capsule.

A heat dump may also consist of a separate heat exchange coil within another storage tank or a coil placed in a large tank or swimming pool. For large systems, the heat dump could be a large radiant system installed underground. The disadvantage of using a heat dump is the additional piping from the top of the collector to the heat dump, resulting in an increase in material and labor cost.

Steamback Systems. A steamback system is a solar water heating system that uses the steam pressure created by high temperatures within a collector to remove the HTF from the exposed areas of a solar loop. In a steamback system, an overtemperature condition within the solar collector causes the HTF to stagnate and reach boiling temperatures. This causes the water within the propylene glycol HTF solution to separate from the solution, boil, and evaporate. As the water evaporates into steam, it expands and pushes the remaining liquid out of the collector and downstream piping and into an expansion tank suitably sized and placed within the solar loop to collect the HTF. The expansion tank must be sized correctly to ensure that the solar loop does not reach pressures that will activate any pressure-relief valves. A steamback system is normally installed as part of an active indirect system. **See Figure 5-20.**

The solar collector(s) in the steamback system must be designed for steamback conditions, as not all collectors will be able to withstand these temperatures and pressures. They must also be installed so that the fluid will easily leave the collector and fill the expansion tank. In order to accommodate the added HTF being pushed into the tank, the expansion tank in a steamback system is significantly larger than a normally sized expansion tank for a solar water heating system. Piping must be installed to avoid sags and must consist of material and fittings designed for high temperatures. The HTF must be a water and propylene glycol solution.

The expansion tank in this system must be installed downstream of the pump or circulator. A check valve must be placed ahead of the branch connecting the expansion tank. This prevents the HTF from pushing through the pump and back into the storage tank, bypassing the expansion tank. The steamback system is used extensively in Europe and is now beginning to be used in North America.

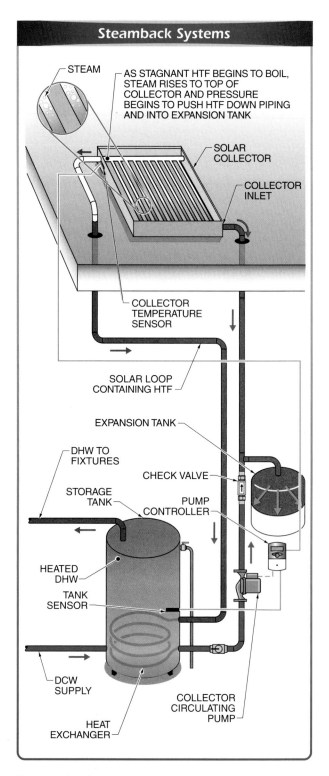

Steamback Systems

STEAM

AS STAGNANT HTF BEGINS TO BOIL, STEAM RISES TO TOP OF COLLECTOR AND PRESSURE BEGINS TO PUSH HTF DOWN PIPING AND INTO EXPANSION TANK

SOLAR COLLECTOR

COLLECTOR INLET

COLLECTOR TEMPERATURE SENSOR

SOLAR LOOP CONTAINING HTF

EXPANSION TANK

DHW TO FIXTURES

CHECK VALVE

STORAGE TANK

PUMP CONTROLLER

HEATED DHW

TANK SENSOR

DCW SUPPLY

COLLECTOR CIRCULATING PUMP

HEAT EXCHANGER

Figure 5-20. A steamback system uses the pressure created by water expanding into steam to push the remaining HTF out of the downstream piping and into an expansion tank. A check valve placed between the expansion tank and collector circulating pump protects the pump from damage due to excessive HTF temperature.

Facts

Steamback technology does not require expensive components or additional HTF that may be discharged during overheating. The system is safe, allowing the recovery of HTF from collectors during overheating or a power failure.

Combination Solar DHW and Space Heating Systems

Solar water heating systems can be designed not only to provide heat for DHW but also to assist in the comfort heating of the building. These systems can use heat generated by the solar DHW system to provide heat for hydronic space heating systems. A *combination system,* also called a combi system, is a solar water heating system that uses the heat harvested from collectors to heat the DHW supply and HTF used for other purposes.

The simplest type of combination system uses a separate heat exchange coil within the solar storage tank to transfer heat or to preheat HTF in a hydronic radiant floor system. **See Figure 5-21.** The second heat exchange coil piping from the solar storage tank connects to the primary loop of the hydronic radiant system. The heat transferred within this coil will either provide all of the heat for the radiant system or it will preheat the HTF and, through a three-way valve, be combined with additional HTF from the radiant system boiler to feed the radiant floor system.

Combination solar water heating systems can be used to heat many other types of fluid heating systems. There are numerous ways to combine systems using various components such as multiple storage tanks, multiple heat exchangers, pumps, and controls. Combination system designers are only limited by the efficiency of the solar water heating systems and their imagination. The goal is to use as much of the sun's radiant energy as possible for the maximum benefit of the consumer and the maximum efficiency of the solar water heating system.

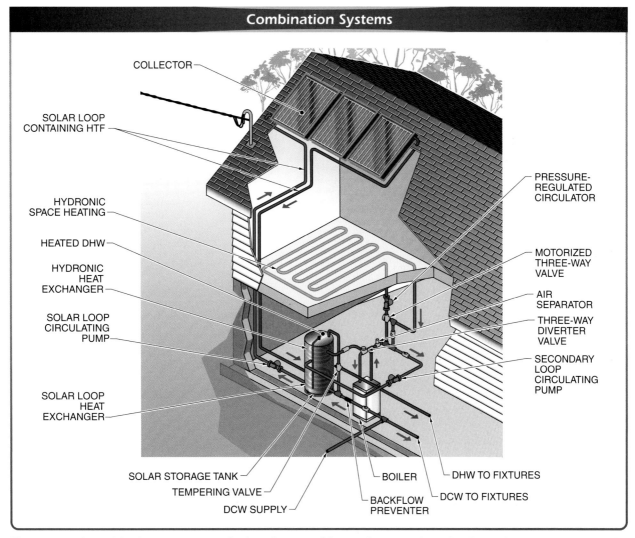

Combination Systems

COLLECTOR

SOLAR LOOP CONTAINING HTF

HYDRONIC SPACE HEATING

HEATED DHW

HYDRONIC HEAT EXCHANGER

SOLAR LOOP CIRCULATING PUMP

SOLAR LOOP HEAT EXCHANGER

PRESSURE-REGULATED CIRCULATOR

MOTORIZED THREE-WAY VALVE

AIR SEPARATOR

THREE-WAY DIVERTER VALVE

SECONDARY LOOP CIRCULATING PUMP

SOLAR STORAGE TANK

TEMPERING VALVE

DCW SUPPLY

BACKFLOW PREVENTER

BOILER

DHW TO FIXTURES

DCW TO FIXTURES

Figure 5-21. A combination system uses the heat harvested from collectors to heat the domestic water supply and HTF used for other purposes such as hydronic space heating.

Chapter 5 Review and Resources

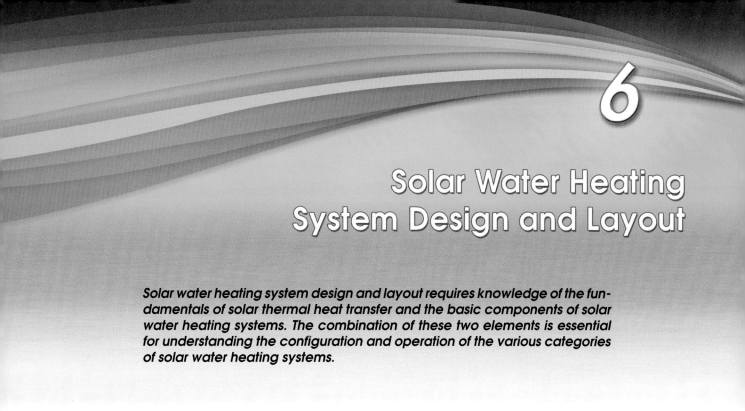

Solar Water Heating System Design and Layout

6

Solar water heating system design and layout requires knowledge of the fundamentals of solar thermal heat transfer and the basic components of solar water heating systems. The combination of these two elements is essential for understanding the configuration and operation of the various categories of solar water heating systems.

SOLAR WATER HEATING SYSTEM DESIGN

Solar water heating systems are designed and configured based on the system operation fundamentals that correspond to parameters at the location site such as amount of insolation, average temperatures, and building structure. Most importantly, they are designed based on customer needs, aesthetic preferences, and cost concerns. These systems may also be designed according to designer preference of system type based on experience and installation ease.

In most cases, the solar water heating system installer will not design the system. Except for the most basic solar water heating systems, the design of these systems should be left to solar thermal, plumbing, or mechanical system design professionals. Therefore, this book is not intended to instruct the solar water heating installer in the design of solar water heating systems. This book is intended to guide the solar water heating installer in the proper installation of solar water heating systems. However, the installer must be able to recognize and understand the various types and categories of solar water heating systems.

Complementing the OG-100 Solar Thermal Collector Certification Program, the Solar Rating and Certification Corporation (SRCC) has created a program that assists the solar water heating industry in organizing and classifying the wide variety of solar water heating system designs. It is essential that the solar water heating installer become familiar with the SRCC OG-300

Rating and Certification Program and Directory of Certified Solar Water Heating Systems. The following information about the SRCC process is from the SRCC website, www.solar-rating.org, on the Solar Facts page. It is reprinted with permission, and the copyright is owned by the Solar Rating and Certification Association.

SRCC OG-300 CERTIFICATION OF SOLAR WATER HEATING SYSTEMS

The OG-300 Rating and Certification Program for solar hot water systems integrates results of collector tests with a performance model for the entire system and determines whether systems meet minimum standards for system durability, reliability, safety, and operation. Factors affecting total system design, installation, maintenance, and service are also evaluated. The purpose of this solar water heating system certification and rating program is to improve performance and reliability of solar products. It gives suppliers the opportunity to submit their solar domestic hot water (SDHW) system designs to an open-ended review, encouraging them to produce the best products possible.

The OG-300 directory contains information about solar water heating systems that have been certified and rated by the SRCC. The information in the directory provides reliable and comparable data for solar water heating systems. The rating information is a helpful tool for comparing the efficiency of the various solar systems on the market. Remember, though, that not all solar energy systems are tested and rated as a package or a system. Many systems may have only the collectors rated by the SRCC. However, collector ratings cannot be compared with system ratings. All systems that have been certified by the SRCC will bear the SRCC label, which is assurance that an independent party has verified the performance and basic durability of the solar product. Copies of SRCC labels are available in the directory.

The directory contains descriptive information about solar systems, including performance information. Performance data relates to the energy output of the system. The SRCC performance information contained in the directory provides a way to compare the relative performance of different solar water heating systems, not the actual performance expected from a given system. This is because the collectors and systems are tested under standard laboratory conditions, which are certain to be different from those in the home.

SRCC ratings can be thought of as the MPG ratings for cars: a benchmark, but not necessarily the same performance that will be experienced. Performance, or energy output, is only one criteria used in choosing a solar water heating system. Quality of installation, cost, availability of service and parts, and the expected life of the equipment are also important points to consider. Equipment that is well-designed and well-built but poorly installed cannot perform according to the manufacturer's specifications.

Under the OG-300 Rating and Certification Program, systems must meet or exceed all criteria required to meet HUD Minimum Property Standards. Ratings include a simple way for consumers to compare expected annual performance and energy savings. Unique features of this certification program include evaluation of system design, safety, installation procedures, operations and maintenance materials, and system performance.

How Systems Are Certified Under the OG-300 Protocol

The process for rating and certifying solar water heating systems under the OG-300 protocol includes the following five steps for each system being rated:

1. The collectors, which are part of the system, must be tested and rated under the OG-100 protocol.
2. Passive systems in which the collector cannot be tested separately must be rated and certified under a system test protocol.

3. The complete specified system of collectors, tanks, pumps, motors, valves, piping, etc., is evaluated for essential elements related to the following:
   - system design
   - projected durability and reliability
   - safety
   - operation and service procedures
   - installation guidelines
   - operation and maintenance manuals

   During this evaluative process, the SRCC reviews the components, assemblies, and materials in the system for compatibility, drawing on nationally recognized resources such as the Recommended Requirements to Code Officials for Solar Heating, Cooling, and Hot Water Systems, which is jointly prepared by the Council of American Building Officials, Building Officials and Code Administrators International, Inc., International Conference of Building Officials, National Conference of States on Building Codes and Standards, Inc., and Southern Building Code Congress International, Inc.
4. Data from the OG-100 collector test, the system test, and review of design and installation guidelines is entered into a computer program called TRNSYS. The program projects system performance.
5. Numerical results of the design and installation review (step 3) and the TRNSYS evaluation are integrated and entered on a certification to the supplier. Data pertinent to solar distributors, retailers, installers, and homeowners is specified for printing on the rating pages. The rating pages that follow are a good means by which to compare systems and select the best one to meet needs.

When installing SRCC-rated systems, properly-trained installers are essential to ensure that the completed system will perform as efficiently as possible.

On-Site Inspection

Compliance with many of the installation requirements of OG-300 can be verified only by an on-site inspection made after installation. The SRCC may randomly inspect installed systems to verify that: (1) approved components have been installed; (2) an approved O&M manual has been provided; and (3) the installation conforms to the approved installation guidelines.

Thermal Performance Rating

The intent of the thermal performance rating is to present to consumers an easily understood comparison between solar domestic hot water systems and conventional hot water systems. Note that the performance any individual consumer will experience may differ due to location and hot water usage. Additional location specific information on the performance of SRCC-certified solar water heating systems is provided in the SRCC Directory of Annual Performance Ratings.

The thermal performance rating is based on the system design and performance projections derived from testing of the collector components used in the system, or from testing and evaluation of the system as a whole. The type of auxiliary system (e.g., gas or electric) used will have a large impact on the overall performance of the system. These differences arise because different types of auxiliary systems have varying standby losses and fuel conversion efficiencies.

Although the auxiliary system may affect solar system performance, in many cases, the solar output is mostly independent of the auxiliary system used. Because gas back-up systems have lower efficiencies and higher standby losses than electric systems, it should be expected that the entire system's (including back-up) performance will be lower, even if the solar output from both system types is equal.

The SRCC uses the solar energy factor (SEF) as its performance rating for solar domestic water heating systems. The SEF is defined as the energy delivered by the system divided by the electrical or gas energy put into the system. The SEF is presented as a number similar to the energy factor (EF) given to conventional water heaters by the Gas Appliance Manufacturers Association (GAMA) but with the exceptions noted in the Rating Parameters Section.

In this context, the solar fraction is the portion of the total conventional hot water heating load (delivered energy and tank standby losses) provided by solar energy. Note that an alternate definition for solar fraction is often used. In this alternate definition, solar fraction is the portion of the total water heating load (losses are not included) provided by solar energy. The alternate method of calculating solar fraction will yield higher solar fractions. Therefore, caution must be used when comparing the solar fraction for specific systems, inputs into energy codes (such as California's Title 24), or outputs from software (such as the F-chart) to ensure that the same calculation procedure for solar fraction has been used.

Fluid Classes

The American Water Works Association (AWWA) has created a classification system for heat transfer fluids (HTFs) that are used in heat exchangers. This classification system is intended to indicate the potential for contamination of the water supply during a heat exchanger failure and as a basis for the types of heat exchangers that should be used with these fluids. The rating system consists of the following three categories:
- Fluid Class I: Potable Heat Transfer Fluid (e.g., water)
- Fluid Class II: Non-Toxic Heat Transfer Fluid (e.g., propylene glycol)
- Fluid Class III: Toxic Heat Transfer Fluid (e.g., ethanol)

Freeze Tolerance

Each system has a freeze tolerance temperature specified by the system supplier. If the air temperature falls below this temperature, fluid in the system is likely to freeze. Unless a system is installed in a nonfreezing climate, every system must have an automatic mechanism, such as automatic draining, antifreeze fluids, or thermal mass, to at least partially protect it from freezing. For systems using water in portions of the system exposed to outdoor air conditions, the mass of the water itself, along with insulation, can provide limited protection from freezing, after which manual intervention may be required. For systems using water in the collector, the freeze tolerance temperature may actually exceed the freezing temperature of water because of radiative cooling by the sky. All systems must be able to be manually drained to protect them from extreme freezing conditions.

SRCC Certification Labels

All solar products certified by the SRCC are required to be labeled with an approved SRCC certification label within sixty (60) days of receipt of certification. The label should be on each system certified under the SRCC OG-300 protocol. **See Figure 6-1.**

SRCC OG-300 Solar Certification Label

	This product certified by: Solar Rating & Certification Corporation 400 High Point Drive, Suite 400 Cocoa, FL 32926 (321) 213-6037 www.solar-rating.org	Sample Solar Corporation P.O. Box 12345 Anytown, CA 97402 System Serial No._____ SRCC Document OG-300
System Model: 1. Super Sample 2B 2. Super Sample 2C 3. Super Sample 2D 4. Super Sample 2E	SRCC Certification Number: 300-2001-078A 300-2001-078B 300-2001-078C 300-2001-078D 300-2001-078E	Solar Energy Factor: 2.0 2.6 1.8 3.7 4.2
The installed system is checked above.		

Figure 6-1. Labels are placed on each system certified by the SRCC.

Types of Solar Water Heating Systems

In general, solar water heating systems fall into one of four categories: forced circulation (pumped), integral collector storage, thermosiphon, and self-pumping. Forced circulation (active) systems use a pump to circulate the water or other fluid from the collector, where it is heated by the Sun, to the storage tank, where it is kept until needed. Integral collector storage (ICS) systems, or batch water heaters, combine the collector and the storage tank into one. That is, the Sun shines into the collector and strikes the storage tank directly, heating the water. Thermosiphon systems have a separate storage tank located above the collector. Liquid, such as water or an antifreeze solution, warmed in the collector rises naturally to the storage tank where it is kept until needed. Self-pumping systems use a phase change (liquid-vapor) or other passive means to cause fluid in the collector to circulate and transport heat from the collector to storage.

The ICS, thermosiphon, and self-pumping systems are often called passive solar systems because they do not use mechanical energy to move the heated water. All four types of solar systems work well, and the performance of one type should be compared to that of the others.

SRCC Directory and Rating Certificate

Certified solar water heating system types are sub-categorized and searchable in the OG-300 directory by system type, backup source (storage/back-up tank type), company (manufacturer or packaged system supplier), and brand name (nomenclature from manufacturer/supplier). The auxiliary backup source types listed on the SRCC website are boiler, electric tank, electric tankless, gas tank, and gas tankless. **See Figure 6-2.** The SRCC OG-300 directory can be found at: https://secure.solar-rating.org/Certification/Ratings/RatingsSummaryPage.aspx.

A search using the parameters above will show all solar water heating systems in that category that have been rated and certified by the SRCC. Each selection will contain the name of the manufacturer, SRCC system certification number, system name, system model number, type, total area (sq ft) of collectors, main storage tank volume, auxiliary storage tank volume, and SEF value. These values can be compared to find the proper system for a specific installation. The SRCC certification number is linked to individual system rating certificates. **See Figure 6-3.**

When selected, the rating certificate will display the supplier's name, address, and phone numbers. It will identify the system name and type. The certificate will give a description for the system including collector type, controller type, freeze tolerance, fluid class, and whether an auxiliary tank is included in the system. More than one system can be submitted by the supplier for a specific category. Each system in the category from that specific supplier will be listed for selection and comparison. Also contained on the certificate are the system and collector name and collector manufacturer so that the OG-100 certificate can be selected. The volume and areas of the collectors and storage and auxiliary tanks are included. The SEF value is also included.

The collectors that are part of an OG-300 certified system must be OG-100 certified.

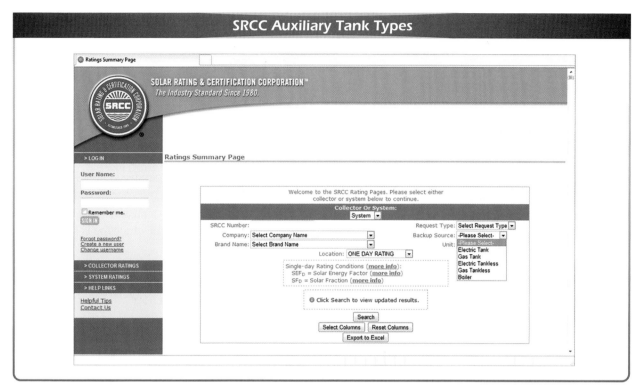

Figure 6-2. The SRCC OG-300 directory is searchable by different characteristics of solar water heating systems such as the auxiliary backup source used, system type, company name, or system name.

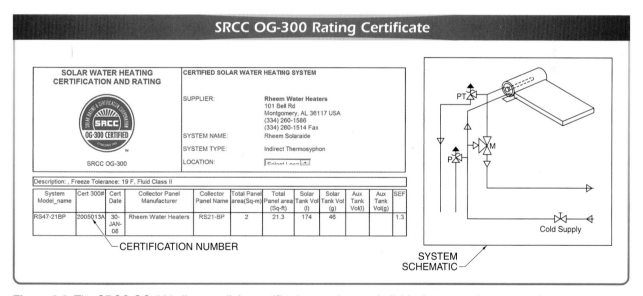

Figure 6-3. The SRCC OG-300 directory links certification numbers to individual systems that are rated.

An additional comparison feature is the ability to select a location, state, and city to customize the SEF value for that location. This will provide values for annual savings and annual solar fraction that are calculated for the location.

A valuable feature of the certificate for the solar water heating installer is the system schematic, which details the proper configuration and installation of the system. This schematic allows the installer to select any certified solar water heating system for installation with the knowledge that there will be no guesswork as to how the system will be configured or how it should be installed. Common plumbing symbols are used for the schematics to make them easy to understand. **See Figure 6-4.**

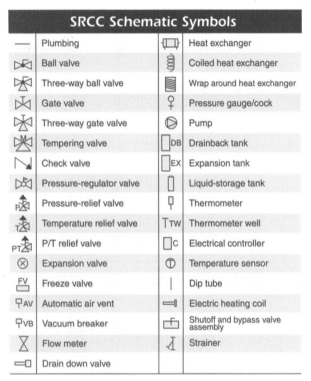

Figure 6-4. Common plumbing symbols are used so that schematics are easy to understand.

In most instances, except for large industrial or commercial purposes, SRCC OG-300 rated systems should be considered for installation rather than custom designs. This will provide the installer with proper installation specifications and the best possible outcomes for building owners especially for warranty and possible subsidy or rebate qualification purposes.

SOLAR WATER HEATING SYSTEMS REVIEW

The general solar water heating system categories can then be subcategorized as direct or indirect and passive or active systems. In all, there are ten system types. **See Figure 6-5.** The system types include the following:
- direct ICS—a passive system that uses an ICS collector in which the HTF is the potable DHW

- direct ICS pumped—an active system that uses an ICS collector in which the HTF is the potable DHW (no systems currently rated)
- direct thermosiphon—a passive system that uses a thermosiphon collector in which the HTF is the potable DHW
- direct pumped—an active system that uses a flat-plate or evacuated tube collector in which the HTF is the potable DHW
- direct self-pumping—a passive system in which the HTF is the potable DHW (no systems currently rated)
- indirect self-pumping—a passive system that uses a heat exchange system in which the HTF is not the potable DHW
- indirect ICS—a passive system that uses an ICS collector and a heat exchange system in which the HTF is not the potable DHW
- indirect ICS pumped—an active system that uses an ICS collector and a heat exchange system in which the HTF is not the potable DHW
- indirect thermosiphon—a passive system that uses a thermosiphon collector and a heat exchange system in which the HTF is not the potable DHW
- indirect pumped—an active system that uses a flat-plate or evacuated tube collector and a heat exchange system in which the HTF is not the potable DHW

Solar Water Heating System Types		
	Passive	**Active**
Direct	Direct Self-Pumping Direct ICS Direct Thermosiphon	Direct ICS Pumped Direct Pumped
Indirect	Indirect Self-Pumping Indirect ICS Indirect Thermosiphon	Indirect ICS Pumped Indirect Pumped

Figure 6-5. The general solar water heating system categories are subcategorized as direct or indirect and passive or active systems.

Rather than examine systems in each category, nine of the most common types of systems an installer might encounter in North America are examined here to explain the operation of the systems and to illustrate the proper placement of components and piping. The systems reviewed are separated into passive systems and active (pumped) systems. For each of the nine systems described, the system type, method of HTF protection, and recommended climate are noted as well as SRCC Collector and System Rating Certification numbers when applicable.

Passive Direct Systems

Passive direct systems use potable DHW as the HTF but do not use pumps or other mechanical means to circulate the fluid through the collector. The passive direct system instead relies on natural gravity flow or convection to create circulation through the system. The two types of passive-direct solar water heating systems are single-tank and multiple-tank ICS systems.

Single-Tank ICS Systems

System Type: Direct ICS

System Name: Sunbather Integral Solar Collector and Storage System

SRCC OG-300 Certification #: None

Solar Collector: AAA Solar Sunbather Integral Solar Collector (batch collector; glazed insulated enclosure; single 40 gal. storage tank)

SRCC OG-100 Certification #: None

Storage Tank: No additional storage tank

HTF: Potable DHW

HTF Protection: Thermal mass and manual drain down

Recommended Climate: Mild; minimum temperature of 40°F (5°C)

A single-tank ICS system is commonly referred to as a batch solar water heating system. The single-tank ICS system is the simplest solar water heating system. Many installation projects for single-tank ICS systems are do-it-yourself projects. The systems are commonly installed by the owners. The Sunbather ICS collector from AAA Solar Supply Inc. is one of the few batch collectors available for purchase in North America. **See Figure 6-6.**

Although the batch collector directly heats the DHW and can act as the sole source of heat for the DHW, this use is recommended only for the mildest of climates in North America. The configuration is also only recommended for buildings with one to four occupants. This

is due to the fact that there may not be enough water heated daily in cooler climates to satisfy the needs of several occupants. In these situations, DHW is needed at higher temperatures and at larger volumes than the system can deliver. Therefore, it is recommended that the batch collector be installed as a preheater for an existing or new water heater.

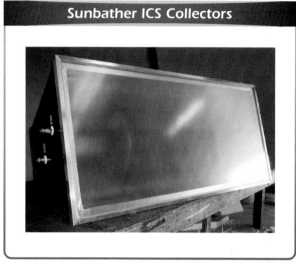

Sunbather ICS Collectors

AAA Solar Supply Inc.

Figure 6-6. The Sunbather ICS collector is one of the few batch collectors available for purchase in North America.

The collector is typically mounted above the storage tank on the building roof. Care should be taken to ensure that the roof can support the weight of the collector fully filled. In this case, the single tank within the collector enclosure is a 40 gal. glass-lined tank. The fully filled collector with glazing and mounting hardware could weigh in excess of 450 lb. The collector should also be mounted to the correct tilt for maximum efficiency. For this model, the manufacturer-recommended tilt is latitude plus 15° from horizontal at true south. Local and national building codes must be followed to ensure the proper support of roof-mounted collectors.

The manufacturer-provided system schematic details the recommended plumbing configuration to ensure proper operation of the system. **See Figure 6-7.** In order for the collector to be used as a preheat source for the water heater, the cold water supply to the water heater must be piped to the collector. With this configuration, the cold water inlet piping is modified by adding a manifold consisting of three tees, three ball valves, and one drain valve to the existing water heating piping downstream of the cold water shutoff valve.

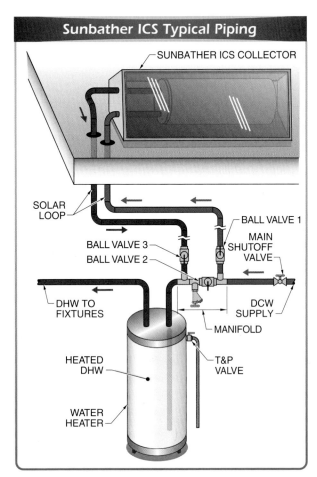

Figure 6-7. Use of the plumbing configuration recommended by the manufacturer ensures the proper operation of the system.

proper air gap per plumbing code requirements. The third downstream tee is installed to connect ball valve 3 and the outlet piping from the collector. Ball valve 3 remains NO while the collector is in use.

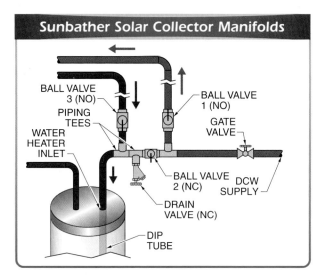

Figure 6-8. A plumbing manifold allows for the installation of piping to the existing system without turning off the cold water service to the building.

The manifold will allow for the installation of piping to the existing system without turning off the cold water service to the building. **See Figure 6-8.** This type of manifold can be used for connecting the solar loop piping in all solar water heating systems.

The first tee and ball valve 1 are installed downstream of the main shutoff valve and connect the cold supply piping to the inlet connection of the collector. Ball valve 1 is normally open (NO) as is the water heater cold water shutoff valve. Ball valve 2 is installed downstream of the first tee. Ball valve 2 is the collector bypass valve and also acts as a drain assist valve. This valve is normally closed (NC) when the collector is in use and opened when draining the collector. The second tee in the manifold is for the installation of the collector drain valve and piping. The collector drain valve is NC until needed to manually drain the collector. The outlet of the valve must be piped to a suitable drain and provided with the

Once the piping is installed and tested, the collector is filled by closing ball valve 2 and opening ball valves 1 and 3. When the collector is being filled, the collector drain valve should be opened to release air from the system, which will prevent the air from entering the water heater storage tank. Once the collector is filled, the drain valve is closed. The hot water fixture fittings should then be turned on to remove any air that may have entered the DHW lines.

Pumps are not used in batch solar water heating systems. Cold-water-line pressure keeps the collector filled and DHW flowing. During the day, solar heat energy is absorbed by the collector, which transfers the heat to the DHW within the collector. Thermosiphoning occurs within the collector as cold water enters the bottom and rises to the top of the tank as it is heated. When hot water is called for by the opening of a hot water fixture fitting (faucet or valve), hot water will flow from the water heater storage tank to the fixture. At the same time, heated water will flow from the collector into the water heater storage tank.

As long as the water temperature within the storage tank stays above the water heater ON setpoint, the DHW will be heated by the collector only. If the temperature within the water heater drops to the ON setpoint, the water heater will turn on to heat the water to its maximum temperature setpoint. Additional temperature controls are not needed for this system.

In this configuration, the batch solar water heater either heats the DHW or acts as a preheater as the cold water supply flows through the collector and is heated by the sun. This heated water then flows through a dip tube into the bottom of the water heater where it is stored or heated to the maximum setpoint of the water heater. The DHW flows to the building's hot water fixture supply when needed.

The HTF in this system, in this case the DHW, is protected by thermal mass and HTF flow. However, when the batch collector is not in use or when prolonged freezing temperatures are expected, the collector and collector piping should be drained. Collector piping must be installed properly with no sags or areas where the HTF may be trapped and freeze. Horizontal piping should be sloped at a minimum of ¼″ per foot to allow for draining.

To drain the collector, the cold water main shutoff valve is closed and ball valves 1, 2, and 3 are opened. The drain valve is then opened until the collector and piping are emptied. To bypass the collector, the drain valve is closed and valves 1 and 3 are closed. Valve 2 can now be opened as well as the cold water main shutoff valve. The collector and piping are now isolated and protected from freezing temperatures. The water heater is now heating the DHW without the assistance of the solar water heating system.

Multiple-Tank ICS Systems

System Type: Direct ICS

System Name: Thermal Conversion Technology Inc. (TCT Solar) ProgressivTube PT-30-CN

SRCC OG-300 Certification #: 1995002A

Solar Collector: ProgressivTube PT-30-CN (multitube glazed insulated ICS; 30.84 gal. capacity)

SRCC OG-100 Certification #: None

Storage Tank: Minimum 40 gal. water heater properly sized per occupants

HTF: Potable DHW

HTF Protection: Thermal mass, manual drain down, and pressure relief

Recommended Climate: Mild; minimum temperature of 40°F (5°C)

The ProgressivTube PT-30-N is a multitube ICS solar thermal collector. It is manufactured using multiple 4″ copper tubes that are welded to interconnecting pipes and provide a series flow pattern through the collector. **See Figure 6-9.** As the HTF flows into the collector from the bottom piping connection, the welded pipe system allows for the lowest tube to fill with HTF and then flow through to the next tube and so on until the heated HTF flows out the top piping connection. This method of thermosiphoning allows for even heat gain as the HTF flows through the collector. The tubing system within the collector is the solar radiation absorber and is treated with a selective coating. The enclosure is insulated, double glazed, and sealed against the elements for maximum efficiency and minimum heat loss.

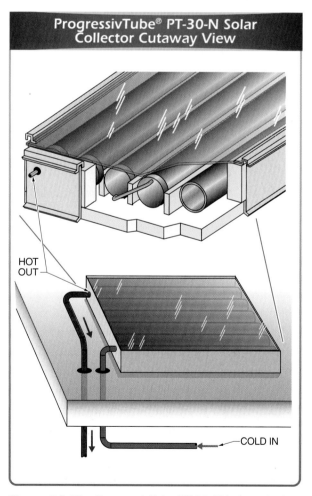

ProgressivTube® PT-30-N Solar Collector Cutaway View

HOT OUT

COLD IN

Figure 6-9. The ProgressivTube PT-30-CN, through thermosiphoning, heats the DHW contained in the 4″ copper tubes located in the collector.

The PT-30-CN collector can be used as a stand-alone solar water heater. However, it is impractical for ICS systems except for those used in the mildest of climates and for only a few occupants. The manufacturer recommends that the collector be used as a solar DHW preheater for an existing conventional electric water heater. **See Figure 6-10.** There are other certified systems available from the same manufacturer for use with gas or tankless water heaters.

Note: If an SRCC-300-certified solar water heating system is not installed per the schematic included on the certification form, the system may not meet warranty or subsidy requirements. The system also may not work as efficiently as stated.

The collector should be mounted on the building roof. **See Figure 6-11.** Care must be taken to properly support the system on the roof. The full or wet weight of the system is in excess of 425 lb. The roof structure may need to be reinforced. Local and national building codes must be followed to ensure the proper support of roof-mounted collectors.

TCT Solar

Figure 6-11. An ICS collector should be mounted on a building roof.

The existing conventional water heater is located below the collector within the building. Thermosiphoning in this passive system occurs within the collector rather than from the collector to the water heater. Flow to and through the collector, to the water heater, and on to the fixtures is accomplished by the potable cold water line pressure. The water heater should be sized according to the number of occupants in the building.

This system is piped differently from the single-tank ICS system, which creates three differences between the two ICS systems. First, two three-way ball valves are placed in the system manifold to allow more flexibility in directing the flow of HTF in the system. The second difference is the addition of a tempering valve on the hot water outlet of the water heater, which is an essential safety measure for system users and is required in all SRCC-certified systems. The third difference is the installation of a pressure-relief valve (PRV) in the collector solar loop, which is an essential safety measure for the solar loop.

Three-way ball valve 1 is installed on the cold water inlet to the water heater after the shutoff valve to the water heater. The center port of the three-way valve

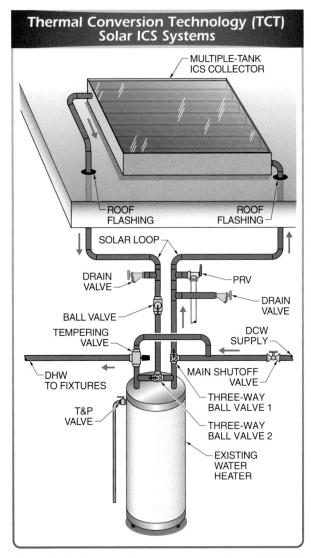

Figure 6-10. The manufacturer recommends the PT-30-CN be used as a solar DHW preheater for an existing electric water heater.

connects the cold water with the upper port connecting the inlet piping to the collector. The bottom port is connected to the water heater cold water inlet piping. The water heater inlet and outlet piping are interconnected by a manifold before any other connections are made. Three-way ball valve 2 is installed in the center of this manifold. The center port of the manifold connects the outlet piping of the collector. The other two ports connect to the inlet and outlet piping of the water heater.

This piping arrangement allows for HTF flow to be directed in one of three ways. **See Figure 6-12.** Operation of ball valve 1 directs HTF, in this case the domestic cold water (DCW), to the solar collector inlet or directly into the water heater. The latter stops the flow of HTF into the solar collector. Operation of ball valve 2 directs the HTF flow from the collector to either the inlet or outlet piping of the water heater. Directing HTF flow to the inlet piping of the water heater allows for the preheating of cold water in the collector before it enters and is stored in the water heater. Directing HTF flow from ball valve 2 into the outlet piping of the water heater essentially eliminates the need for the water heater. HTF will then flow from the collector directly to the fixtures and allow for the use of the collector as the sole heating system for the DHW.

These changes in the HTF direction of flow from the collector are done on a seasonal basis. For example, during the fall or winter months when the temperatures in the collector and amount of insolation are low, ball valve 2 would be adjusted to direct HTF flow into the water heater. The HTF, preheated in the collector could then be heated to proper temperatures for use within the water heater and flow out to the fixtures. During the spring and summer months when there is high insolation and high HTF temperatures, ball valve 2 would be adjusted to direct HTF flow to the outlet piping of the water heater and to flow directly to fixtures.

During winter months when use of the collector is impractical due to low temperatures, this piping arrangement also allows for the isolation of the collector. Ball valve 1 is adjusted to direct cold water flow into the water heater. Ball valve 2 is adjusted to direct flow to the outlet piping of the water heater, and the two-way ball valve is closed. This isolates the collector so that HTF does not flow through it. The collector should then be drained using the boiler drain valves on the collector supply and return piping. The system collector piping must be installed properly with no sags or areas where HTF may be trapped and freeze. Horizontal piping should be sloped at a minimum of ¼″ per foot to allow draining.

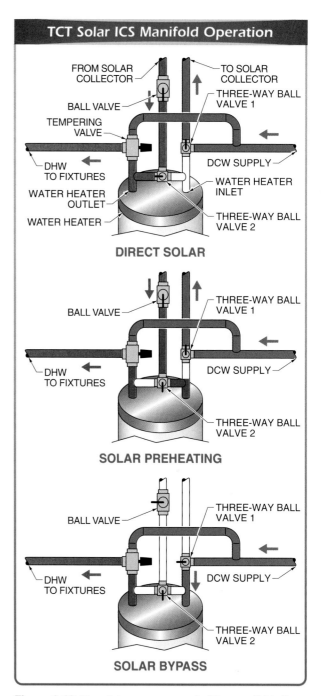

Figure 6-12. The piping arrangement of the manifold allows the HTF flow to be directed in one of three ways.

This multiple-tank ICS system is a direct solar water heating system. The HTF in this system is the DHW. The ICS collector has the capability of heating the HTF to at least 350°F (177°C). If DHW at this temperature were allowed to continue to the fixture supplies, severe scalding of fixture users could occur. In order to protect users and for the system to meet SRCC OG-300 certification standards, antiscalding measures must be taken to lower the temperature of the DHW leaving the water heater to a safe temperature.

The most cost effective and safe method of lowering the DHW temperature is to temper or cool the DHW by mixing it with cold water. With the multiple-tank ICS system, tempering the DHW is accomplished by installing an automatic tempering valve at the hot water outlet of the water heater. **See Figure 6-13.** This valve is also connected to the cold water piping between the cold water shutoff valve and three-way ball valve 1. This connection allows for cold water to mix with the DHW before it flows to the fixtures.

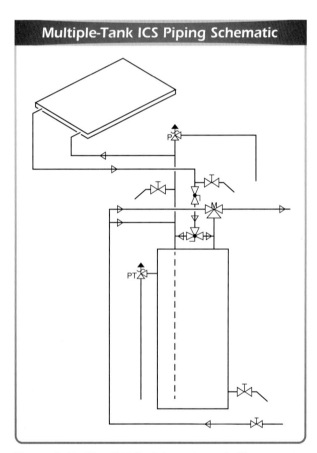

Multiple-Tank ICS Piping Schematic

Figure 6-13. The SRCC piping schematic illustrates the position of system components such as collectors, piping, and valves.

The tempering valve automatically maintains the outgoing DHW at a safe temperature. The tempering valve is normally set to maintain a temperature of about 120°F (49°C). In addition to the tempering valve, many local and national plumbing codes require that antiscald valves be installed at shower or bath fixtures when the DHW can

reach scalding temperatures. Local building codes must be consulted for proper compliance.

The collector piping in this system also includes valves to isolate and drain the ICS collector. The two-way ball valve on the piping connected to the collector output must be closed. Three-way ball valve 2 must be adjusted for full flow to the outlet piping of the water heater. Three-way ball valve 1 must be adjusted for full flow of the DCW to the water heater inlet piping. The boiler drain valves can be opened to allow for the collector to be drained and thus protected from freezing temperatures. However, this is a manual operation, which requires the building occupants to remember to isolate and drain the system when necessary.

Isolating and draining the collector and vulnerable solar loop piping is the primary method of protecting HTF from freezing temperatures. The collector is also protected by the thermal mass of 4″ of HTF in each collector tank or line. It would take several days of low temperatures to freeze the HTF in the collector. However, it would only take one night of freezing temperatures to freeze the collector piping.

The collector and solar loop in this system are also protected from dangerous pressure conditions that could occur by the installation of a PRV in the collector inlet line. The PRV for this system is factory set at 150 psi. If that pressure is reached within the system, the valve will release HTF from the system until safe pressures are reached and the PRV closes. Until then, HTF will be discharged out of the PRV. The PRV outlet must be properly piped and drained per plumbing code requirements.

Under normal operation conditions, this ICS system is designed to have the DCW flow through the collector where it will be heated by solar radiation and then enter the water heater. It will then be ready for use by the building occupants at the various DHW-supplied plumbing fixtures. The DHW flow will also be tempered if necessary by the cold water supplied to the tempering valve. The valve settings for a multiple-tank ICS system must be as follows:

* Both boiler drain valves located on the solar loop piping must be fully closed.
* The ball valve must be fully opened.
* Three-way ball valve 1 must be adjusted for full flow to the collector.
* Three-way ball valve 2 must be adjusted for full flow to the inlet piping of the water heater.
* The DCW shutoff valve should be opened to fill the system once the drain valves and ball valves are properly set.
* The DHW fixture supply valves should be opened to eliminate air from the system.

HTF flow through the system is initiated by the use of a DHW-supplied fixture. This causes flow that is created by the DCW line pressure. DCW will then flow to the collector where it will be heated as it flows through the collector. When the HTF in the collector is not flowing, it is being heated and moving by convection due to thermosiphoning that occurs within the ICS collector. When HTF flow is again initiated by fixture use, the HTF will flow through the water heater dip tube to the bottom of the tank. While in the water heater, the HTF, which is now the DHW, will be stored and heated to the water heater maximum temperature setpoint if needed.

Note: The DCW line pressure must be high enough to provide flow up to the ICS collector and be a minimum of 15 psi at the farthest fixture from the water heater. If not, a pump or circulator must be used.

Passive Indirect Systems

Passive indirect systems use thermosiphoning or self-pump methods to circulate an HTF that is not the DHW through the solar loop containing a heat exchange system. In most passive indirect systems, the storage tank is mounted above the solar collector. However, this is not the case in the self-pumping system.

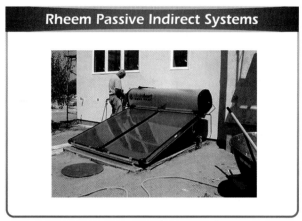

Rheem Manufacturing Company

Figure 6-14. The Rheem Solaraide water heating system uses thermosiphoning to circulate the HTF that is not DHW through the solar loop.

The Rheem Solaraide is a stand-alone solar water heater with a back-up electric heating element located in the storage tank. It is a versatile system designed to be installed as the primary water heater in newly built homes, a replacement for an existing water heater, or a preheater for a larger water heating system. As a preheating system, it can be used with an electric, gas, or tankless water heater or as a preheater in a boiler system.

The 80 gal. water heater or heat exchanger has three functions in this solar water heating system. The first function is as a storage tank to store and heat the DHW. The second function is as a back-up water heater for the solar water heating system. The storage tank is a horizontal electric water heater with a single copper heating element capable of heating the DHW to 140°F (60°C) when there is not enough insolation for the flat-plate collectors to properly heat the DHW. The third function of the tank system is as a heat exchanger.

A single-wall heat exchanger completely surrounds the inner storage tank, providing the method of heat transfer from the HTF within the heat exchanger to the DHW in the storage tank. The two glazed flat-plate solar collectors are attached below the horizontal tank and piped to the tank's heat exchanger connections. **See Figure 6-15.** The two harp-style collectors are piped together at the upper and lower manifold outlets using supplied collector unions. The heat exchanger outlet is piped to the collector inlet at the bottom right-hand side of the lower collector manifold. The collector outlet at the top left-hand side of the upper collector manifold is piped to the inlet of the heat exchanger. This piping comprises the solar loop of the system.

Thermosiphon Solar Water Heating Systems

System Type: Indirect Thermosiphon

System Name: Rheem Solaraide RS80-42BP

SRCC OG-300 Certification #: 2005013B

Solar Collector: Rheem RS21BP (glazed flat-plate)

SRCC OG-100 Certification #: 200501A

Storage Tank: Nonseparable electric water heater with heat exchanger (supplied with system); 80 gal. capacity

HTF: Antifreeze

HTF Protection: Antifreeze

Recommended Climate: Mild; minimum temperature of 40°F (5°C)

The Rheem Solaraide RS80-42BP is a thermosiphon solar water heating system. It is a passive indirect system that uses two glazed flat-plate solar collectors to harvest solar radiation and an attached 80 gal. water heater with a heat exchanger for storage. **See Figure 6-14.**

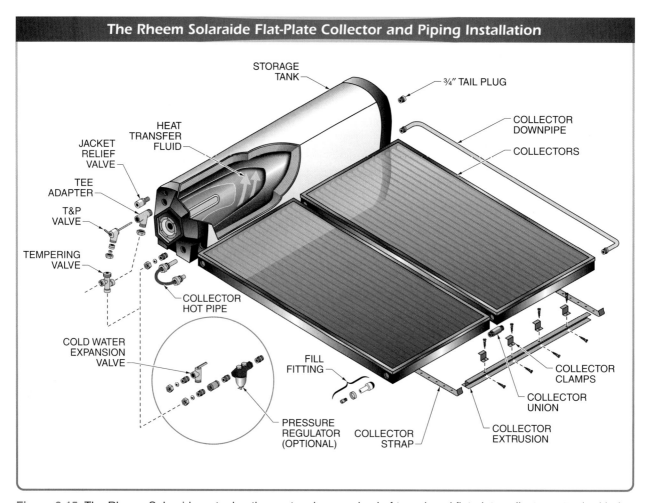

The Rheem Solaraide Flat-Plate Collector and Piping Installation

Figure 6-15. The Rheem Solaraide water heating system is comprised of two glazed flat-plate collectors attached below the horizontal tank and piped to the tank heat exchanger connections.

The HTF used in this system is a nontoxic antifreeze solution. The HTF is heated in the collectors and flows by thermosiphoning up and around the tank within the heat exchanger. The HTF transfers its heat to the DHW contained within the tank. The HTF becomes heavier as it cools, and it flows through the collector down pipe and back into the collector lower manifold. As the HTF is heated within the collectors, the HTF flows up and out the collector outlet, to the heat exchanger, and through the collector hot pipe. This process continues, without the use of pumps or electricity, while the system is absorbing heat. Thus, it is a passive and indirect solar heat transfer system. Because the HTF runs through the heat exchanger and does not come in contact with the DHW in the water heater storage tank or the DHW piping, it is considered a closed loop system.

The tank and collectors are typically mounted on the building roof using proprietary mounting hardware from the manufacturer. **See Figure 6-16.** Care must be taken in supporting the system on the roof. The full or wet weight of the system is in excess of 1100 lb. The roof structure must be reinforced. Manufacturer mounting instructions as well as local and national building codes must be followed to ensure the proper support of the system.

Facts

A roof-mounted stand-alone solar water heating system includes a back-up heater. It combines simplified installation with versatile operation. The attached storage tank can replace an existing water heating system.

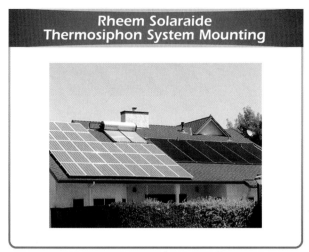

Figure 6-16. System tank and collectors are typically mounted on a building roof using proprietary mounting hardware from the manufacturer.

The tank must be piped according to local and national building mechanical codes. The DCW is piped to the water heater inlet, and the DHW is piped to the water heater outlet. The DCW and DHW lines must be connected to the building DCW and DHW lines. A shutoff valve must be installed in the DCW line to the water heater. The shutoff valve is used to turn off the DCW to the water heater. **See Figure 6-17.**

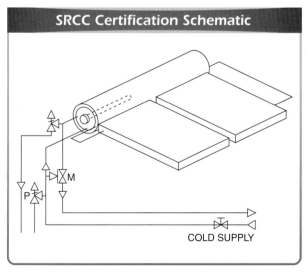

Figure 6-17. The tank must be piped according to local and national building mechanical codes.

A temperature and pressure (T&P) relief valve is mounted to the tank at the tank outlet. The T&P discharge line must be piped to an appropriate drain. If the DCW operating pressure can reach over 60 psi, a PRV should also be installed after the DCW shutoff valve to the water heater. To protect against high temperatures at fixture supplies, a tempering valve should be installed in the DHW line and piped to the DCW before any DHW fixture supply branches.

Note: All plumbing and discharge lines must be installed per local plumbing code requirements. All plumbing lines within attic areas and exposed roof areas must be graded at ¼″ per foot to drain properly and for protection against freezing. All valves should be installed where they are readily accessible.

Under normal operating conditions, the DCW shutoff valve to the water heater is open. The tempering valve is set to the proper temperature for DHW per the local plumbing or health codes. The water heater thermostat on the Rheem Solaraide should be set to the desired temperature, typically 120°F (49°C), for the DHW.

A thermosiphon system is simple in operation. The HTF in the flat-plate collectors absorbs heat throughout the day and flows up and into the heat exchanger surrounding the water heater tank. As the heat is transferred to the DHW within the tank, the HTF cools and returns to the collector inlet and repeats this thermosiphoning process. There are no pumps, valves, or controls in this solar loop.

The HTF protection in the thermosiphon system is an antifreeze solution. The manufacturer specifies that the solution in the system be 20% antifreeze and 80% water. The operating temperature range of the solution is –40°F to 300°F (–40°C to 149°C), with freezing to occur at –50°F (–46°C). However, since the system also contains DHW in the interior storage tank, the recommended local area minimum temperature for the system is 40°F (5°C).

The water heater section of this system works as any other electric water heater as should be wired per local and national electrical codes. DHW flow to the fixtures is maintained by DCW line pressure. If there is not enough insolation during the day for the solar collectors to provide enough heat to keep the DHW at 120°F (49°C), the electrical heating element within the tank will come on and heat the DHW. The DHW in the tank is protected against freezing by the heating element and also by its thermal mass. Since the water heater is mounted in an exposed area, the system is only recommended for mild temperatures.

As a stand-alone solar water heating system, the thermosiphon system is very simple to pipe and operate. However, if the system is used with an additional storage tank or as a preheating system, the piping will be more complicated. **See Figure 6-18.**

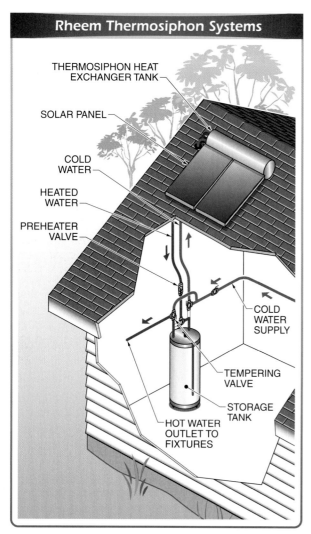

Figure 6-18. The piping configuration for a Rheem thermosiphon system used with an additional storage tank or as a preheating system is more complicated than for a stand-alone system.

Self-Pumping Systems

A *self-pumping system* is a passive solar water heating system that uses a phase change (liquid to vapor) or other passive means to cause the fluid in a collector to circulate and transfer heat from the collector to the storage system. A direct self-pumping system uses the DHW as the HTF. An indirect self-pumping system uses a heat exchange system to transfer heat to the DHW.

SELF-PUMPING SOLAR WATER HEATING SYSTEMS

System Type: Indirect Self-Pumping

System Name: Sunnovations S8080-EICa

SRCC OG-300 Certification #: 2011137H

Solar Collector: SunEarth, Inc. Empire EC-40 (glazed flat-plate)

SRCC OG-100 Certification #: 2006024E

Storage Tank: Bradford White EcoStor2™ S-DW2-75R6DS (indirect electric water heater system with single-coil double-wall heat exchanger); 75 gal. capacity

HTF: 50% propylene glycol and 50% distilled water solution

HTF Protection: Antifreeze, overtemperature protection is a form of steamback

Recommended Climate: Moderate, minimum temperature −22°F (−30°C)

The Sunnovations S8080-EICa is a passive, indirect, self-pumping solar water heating system. **See Figure 6-19.** This system functions differently from other solar water heating systems. The Sunnovations system uses what the manufacturer describes as "geyser pumping" for circulation of HTF.

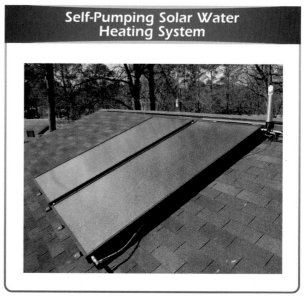

Sunnovations, Inc.

Figure 6-19. The Sunnovations S8080-EICa is a passive, indirect, self-pumping solar water heating system that uses a geyser pumping principle for circulation of HTF.

"Geyser pumping" refers to the principle in which geysers are formed such as those in Yellowstone National Park. The principle used in this solar water heating system is the same as those natural geysers. Heat in the form of solar radiation brings HTF to boil within the collector and creates steam. That powerful steam pushes the HTF up the collectors and into system piping, which creates the geyser effect. After the steam is released, the process starts again. It continually pumps until there is no insolation to create steam.

This packaged system consists of two glazed flat-plate harp-style solar thermal collectors and a 75 gal. electric water heater with a single double-wall immersed heat exchange coil. The system is also comprised of several proprietary components only used in the Sunnovations system. These are the passive pump assembly and the overflow reservoir. The system also is supplied with the solar loop piping consisting of insulated ¾″ oxygen barrier PEX tubing (Uphonor HePEX). Valve manifolds for the heat exchanger connections are also supplied.

The flat-plate collectors should be mounted on the roof and properly oriented. The manufacturer recommends that the collectors be oriented at a minimum of ±45° from true south and tilted to the location's latitude. Only harp-style collectors are used in the Sunnovations system, and they must be set level so that the HTF will flow freely through the inlet piping and collector to the passive pump.

The passive pump is mounted to the collector above the collector outlet. **See Figure 6-20.** The pump comes with the collector inlet and outlet connections. The collector inlet piping is soft copper connected to the pump reservoir (R2), rolled down to the collector inlet, and soldered to the collector. The collector outlet piping is connected to the pump riser. The solar loop piping from the heat exchanger penetrates the roof at the same point next to the pump and attaches to the proper connection points on the pump assembly.

The overflow reservoir is mounted to the second collector. **See Figure 6-21.** It contains a PRV that is connected back to the collector inlet piping. The PRV, or overflow orifice, relieves any over pressure within the system. The overflow reservoir (R3) collects the HTF steam, cools and condenses it, and delivers it back to the system.

The solar loop piping is connected to the storage tank heat exchanger through the supplied valve manifolds. The collector outlet piping is connected to the top/inlet heat exchanger manifold. The collector inlet piping is connected to the bottom/outlet heat exchanger manifold, which also contains the system fill and drain valve. **See Figure 6-22.**

Sunnovations, Inc.

Figure 6-20. The passive pump is mounted to the collector above the collector outlet. The solar loop piping penetrates the roof at the same point next to the pump and attaches to the proper connection points on the pump assembly.

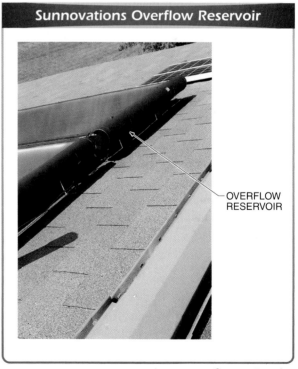

Sunnovations, Inc.

Figure 6-21. The overflow reservoir (R3) connected to the second collector collects the HTF steam, cools and condenses it, and delivers it back to the system.

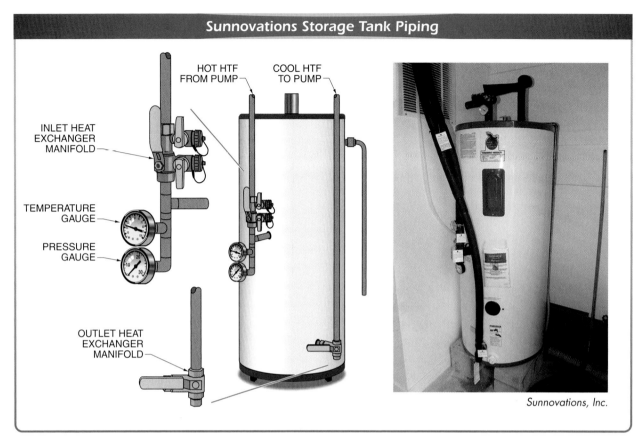

Sunnovations Storage Tank Piping

HOT HTF FROM PUMP

COOL HTF TO PUMP

INLET HEAT EXCHANGER MANIFOLD

TEMPERATURE GAUGE

PRESSURE GAUGE

OUTLET HEAT EXCHANGER MANIFOLD

Sunnovations, Inc.

Figure 6-22. Solar loop piping is connected to the storage tank heat exchanger through the supplied valve manifolds.

Oxygen barrier PEX tubing is used for the solar loop because the working pressure and temperatures are lower than they are in other solar water heating systems. The HTF temperature range on high-insolation days will range from 100°F (38°C) to 185°F (85°C). This is well within the operating temperatures of the supplied PEX tubing.

The DCW and DHW should be piped to the storage tank per local plumbing codes. A tempering valve should be installed on the outlet of the DHW and interconnected to the DCW. The tempering valve should be set to normal DHW temperature, which is usually 120°F (49°C). This self-pumping system has no other valves, pumps, controls, sensors, wires, or expansion tank and uses no electrical energy.

The installation, filling, and startup procedures provided by the manufacturer should be followed closely. The manufacturer recommends using a filling station to fill the system. Once the system is filled, the fill station pump is reversed and 1 gal. of HTF must be removed. This process creates a slight vacuum in the system. This step is crucial for the proper operation of the system. The

system is specially designed and will not operate as other solar water heating systems. Installing part of the system incorrectly will cause the system to be inoperative.

Under normal operating conditions, the isolation valves at the heat exchange manifolds are open. Steam bubbles are formed inside the risers inside the collectors once the fluid in the collector and pump riser reaches its boiling point. **See Figure 6-23.** The initial boiling point temperature is low, approximately 100°F (38°C), because the system is under a vacuum.

The rapid volume expansion of the steam bubbles forces HTF up through the collector, into the pump riser, and to reservoir R1. This creates a difference in fluid levels between R1 and the collector input reservoir R2. Due to the principle of fluid gravity balance, the HTF level of the hot "leg" becomes higher than the HTF level in the cold "leg" of the solar loop. This causes the HTF to rise in the cool leg and flow to the heat exchanger in the tank. Circulation is thus created throughout the solar loop. Heat is transferred to the solar storage tank and cool fluid returns to R2.

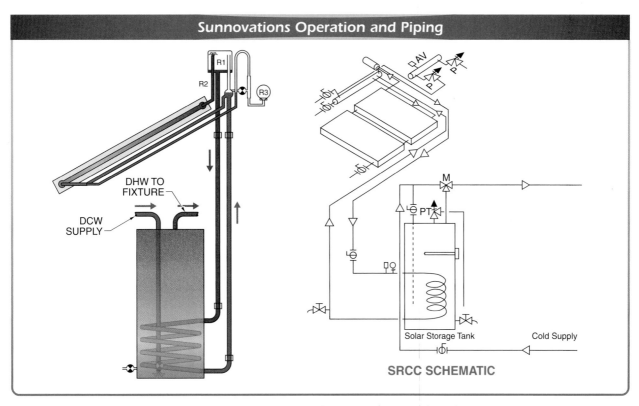

Figure 6-23. Circulation through the solar loop is created by the formation of steam bubbles in the riser tubes inside the collector.

The cooler HTF in R2 keeps a low dew point, which maintains a vacuum in the system. This also maintains a low temperature in the system. The condensation of steam in R2 preheats the HTF returning to the collector and effectively recovers the heat energy used for geyser pumping.

The pressure in the system rises with increased insolation. This results in adaptive flow rates and operating temperatures. In other words, as temperatures increase, the flow rate increases. As temperatures decrease, the flow rate decreases until stopping under no insolation conditions (nighttime). This effectively controls the system without electrical controls or pumps. Also because the hot and cold fluid sections are separated from each other, there is no risk of reverse thermosiphoning and no need for check valves.

The system is also automatically controlled under high-insolation conditions. If insolation is very high, system pressure will rise. Once the pressure in the system reaches ambient pressure, the relief valve opens to vent overpressure via the overflow orifice. Steam will condense to HTF in overflow reservoir R3. The system will continue to operate while it limits the pressure and operating temperature. The maximum temperature to the storage tank is 185°F (85°C).

If the temperature of the HTF returned from the heat exchanger exceeds 140°F (60°C) because the tank is charged, the pressure in the system will rise rapidly. The excess steam cannot be vented quickly enough through the temperature-limiting orifice. This causes HTF from the collectors to be purged to reservoir R3. After all of the HTF is purged, the system comes to rest without any HTF in the collectors. Once the system cools down, all HTF is drawn back into the system and a strong vacuum is restored.

The system can be drained for the vacation or replacement of HTF. The lower heat exchanger manifold contains a drain valve. Both heat exchanger isolation valves and the drain valve should be opened to drain the system.

Maintaining this system is relatively easy. Since there are no electrical or mechanical components and the glycol never reaches the breakdown temperature the system is virtually maintenance free. A periodic check of the pressure and temperature gauges is recommended especially to determine whether a vacuum is present in the system. This is the key health indicator for the system. The pressure gauge at the tank should read between 0 psi and 15 psi. This positive pressure is due to the head pressure created by the height of the vertical HTF column. The temperature gauge at the heat exchanger inlet should read 120°F (49°C) during good insolation days.

Normally, the propylene glycol solution should not suffer chemical degradation from overheating because temperatures remain relatively low. Inhibitors in the propylene glycol should prevent corrosion. Since the Sunnovations system operates under a vacuum, there is no oxygen inside the system to form corrosion. Six years after installation, an annual check of the storage tank is recommended.

Active Direct Systems

An active direct solar water heating system uses domestic potable water as the HTF and employs a pump or circulator to circulate the fluid through the solar collector. Because the active direct system does not depend on thermosiphon action to move the HTF, the solar collector is usually installed above the storage tank and fixtures.

PUMPED SOLAR WATER HEATING SYSTEMS

System Type: Direct Pumped

System Name: Alternate Energy Technologies (AET) Eagle Sun D-80-40

SRCC OG-300 Certification #: 2011020A

Solar Collector: AET American Energy AE-40-E (glazed flat-plate)

SRCC OG-100 Certification #: 1999001C

Storage Tank: American Water Heater SE62-80H-045S or Richmond (Rheem) Water Heater 81VR80-TC-1

HTF: Potable DHW

HTF Protection: Pump circulation, freeze prevention valve, and manual drain down

Recommended Climate: Mild; minimum temperature of 40°F (5°C)

The Alternate Energy Technologies (AET) Eagle Sun D-80-40 system is an active direct, or forced circulation, solar water heating system. The system is comprised of a glazed flat-plate solar collector, an 80 gal. storage tank / water heater, and a pump for circulation through the solar loop. The HTF for the system is the DHW.

The solar collector in this solar water heating system primarily functions as a DHW preheater for the recommended storage tank / water heater. On good insolation days, the single collector is able to supply enough heat energy to provide all the DHW heating. However, on cloudy days or days of high DHW usage, the water heating elements of the tank provide the additional heat energy needed to heat the DHW to the proper temperature.

Alternate Energy Technology
High demand for DHW may require the use of more than one collector to meet the DHW requirements of a household.

The glazed flat-plate collector is typically mounted on the roof and piped to the storage tank / water heater below. The recommended storage tank / water heater is a multiple port tank, which allows for separate connection of the collector inlet and outlet piping. For this system, either top or side ports can be used. **See Figure 6-24.**

The collector inlet piping should connect to the solar outlet port on top of the tank or to the side port bottom outlet. The top outlet port should have a dip tube that extends to approximately 6″ from the bottom of the tank. The use of either connection ensures that only the coolest water, which has sunk to the bottom of the tank, is circulated up to the collector.

The collector outlet piping should connect to the top solar inlet port or the top side inlet port. Care must be taken to ensure that if the solar inlet top port is used that it has a dip tube that extends approximately halfway to the bottom of the tank. This will allow for the stratification of DHW temperatures within the tank. It will also eliminate the possibility of circulating only the bottom portion of DHW in the tank. If this is not done and both dip tubes extend to the bottom of the tank, only the solar-heated water will be circulated back up to the collector.

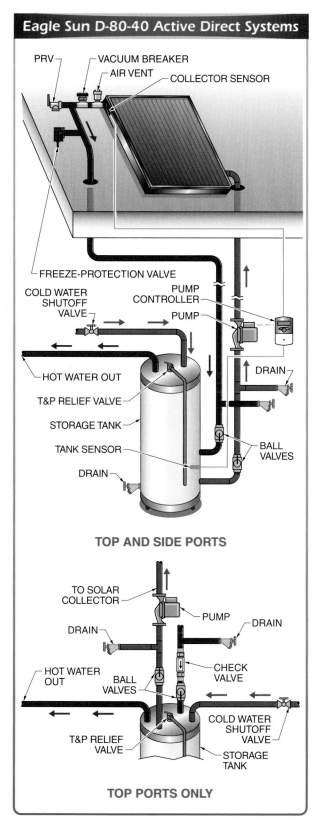

Eagle Sun D-80-40 Active Direct Systems

PRV — VACUUM BREAKER
AIR VENT — COLLECTOR SENSOR

FREEZE-PROTECTION VALVE
COLD WATER SHUTOFF VALVE
PUMP CONTROLLER
PUMP
HOT WATER OUT
T&P RELIEF VALVE
STORAGE TANK
TANK SENSOR
DRAIN
DRAIN
BALL VALVES

TOP AND SIDE PORTS

TO SOLAR COLLECTOR
PUMP
DRAIN
DRAIN
HOT WATER OUT
BALL VALVES
CHECK VALVE
T&P RELIEF VALVE
COLD WATER SHUTOFF VALVE
STORAGE TANK

TOP PORTS ONLY

Figure 6-24. Depending on the available area and location of existing piping, top or side ports may be used.

Whether using the top or side ports of the tank, the pump and valve configurations are the same. **See Figure 6-25.** Isolation ball valves should be placed after the piping connection to the tank. A boiler drain valve should be installed on the collector inlet piping after the isolation valve and then the circulation pump with integral check valve. If the pump assembly does not also have integral isolation valves, two full-flow ball valves should be installed before and after the pump to facilitate repair or replacement. The piping should then be connected to the inlet of the collector.

After the isolation ball valve on the tank solar inlet piping, a check valve and a boiler drain valve should be installed. The check valve is oriented so that there is no flow up the collector outlet piping. This check valve plus the integral check valve in the pump eliminate the possibility of reverse thermosiphoning. Reverse thermosiphoning is the movement of warm water up to the collector when ambient temperatures are low enough to cool the HTF in the collector. It creates a reversal of intended flow.

Facts

Thermosiphoning, or natural circulation, is caused by the action of gravity on the heavier and cooler HTF displacing the lighter and hotter HTF. Natural circulation occurs when the temperature difference is enough to create a sufficient pressure difference to overcome the friction caused by the flow.

At the collector outlet, an air vent that allows air to be removed from the system should be installed. A freeze-protection valve is also installed at this point along with a PRV. The freeze-protection valve opens at near freezing temperatures to discharge the HTF, in this case the DHW. The discharged HTF flows through the collector, preventing the HTF within the collector from freezing. The PRV should be set for 60 psi in order to protect the system from excessive pressures.

Note: All plumbing and discharge lines must be installed per local plumbing code requirements. All plumbing lines within attic areas and exposed roof areas must be graded at ¼″ per foot back to the drain valves to drain properly and be protected against freezing. All valves should be installed where they are readily accessible.

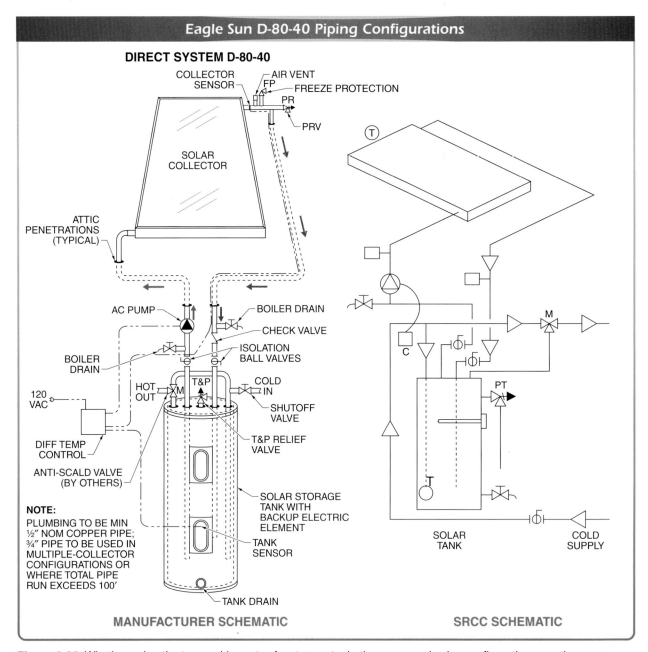

Eagle Sun D-80-40 Piping Configurations

DIRECT SYSTEM D-80-40

COLLECTOR SENSOR
AIR VENT
FP
FREEZE PROTECTION
PR
PRV

SOLAR COLLECTOR

ATTIC PENETRATIONS (TYPICAL)

AC PUMP
BOILER DRAIN
CHECK VALVE
ISOLATION BALL VALVES
BOILER DRAIN

HOT OUT
T&P
COLD IN
SHUTOFF VALVE
T&P RELIEF VALVE

120 VAC

DIFF TEMP CONTROL

ANTI-SCALD VALVE (BY OTHERS)

SOLAR STORAGE TANK WITH BACKUP ELECTRIC ELEMENT

TANK SENSOR

NOTE:
PLUMBING TO BE MIN ½" NOM COPPER PIPE; ¾" PIPE TO BE USED IN MULTIPLE-COLLECTOR CONFIGURATIONS OR WHERE TOTAL PIPE RUN EXCEEDS 100′

TANK DRAIN

MANUFACTURER SCHEMATIC

T

M
C
PT
T

SOLAR TANK
COLD SUPPLY

SRCC SCHEMATIC

Figure 6-25. Whether using the top or side ports of a storage tank, the pump and valve configurations are the same.

This system is supplied with a differential temperature controller. A differential temperature controller senses the difference in temperature between two temperature sensors and activates or deactivates the system pump depending upon the action required. One sensor is placed on the water heater tank near the bottom of the tank, and the other sensor is placed on the collector outlet piping. **See Figure 6-26.** These sensors are wired to the system controller, which should be mounted close to the storage tank / water heater. The controller is wired to the electrical system and also to the pump. The manufacturer recommends that the controller be set at 135°F (57°C) maximum temperature for the storage tank sensor. A 10° differential temperature is the default setting on this controller with a 5° differential shutoff setting.

Temperature Sensor Connections to Collector Outlet Piping

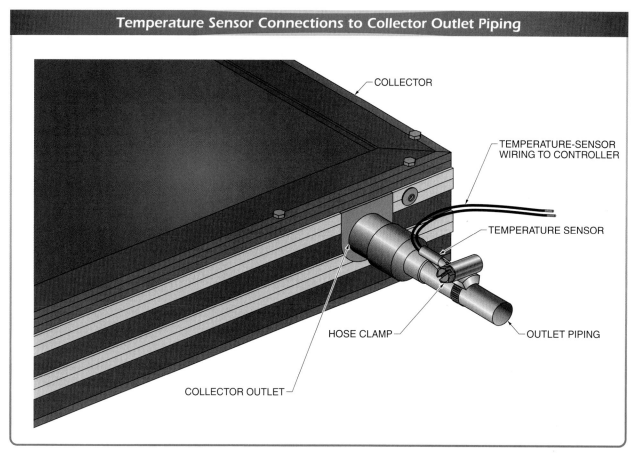

COLLECTOR

TEMPERATURE-SENSOR
WIRING TO CONTROLLER

TEMPERATURE SENSOR

HOSE CLAMP

OUTLET PIPING

COLLECTOR OUTLET

Figure 6-26. For this system, the collector temperature sensor is placed on the collector outlet piping.

Under normal operating conditions, the boiler drain valves are closed and the collector isolation valves are open. When the controller senses a 10° differential in temperature between the storage tank and the collector sensor, the controller will activate the pump. For example, if the collector sensor reads 100°F (38°C) at the storage tank and 110°F (43°C) at the collector, the pump will be activated to provide HTF flow through the collector.

When the pump is activated, DHW or HTF circulates through the solar loop to transfer heat from the solar collectors to the DHW within the tank. This process continues until the sensor temperatures are within 5° of each other, at which point the pump will be deactivated. If the temperature in the storage tank reaches 135°F (57°C), the pump will also be deactivated. If temperature within the tank is low or if there is increased usage of DHW for an extended period of time, the water heater electrical elements will be energized to bring the tank up to temperature and function as a normal water heater. This can be customized for varying time and temperature parameters.

The DCW inlet and DHW outlet tank piping along with a T&P relief valve are installed as normally required. DCW line pressure provides for the flow of DHW to plumbing fixtures. A tempering valve must be installed in the DHW outlet piping. The tempering valve is also connected to the DCW line to temper the DHW and provide a form of antiscald protection for system users. The primary antiscald protection is installed at the fixture itself in the form of thermostatic-, pressure-, or thermostatic-pressure-sensitive antiscald valves. Local plumbing codes should be consulted for antiscald requirements.

The HTF in the solar loop of this system is protected using three methods. The first method of freeze protection involves a freeze protection setting that activates the pump if the collector sensor reaches a set temperature. A temperature of 40°F (4°C) is recommended. This will circulate warmer HTF contained within the tank to the collector, protecting it and the vulnerable collector piping against freezing.

The second method of freeze protection involves using a freeze-protection valve that can be set to drain a small amount of HTF out of the system. Therefore, HTF flow is created through the collector until the ambient temperature rises and the valve closes. The setpoint for a freeze-protection valve is usually between 35°F (2°C) and 45°F (7°C). This valve acts as freeze protection backup if power to the system is lost for an extended period of time in cold weather.

The final method of freeze protection involves manually isolating and draining down the collector and the solar loop piping. This is accomplished by closing the two solar loop isolation valves and opening the boiler drains. All HTF in the collector and piping must be removed, or any remaining HTF will freeze and possibly damage the collector or piping. *Note:* To successfully drain down the solar collector and remove all HTF, the collector should be installed with a grade of ¼″ per foot sloped toward the collector inlet. **See Figure 6-27.**

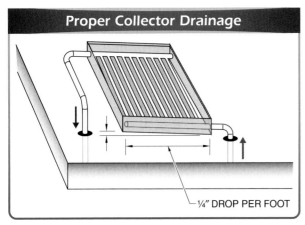

Proper Collector Drainage

¼″ DROP PER FOOT

Figure 6-27. The collector should be installed with a grade of ¼″ per foot sloped toward the collector inlet to ensure complete drain down.

Since this system uses DHW as the HTF, it is recommended for mild temperatures only. Freeze protection for this system should be considered only as protection against extreme conditions and not for normal operation.

Active Indirect Systems

An active indirect system uses a pump or circulator to force an HTF that is not the DHW through the solar loop. Heat transfer takes place in a heat exchanger. The solar loop consists of the solar collector and piping connected to the inlet and outlet of the heat exchanger. The secondary loop consists of the potable cold water inlet, storage tank, and outlet piping to the DHW fixtures.

PUMPED HTF HEATING SYSTEMS

System Type: Indirect Pumped

System Name: A. O. Smith Cirrex® Solar Water Heating System SAC102 200402-80

SRCC OG-300 Certification #: 2010123B

Solar Collector: Chromagen Solar Energy Systems CR-130-A-P (glazed flat-plate)

SRCC OG-100 Certification #: 2009061D

Storage Tank: A. O. Smith SUNX-80 (residential electric indirect solar-booster water heater with integral single-wall heat exchanger); 80 gal. capacity

HTF: 50% propylene glycol and 50% distilled water solution

HTF Protection: Antifreeze and circulation

Recommended Climate: Moderate; −30°F (−34°C) HTF freeze point

The A. O. Smith Cirrex® SAC102 200402-80 solar water heating system is a complete packaged system that includes solar collectors, a solar storage tank / water heater with integral heat exchanger, an expansion tank, and a pump control module. **See Figure 6-28.** It is an active indirect system designed specifically for residential installations. Because the HTF runs through the heat exchanger and does not contact the DHW in the storage tank / water heater or the DHW piping, it is considered a closed loop system.

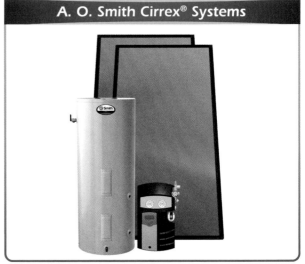

A. O. Smith Cirrex® Systems

A. O. Smith

Figure 6-28. The A. O. Smith Cirrex® solar water heating system includes solar collectors, a solar storage tank/water heater with internal heat exchanger, an expansion tank, and a pump control module.

The manufacturer of this system made it simple to choose the proper system for most installations. By using the A. O. Smith Residential Solar Sizing Guide, it is easy to determine that this particular system can be used by a family of four in almost three-quarters of the United States. **See Figure 6-29.** Finding the right system to install is made easy with this guide and similar guides. However, the guides are only designed for use with the systems of the specific manufacturer. Although many solar water heating systems are similar, each manufacturer has sizing parameters for its systems, and only those parameters should be followed.

This particular system uses two glazed flat-plate solar collectors to harvest solar radiation. The collectors are typically mounted on the roof of the residence and piped to the pump control module. The module for this system is referred to as the pump station and is mounted close to the storage tank / water heater. Care should be taken to ensure that the length of system piping is as short as possible. **See Figure 6-30.** Per the manufacturer specifications, the system configuration is limited to a maximum of 200′ of solar loop piping. Longer lengths of piping may require the use of a different pumping system or configuration due to friction losses. Friction losses reduce the benefits of the ease of design and installation and increase costs.

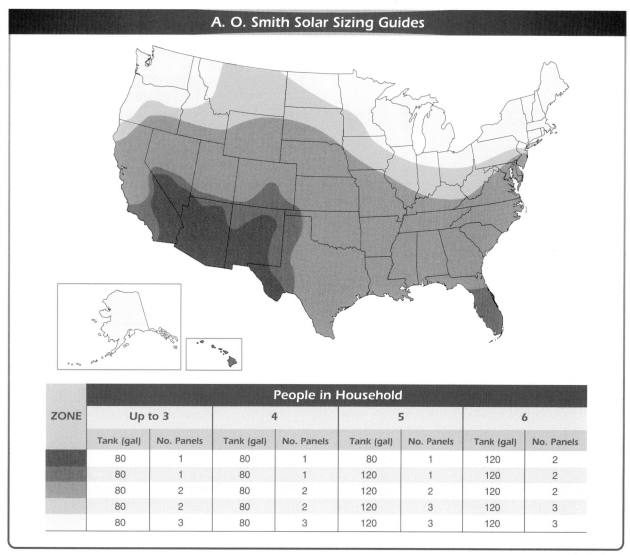

A. O. Smith Solar Sizing Guides

ZONE	People in Household							
	Up to 3		4		5		6	
	Tank (gal)	No. Panels	Tank (gal)	No. Panels	Tank (gal)	No. Panels	Tank (gal)	No. Panels
	80	1	80	1	80	1	120	2
	80	1	80	1	120	1	120	2
	80	2	80	2	120	2	120	2
	80	2	80	2	120	3	120	3
	80	3	80	3	120	3	120	3

A. O. Smith

Figure 6-29. The A. O. Smith Residential Sizing Guide makes it easy to determine which residential solar water heating system meets the needs of occupants based on location and number of occupants.

A. O. Smith Cirrex® System Components

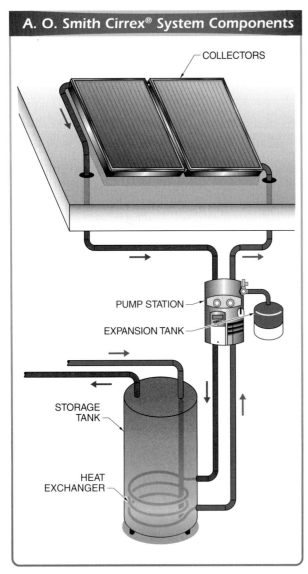

Figure 6-30. Per the manufacturer specifications, the A. O. Smith Cirrex solar water heating system is limited to 200 ft of solar loop piping.

A. O. Smith Pump Stations

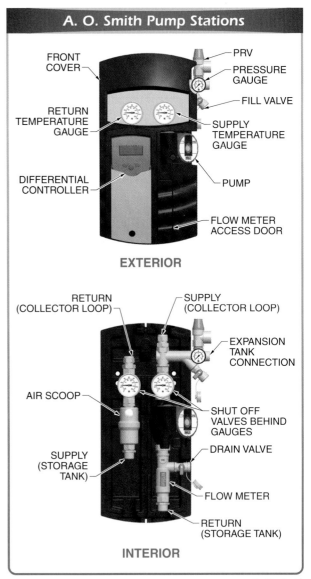

A. O. Smith

Figure 6-31. A compact pump station includes the pump, valves, gauges, and piping connection needed to operate the system.

The ease of system installation is made possible by the pump station. The pump station contains most of the manifold and necessary valves. Isolation valves, check valves, PRVs, pressure gauges, the flow meter, the pump, and piping connections are all contained in this relatively small pump station. **See Figure 6-31.** There is no need to create manifolds or solder valves and fittings for this system. It has already been done for the installer. Therefore, a small amount of space is needed for the system, and the cost of installation is low.

The pump station also contains the differential controls and wiring connections for the system. Two temperature sensors are required to operate the system. One temperature sensor must be installed on the collector outlet piping and one sensor on the storage tank at the lower door opening. The sensor wiring is then connected to the electrical terminals on the system controller, which is part of the pump station. The power supply wiring is also connected to the terminal. **See Figure 6-32.**

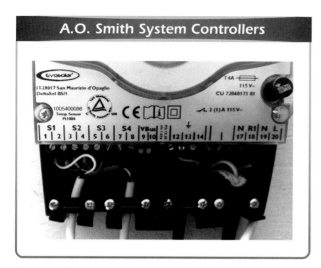

A.O. Smith System Controllers

Figure 6-32. Sensor and power supply wiring are connected to terminals on the system controller.

The pump station has a programmable digital system controller that can be used with the system default parameters or be customized by the installer or building owner. Other control functions are available if additional sensors are installed and wired to the system. The default settings for the pump station are as follows:

- differential temperature at 10°
- switch-off temperature differential at 4°
- maximum storage tank temperature at 160°F (71°C)
- maximum collector temperature at 285°F (141°C)
- freeze protection switch-on at 40°F (5°C)

This type of pump station is a very attractive option compared to a complicated customized system especially for a user familiar with HVACR thermostat LCD screens. The digital controller screen allows the user to easily program the system and monitor system temperatures.

Once the collectors and the pump station are mounted and the storage tank is located, the inlet and outlet piping from the collector are piped to the pump station piping connections at the top of the station. The lower connections of the station are piped to the storage-tank heat exchanger inlet and outlet. **See Figure 6-33.** The collector outlet piping connects through the station to the upper or inlet port of the internal heat exchanger. The collector inlet piping connects from the lower heat exchanger port through the station to the collector inlet. This is the solar loop of the system.

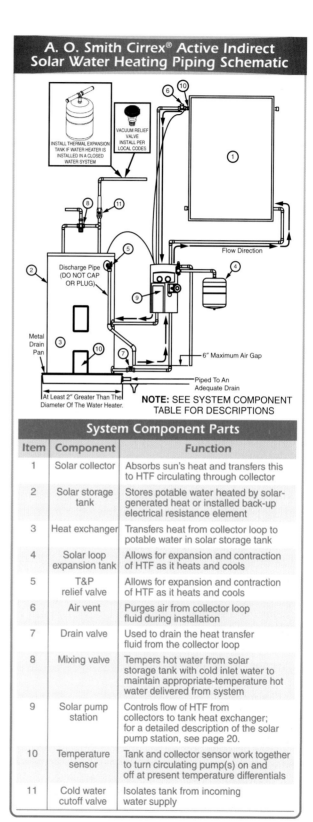

A. O. Smith Cirrex® Active Indirect Solar Water Heating Piping Schematic

NOTE: SEE SYSTEM COMPONENT TABLE FOR DESCRIPTIONS

	System Component Parts	
Item	Component	Function
1	Solar collector	Absorbs sun's heat and transfers this to HTF circulating through collector
2	Solar storage tank	Stores potable water heated by solar-generated heat or installed back-up electrical resistance element
3	Heat exchanger	Transfers heat from collector loop to potable water in solar storage tank
4	Solar loop expansion tank	Allows for expansion and contraction of HTF as it heats and cools
5	T&P relief valve	Allows for expansion and contraction of HTF as it heats and cools
6	Air vent	Purges air from collector loop fluid during installation
7	Drain valve	Used to drain the heat transfer fluid from the collector loop
8	Mixing valve	Tempers hot water from solar storage tank with cold inlet water to maintain appropriate-temperature hot water delivered from system
9	Solar pump station	Controls flow of HTF from collectors to tank heat exchanger; for a detailed description of the solar pump station, see page 20.
10	Temperature sensor	Tank and collector sensor work together to turn circulating pump(s) on and off at present temperature differentials
11	Cold water cutoff valve	Isolates tank from incoming water supply

Figure 6-33. The A. O. Smith Cirrex piping schematic contains all the information on connecting the components of the system.

HTF is drawn from the lowest part of the integral heat exchanger coil after it has transferred its heat to the DHW in the tank. The HTF is then pumped to the solar collector inlet and through the collector to harvest heat from the sun. The HTF then flows out of the collector, down to the upper inlet of the heat exchanger coil, and through the coil where it will transfer its heat to the DHW.

As do other systems, this solar water heating system has DHW and HTF protection devices. The storage tank / water heater must have a T&P relief valve to protect against high temperatures and pressure. The DHW outlet piping from the storage tank must be piped to a tempering valve, which is also connected to the DCW to regulate the outgoing DHW to a safe temperature. Check valves are included in the pump station to prevent reverse thermosiphoning in the solar loop. The pump station also includes a PRV that releases the HTF if pressures build up within the solar loop. A boiler drain valve must be installed in the lower heat-exchanger outlet piping to provide a means for draining the HTF from the collectors. *Note:* Some plumbing codes limit the pressure within a single-wall heat exchanger in contact with the potable water to 30 psi.

The solar loop is protected by an additional safety device, which is an expansion tank. An expansion tank is equipped with an integral bladder. The bladder is cushioned with air and absorbs the increased volume of HTF created by thermal expansion when the water is heated. **See Figure 6-34.** When the solar loop and collectors are filled and the HTF is heated, there will be an increase in volume and pressure inside the piping system. Because it is a closed loop system, there is no room for thermal expansion. *Note:* In open loop systems, the overflow from thermal expansion can be spread throughout the DHW or DCW system. The DHW system is required to have its own thermal expansion protection, which is an expansion tank located close to the water heater in most cases.

Without taking measures to address the added pressure and volume, the PRV would continually relieve HTF from the system. The expansion tank absorbs the increased pressure and volume from thermal expansion. The expansion tank also accepts additional HTF that may be pushed into the tank under steamback conditions within the solar collector.

The expansion tank supplied with the system is attached to the proper port on the pump station. In other systems that may not use a pump control module, the expansion tank must be mounted on the collector

inlet piping with the pump and check valve mounted below the tank branch. The tank must be installed in the downward position. This keeps the bladder wet and also facilitates the collection and holding of HTF that flows back to the tank under steamback conditions. The air reservoir of the tank should be filled to the manufacturer-recommended air pressure. In this case, the recommended air pressure is 25 psi.

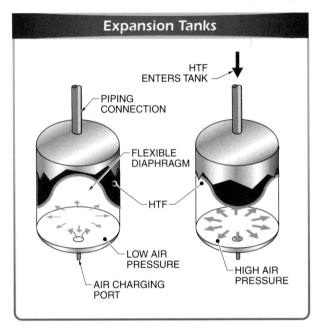

Figure 6-34. An expansion tank is required to prevent thermal expansion damage in a closed loop system.

Once the piping and wiring are complete, the system should be filled with the recommended solution of 50% supplied solar antifreeze and 50% distilled water. An air vent should be installed at the outlet of the solar collector to facilitate removal of air from the system. Flow through the solar loop should be set to the manufacturer-recommended flow rate. In this case, the recommended flow rate is 2 gpm. This rate can be verified by the flow meter within the pumping station. The pump in the station has three speeds. The proper pump speed should be selected to produce the 2 gpm needed. The flow rate can be further adjusted by the flow restrictor on the station mounted just before the pump.

Under normal operating conditions, the isolation valves in the pump station are open. The pump is activated when there is a temperature differential of 10° between the storage tank sensor and the collector outlet temperature. The system will run until there is a 4° temperature differential between the collector sensor and the tank sensor when the pump is deactivated. The pump also deactivates if the tank sensor reaches the maximum tank temperature of 160°F (71°C).

There are other system activation and deactivation parameters available with this controller. For example, a freeze protection parameter can be set to activate the pump when temperatures at the controller outlet drop to 40°F (5°C). HTF will then circulate through the system and warm the collector and vulnerable piping. *Note:* Manufacturer instructions for system and pump station installation and control parameters must be followed for the solar water heating system to function properly and efficiently.

The storage tank / water heater should be installed and piped as required by local plumbing or mechanical codes. The thermostat should be set at the proper hot water temperature, normally 120°F (49°C). The water heating elements, which act as the backup for the solar water heating system, should be left on. If the solar system cannot provide enough heat to create 120°F (49°C) within the tank, the water heating element will provide additional heat to bring the DHW to the proper temperature. DHW is supplied to fixtures through the DCW line pressure.

This type of active indirect system is the most popular type of system currently being installed in North America. The antifreeze protection provided by the HTF solution allows the system to be installed in most areas with moderate freezing temperatures. The solution prevents the system from freezing in temperatures down to –30°F (–34°C). However, as an added safety precaution, the system should be installed only in areas that never reach below 0°F (32°C). For continuous freeze protection, the system must be maintained properly and the HTF must be checked for quality annually.

To ensure system freeze protection, all plumbing and discharge lines must be installed per local plumbing code requirements. To successfully drain down the solar collector and remove all HTF, all plumbing lines within attic areas and exposed roof areas must be graded at a minimum of ¼″ per foot back to the drain valves. All valves should be installed where they are readily accessible. In addition, the collector should be installed with a grade of ¼″ per foot tilted toward the collector inlet.

Drainback Systems

The SunEarth Inc. Cascade Drainback™ solar water heating system model EPRD48-80-2 is an active indirect drainback system. Although it is not a packaged system, the solar collector manufacturer, SunEarth, has specific recommendations for other system components and installation parameters. These components and parameters were selected by the manufacturer to achieve maximum efficiency of the solar water heating system. Therefore, manufacturer recommendations should be followed to ensure maximum efficiency and compliance with the SRCC OG-300 certification requirements.

PUMPED HTF DRAINBACK HEATING SYSTEMS

System Type: Indirect Pumped

System Name: SunEarth Inc. Cascade Drainback™ Solar Water Heating System EPRD48-80-2

SRCC OG-300 Certification #: 2001027E

Solar Collector: SunEarth Inc. Empire EP-24 (glazed flat-plate)

SRCC OG-100 Certification #: 2007032C

Storage Tank: Rheem Solaraide 81V80HE-1 (double-wall heat exchanger and electric water heater); 80 gal. capacity

HTF: Water

HTF Protection: Drainback system

Recommended Climate: Cold; minimum temperatures of –50°F (–46°C)

This active indirect system is designed as a drainback system. A drainback system allows for the HTF to be drained out of the system and into a reservoir or tank every time the pump is deenergized such as when the ambient air approaches freezing temperatures. This will also act as the freeze protection method which allows for the system to be installed in cold temperature locations. The installation manual for this system states the following: "The Cascade Drainback™ solar water heating system may not be installed in areas within the continental United States where the annual ambient temperature has ever fallen below –50° Fahrenheit (–46° Celsius). The Cascade Drainback™ system must be installed as specified in this manual to have effective freeze protection at these low temperatures."

There are several different systems manufactured by SunEarth. The Cascade Drainback consists of two glazed flat-plate solar collectors, a storage tank with integral double-wall heat exchanger, and a back-up electric water heater. This tank configuration is a two-tank system. One tank stores solar heat or acts as a preheater, and the other acts as a back-up water heater. The system also consists of a pump or circulator, a control module, and a drainback tank or reservoir. **See Figure 6-35.**

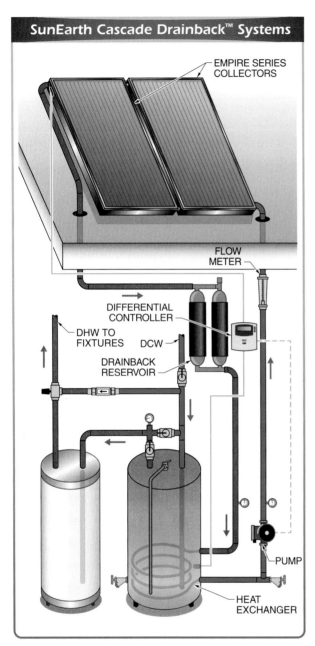

SunEarth Cascade Drainback™ Systems

EMPIRE SERIES
COLLECTORS

FLOW
METER

DIFFERENTIAL
CONTROLLER

DHW TO
FIXTURES   DCW

DRAINBACK
RESERVOIR

PUMP

HEAT
EXCHANGER

Figure 6-35. The SunEarth Cascade drainback system uses one tank for solar heat storage and another as a backup water heater.

The solar collectors must be properly mounted above the system, typically on the roof. In drainback systems, the collectors must be designed for complete emptying of the HTF. The collectors should also be installed with at least a 20° tilt from horizontal to facilitate draining of the HTF. A harp-style solar collector is ideal. The collectors should also be graded to the collector inlet to

further aid draining of the collector. An air vent is not required at the collector outlet for this installation and, in fact, it would be detrimental to system operation. Air must be retained in the solar loop and not eliminated.

The two tanks should be located as close as possible to the collectors. This is necessary to allow the HTF to quickly flow out of the collectors by gravity. Long lengths of horizontal piping and too many 90° elbows in the solar loop will slow the flow of HTF. Horizontal offsets in the solar loop should be graded at least ¼″ per foot. When possible, 45°elbows and offsets should be used to maintain the proper flow.

The collector inlet should be piped to the lower or outlet port of the storage tank heat exchanger. The outlet of the collector is piped to the upper port or inlet of the storage tank heat exchanger. A boiler drain should be installed in the lowest portion of the collector inlet piping between the heat exchanger outlet and the system pump to facilitate manual draining of the system.

A properly sized drainback tank or reservoir should be mounted close to the storage tank. **See Figure 6-36.** It is recommended by the manufacturer that the Copperstore drainback reservoir be used rather than a tank. The reservoir is designed for ease of installation. It is mounted in a conditioned space with a wall-mounted bracket. This reservoir is much thinner than a shelf-mounted tank and can save space in the storage/mechanical room. Because the reservoir is constructed of copper, the piping connections can be soldered or brazed rather than joined using pipe and dielectric unions. It also has a threaded female adapter for the installation of a PRV at the top of the reservoir manifold.

The drainback reservoir is sized according to the volume of HTF in the solar collectors and solar loop piping above the drainback tank. As a precaution, it is recommended that volume of up to 1 gal./collector be added to the capacity of the drainback tank. Manufacturer recommendations should be followed for proper drainback tank sizing. The manufacturer may provide an easy method for calculating the tank size and the line sizes to the collectors. **See Figure 6-37.** Tables may be used to determine the proper size of the reservoir and the collector inlet and outlet piping. For this installation, the third line in either Table 1 or 2 could be used. For longer runs of piping, Table 1 could be used in which a ½″ collector inlet piping and ¾″ outlet piping with maximum line distances ranging from 73′ to 185′ are required. The longer the distance, the more volume of HTF is contained in the piping and, consequently, the larger the reservoir.

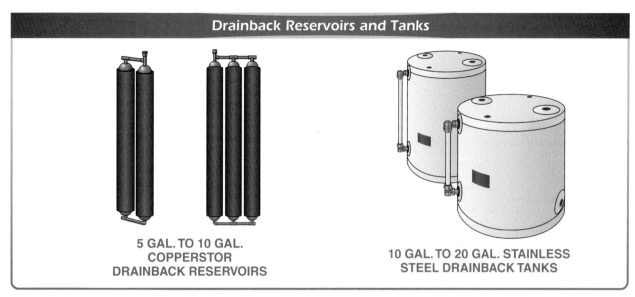

Figure 6-36. SunEarth manufactures drainback reservoirs and drainback tanks.

Pipe Run Lengths for ½″ Supply and ¾″ Return Lines			
	SunEarth Drainback Reservoir		
Array	DB5.0	DB7.5	DB10.0
	Total Run of Solar Supply and Return Loop Piping*		
1 EP/EC-40	82	138	194
2 EP/EC-21	76	132	188
2 EP/EC-24	73	129	185
2 EP/EC-32	62	118	174

Pipe Run Lengths for ¾″ Supply and ¾″ Return Lines			
	SunEarth Drainback Reservoir		
Array	DB5.0	DB7.5	DB10.0
	Total Run of Solar Supply and Return Loop Piping*		
1 EP/EC-40	61	103	144
2 EP/EC-21	57	98	140
2 EP/EC-24	54	96	138
2 EP/EC-32	46	88	130

* in ft

Figure 6-37. Manufacturer piping charts make it easy to choose the proper drainback reservoir for a specific system.

The solar loop in a drainback system does not need to be pressurized. This is because it is not completely filled with HTF as are other solar water heating systems. Air is entrained within the system, which allows the solar loop to drain down when the pump is deactivated. Once flow has stopped within the piping, the air will begin to rise to the upper levels of the solar loop and replace the

HTF. To accomplish this, the drainback tank or reservoir must be installed in the correct location within the solar loop per manufacturer recommendations. In this case, it is installed in the collector outlet piping with the bottom of the drainback tank or reservoir 18″ above the heat exchanger inlet fitting of the storage tank and 48″ above the pump suction inlet. **See Figure 6-38.**

Note: These are minimum requirements. The drainback tank or reservoir can be mounted higher. However, it should be accessible for maintenance and be visible if it includes a sight glass. The sight glass allows for the monitoring of the HTF level when the system is not in operation.

Alternate Energy Technology
Some drainback tanks contain an internal heat exchanger that can be used to transfer heat from the HTF to another fluid such as the DHW.

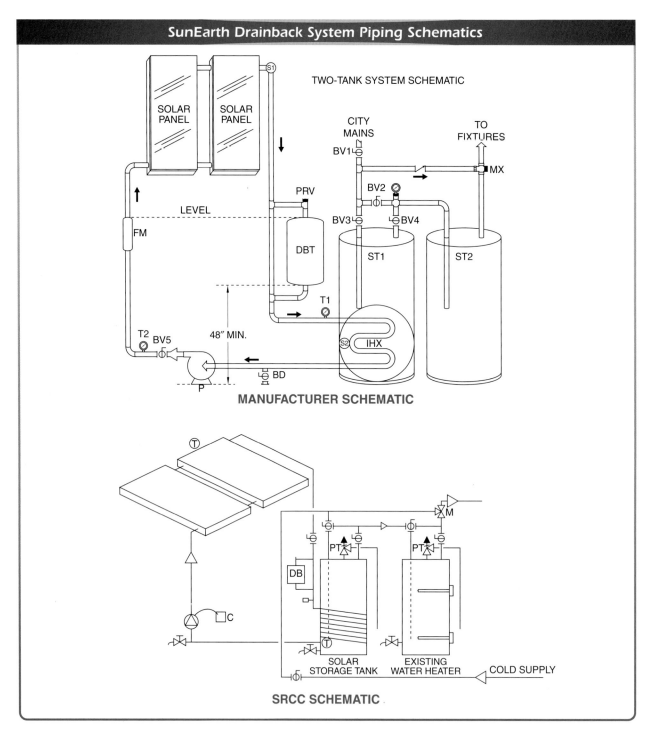

Figure 6-38. The proper piping configuration is provided by the manufacturer. The SRCC also supplies a system schematic for each rated system.

Regardless of whether the drainback tank has a sight glass, a flow meter should be installed in the collector inlet piping at the same level as the drainback tank or reservoir. The center of the flow meter should be at the HTF level in the drainback tank when the HTF is not circulating. This provides a method of monitoring the HTF level in the system and checking the flow rate. The manufacturer of this system recommends the Blue White F-450 flow meter or its equivalent. **See Figure 6-39.**

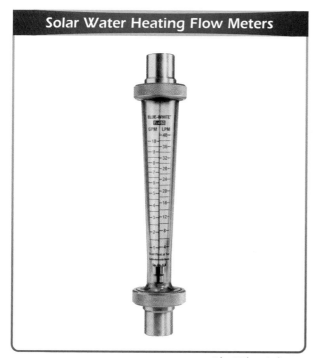

Figure 6-39. A flow meter should be installed in a system and used to monitor HTF level and flow rate.

Isolation ball valves should be installed before and after the pump and on the collector outlet piping before the drainback tank to facilitate maintenance and repair of the system components. It is recommended by the manufacturer that the collector outlet line be teed off to install the drainback tank. However, many systems allow for the collector outlet piping to be connected directly to the drainback tank inlet and outlet. Temperature gauges should be installed on the collector outlet piping before the heat exchanger inlet and on the outlet side of the pump on the collector inlet piping. Sensors should also be attached to the collector outlet piping and on the storage tank at the lower tank opening.

Sensor wiring should then be connected to a temperature differential controller. Several different models can be used. However, the controller should be designed for use on a drainback system. The activation-differential temperature setting should be set at 20°. The deactivation differential should be set at 4°. The maximum storage tank temperature should be set from 150°F (66°C) to 160°F (71°C).

Two three-way ball valves are installed at the storage tank DCW inlet 1 and the back-up water heater

inlet 2. A manifold connects the outlet from the storage tank three-way ball valve to the storage tank DHW outlet, the back-up water heater DHW inlet three-way ball valve, and the back-up water heater outlet piping. Isolation valves are installed on the DHW outlet of the storage tank and the back-up water heater. The DCW connects to a tempering valve at the outlet of the back-up water heater installed after the connection to the three-way valve manifold.

This configuration allows for increased flexibility of the solar water heating system. Under normal operating conditions, three-way ball valve 1 is set to feed DCW into the storage tank where it is heated by the solar-heated HTF in the heat exchanger. That heated water then flows out of the storage tank and to three-way ball valve 2. There, it is set to flow into the back-up water heater where it can be stored or heated to the proper DHW temperature by the water heater electrical elements. The DHW then flows out of the back-up water heater to the tempering valve where it is mixed with cold water to achieve the proper temperature. The water then flows to the fixture supplies.

The three-way ball valves and isolation valves allow for either the storage tank or the water heater to be bypassed if necessary. Check valves should not be installed in this system, including the internal check valve that is installed in many pumps and circulators. The internal check valve must be removed from the pump if present. Check valves stop the flow of air to the collectors when the pump is deactivated.

Once the piping is installed, tested, and flushed, the system can be filled. The storage tank and back-up water heater are filled with DCW. The solar loop can also be filled with water. The solar loop HTF in this system is designed to drain out the collector and vulnerable piping to prevent freezing, thus normal household water can be used. However, distilled water is preferable due to its lack of minerals and added chemicals. An antifreeze solution of 10% to 20% could be used during severe cold conditions.

The solar loop should be filled from the boiler drain on the heat exchanger outlet piping. The loop should be filled to 1″ below the top of the flow meter. On most drainback tanks, a PRV set to a maximum of 35 psi is installed on the side of the tank. With the Copperstore reservoir, the PRV is installed on the top of the unit. While filling the solar loop, the PRV should be left off until the excess air is removed. However, air will remain in the solar loop piping above the drainback tank or reservoir after filling the solar loop.

Under normal operating conditions when there is a 20° differential between the solar collector temperature and the storage tank temperature, the pump will activate and force HTF up and through the solar collector to harvest solar radiation. The HTF then flows down to the drainback tank or reservoir. The entrained air in the loop collects within the tank or reservoir. The HTF drains out of the drainback tank by gravity and pump suction and flows down through the heat exchanger to the pump.

The manufacturer-recommended HTF flow rate through the solar collectors should be used. Manufacturers may provide tables for determining the proper flow rates for their collectors. **See Figure 6-40.** For two EP-24 collectors, the recommended flow rate is 1.2 gpm. The proper pump that will provide this flow should be selected. The flow can be adjusted by throttling the ball valve downstream of the pump if necessary.

SunEarth Recommended Flow Rates	
Collector Array	Flowrate (GPM)
1 EP/EC-40	1.0
2 EP/EC-21	1.1
2 EP/EC-24	1.2
2 EP/EC-32	1.7

Figure 6-40. Manufacturers usually provide tables for determining the proper flow rates through their collectors.

When the temperature differential between the collector outlet and the storage tank reaches 4°, the pump deactivates. The HTF within the collector and vulnerable piping stops flowing, which allows the entrained air to rise in the system. HTF then drains down on both sides of the solar loop down through the drainback tank or reservoir and into the collector inlet piping. The HTF level should be visible in the flow meter. When the storage tank temperature reaches the maximum set temperature, the controller will deactivate the pump, and the solar loop will drain back.

A minimum ambient air temperature can also be set on some controllers. When the minimum ambient air temperature is reached, the system will deactivate and the HTF within the collector will drain back to the reservoir. The controller should also be set so that if there is a loss of power, the system will deactivate and the HTF will drain back. This provides fail-safe protection of the collector and vulnerable piping from freezing conditions.

The back-up water heater should be set for the proper DHW temperature and left on. If the solar water heating system cannot provide enough heat to bring the DHW up to the set temperature, the water heating elements provide the necessary heat.

When properly installed and maintained, the drainback solar water heating system is ideal for mild and cold weather conditions. It is also fairly simple for a homeowner to operate and maintain. There is no need for antifreeze solutions, and the system does not need to be drained down in freezing weather.

Facts

Radiant heating systems have much larger heating surfaces than conventional systems. Much lower temperature is needed to provide the same level of heat transfer.

Combination Systems

A combination system, also called a combi system, is a solar water heating system that uses the heat harvested from collectors to heat the DHW supply and HTF used for other purposes. With the SunMaxx Solar™ HelioMaxx Combi System, the DHW and a radiant space heating system use the solar-heated HTF in an active indirect solar water heating system. **See Figure 6-41.** This particular packaged system is recommended by the manufacturer for a 2000 sq ft home.

PUMPED COMBINATION HEATING SYSTEMS

System Type: Indirect Pumped

System Name: SunMaxx Solar™ HelioMaxx Combi System

SRCC OG-300 Certification #: None

Solar Collector: SunMaxx Solar ThermoPower VHP 30 (evacuated-tube heat-pipe)

SRCC OG-100 Certification #: 2006011D

Storage Tank: StorMaxx CTec 211-3HX (with three integral heat exchangers; 211 gal. capacity)

HTF: Antifreeze

HTF Protection: Antifreeze and circulation

Recommended Climate: Moderate to cold; minimum temperatures of −30°F to −50°F (−34°C to −46°C)

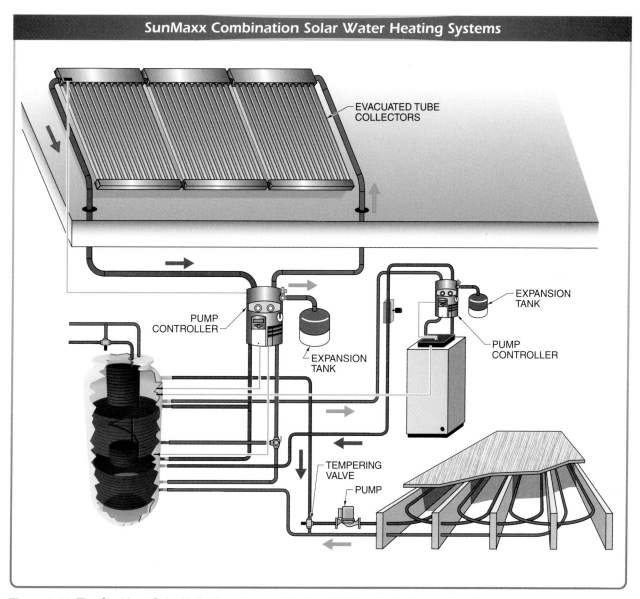

SunMaxx Combination Solar Water Heating Systems

EVACUATED TUBE COLLECTORS

EXPANSION TANK

PUMP CONTROLLER

EXPANSION TANK

PUMP CONTROLLER

TEMPERING VALVE

PUMP

Figure 6-41. The SunMaxx Solar HelioMaxx is a combination DHW and radiant space heating system.

Because the water heating and space heating systems are combined, combination systems are larger than single-use systems. The SunMaxx Solar HelioMaxx combination system includes three 30-tube evacuated-tube heat-pipe solar collectors and a large storage tank. The two systems, a solar DHW system and a radiant space heating system, use the same storage tank. The storage tank is a StorMaxx CTec 211-3HX, which is a 200 gal. storage tank with three integral-coil heat exchangers. **See Figure 6-42.**

Two heat exchanger coils are used for heat transfer from the collectors to the tank, and one heat exchanger coil is used to heat the DHW.

The storage tank interior reservoir contains the HTF used by the radiant space system. The storage tank is heated by a boiler system if the solar water heating system cannot produce enough heat for the entire system. In most cases, the solar water heating system acts as a preheater for the boiler system especially during severe cold weather.

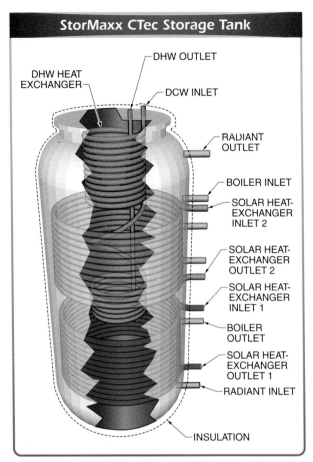

StorMaxx CTec Storage Tank

DHW OUTLET

DHW HEAT
EXCHANGER

DCW INLET

RADIANT
OUTLET

BOILER INLET

SOLAR HEAT-
EXCHANGER
INLET 2

SOLAR HEAT-
EXCHANGER
OUTLET 2

SOLAR HEAT-
EXCHANGER
INLET 1

BOILER
OUTLET

SOLAR HEAT-
EXCHANGER
OUTLET 1

RADIANT INLET

INSULATION

Figure 6-42. The SunMaxx system storage tank has a 211 gal. capacity and three heat exchangers.

The configuration or installation requirements for the solar loop piping are the same for combination systems as for other systems. The solar loop consists of three evacuated-tube heat-pipe solar collectors and a UniMaxx Plus solar pump station containing a pump, a temperature differential controller, an expansion tank, and isolation valves. A preinsulated line set of various lengths may also be supplied with the kit. The line set contains insulated corrugated stainless steel piping and wiring for the collector outlet sensor. **See Figure 6-43.** The line set should be installed with the appropriate supports and the proper horizontal and vertical alignment according to local and national plumbing and mechanical codes.

Evacuated tube collectors are normally mounted and properly oriented on the building roof. They are coupled together in parallel with an air vent installed on the outlet piping. The collector inlet piping connects to the top

of the pump station and then to a three-way valve that connects to the bottom outlet of heat exchanger coils 1 and 2. The collector outlet piping connects to the top of the pump station and then tees to the upper outlet of heat exchanger coils 1 and 2. Isolation valves can be installed at the heat exchanger outlet and a boiler drain piped in the collector inlet piping after the storage tank isolation valve. An expansion tank and a PRV also connect to the proper ports on the pump station.

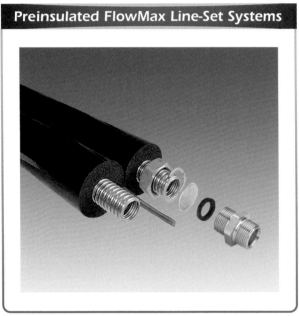

Preinsulated FlowMax Line-Set Systems

SunMaxx Solar

Figure 6-43. Manufacturers may supply corrugated stainless steel line sets with their prepackaged systems.

The solar loop functions as any other solar loop in an active indirect system. HTF is pumped from the lower outlets of heat exchanger coils 1 and 2 and flows to the solar collectors. It harvests the solar radiation absorbed by the collectors and flows back to the inlets of the two heat exchanger coils. The solar-heated HTF then transfers that heat to the HTF contained within the storage tank. The three-way modulating valve allows solar-heated HTF into just one heat exchanger coil, or both as needed. Both coils are used when there is a high temperature differential. The lower coil is used when there is a low temperature differential. The HTF recommended for the solar loop is a 50% water and 50% propylene glycol solution. Manufacturer-recommended HTF should always be used in radiant heating systems.

Temperature sensors are placed on the collector outlet and the upper and lower levels of the storage tank. The pump station differential controller operates the same as the other systems. It activates the pump when it senses the set 10° to 20° difference between the outlet temperature of the collectors and the storage tank temperature. The pump is deactivated when the temperatures are within 4°.

DHW storage is unique in that the DHW is not held in the storage tank but in the center heat exchanger coil. DCW enters the top heat exchanger coil inlet and flows through the heat exchanger coil from bottom to top. Heat is transferred from the HTF to the DCW within the storage tank and exits as DHW at the top outlet. The DHW is tempered at the tempering valve also connected to the DCW. The DHW then flows to the fixture supplies by DCW line pressure.

The boiler supply and return piping connect to two side storage-tank ports. The boiler system has a separate pump station to move HTF from the storage tank to the boiler and back to the tank. It acts as either the primary or back-up heater depending on the amount of solar insolation. The boiler activates if the solar coils cannot heat the HTF in the tank to the proper temperature. The boiler is not included in this packaged system. Once the boiler is selected, manufacturer recommendations for the pump station parameters should be followed.

The radiant heating system is combined with the solar system at the storage tank. The radiant heating system inlet and outlet piping is connected to additional side ports on the storage tank. The radiant system uses the HTF held within the tank for use in the radiant floor-heating system. The radiant hot outlet or supply piping connected at the top portion of the tank connects to a three-way ball valve, which also connects to a tee in the radiant return line. A pump or circulator is placed in the radiant supply line to push HTF through the radiant system. Radiant return water either flows back to the tank or immediately returns to the radiant supply piping through the three-way valve if its temperature is high enough.

Under normal operating conditions, the solar-heated HTF flows through heat exchanger coils 1 and 2 and transfers heat to the HTF within the tank. The HTF within the tank transfers heat to the DHW heat exchanger, which supplies the DHW to fixtures. HTF within the tank is used in the radiant floor-heating system. The boiler, supplied with its own control station, heats the HTF within the tank when the solar-heated HTF cannot heat both the DHW and the radiant HTF.

The HTF in this system is protected from freezing by the antifreeze and water solution. It is also protected from overtemperature conditions by steamback. The expansion tank used in this system is sized to contain the added HTF from steamback conditions. All piping should be installed to drain back to the pump station. The PRV installed at the pump station protects this particular combination system from high pressures. A T&P relief valve must also be installed at the storage tank.

There are hundreds of different configurations of combination solar water heating systems, which use various types of collectors, storage tanks, and piping designs. The uses for solar-heated HTF are many and varied, and solar water heating systems can offer significant cost savings for both residential and commercial building owners.

Active Direct Solar Pool Heating Systems

The residential solar pool heating system is the most commonly installed solar water heating system in North America. The U.S. Energy Information Administration reported in 2010 that over 20,000,000 sq ft of pool heating solar collectors were shipped in the United States during 2008 and 2009. This is approximately 77% of all the solar water heating collectors shipped during those two years. **See Figure 6-44.**

FAFCO

Figure 6-44. Pool heating solar collectors consist of the majority of solar water heating collectors in North America.

There are a few reasons for these facts. The solar pool heating system is the most cost-effective solar water heating system. The unglazed flat-plate solar collector used in nearly all systems is less expensive and much easier to install than other solar collectors. Other system components are also much less expensive than those used in other systems. The system may include a few simple valves and some PVC pipe or a complicated multivalve automatic system with a controller and temperature sensors. Either way, the system is still far less expensive, and thus more popular, than other solar water heating systems.

The solar pool heating system is a low-temperature system that, in most cases, heats pool water from 20°F to 30°F above normal pool temperature. Most pool usage occurs in mild temperatures during the late spring, summer, and early fall. Pool water is normally heated from an existing 60°F (16°C) to 70°F (21°C) or higher to a temperature of 80°F (27°C) to 85°F (29°C). During these mild temperatures, the solar pool heating system can usually provide all of the pool heating needs. Therefore, significant fuel-cost savings result when compared to the fuel usage of a fossil-fuel pool heater. This allows the payback period for the system to be just a few years rather than the 10 to 20 years of other solar water heating systems.

PUMPED POOL HEATING SYSTEMS

System Type: Direct Pumped

System Name: Fafco Solar Pool Heating System

SRCC OG-300 Certification #: None

Solar Collector: Fafco Sunsaver (unglazed flat-plate)

SRCC OG-100 Certification #: 2007051A

Storage Tank: None

HTF: Pool water

HTF Protection: Drain down

Recommended Climate: Nonfreezing conditions

Most solar pool heating systems are active direct nonpressurized systems. Active indirect systems are rare and are usually commercial indoor pools using glazed flat-plate collectors. For example, the Fafco Solar Pool Heating System is a typical residential pool heating system. The solar loop of a typical solar pool heating system has an unglazed flat-plate collector, a three-way valve, a check valve, and PVC pipe and fittings. A more complicated system may include an automatic three-way valve, controls, temperature sensors and gauges, and isolation valves.

Normally pool heating systems use unglazed flat-plate collectors such as the Fafco Sunsaver. **See Figure 6-45.** The collector is lightweight and relatively easy to install. It is normally installed directly on the roof without spacing or support underneath. The roof should be sloped at least 6° to 10° to facilitate the drain down of the panel. Flat roofs do not allow the plates to drain completely for freeze prevention. There are some unglazed flat plates that are designed for flat mounting. It must be ensured that the correct unglazed flat plates are used.

FAFCO

Figure 6-45. Most pool heating systems use unglazed flat-plate collectors.

The collector manifolds at the top and bottom should be attached to the roof with tie-downs rather than bolts and clips as with other solar collectors. Also, if panels are used, the panels should be strapped down midpanel using nylon strapping if the system is installed in high-wind areas. **See Figure 6-46.**

The Fafco collector panels are 48″ wide and come in 8′, 10′, or 12′ lengths. The panels can be coupled together with flexible unshielded couplings. The number of panels needed for a system is based on the surface area of the pool. The panel requirement can be anywhere from 50% to 75% of the pool surface area depending on the local average ambient temperature. Cooler ambient temperatures may require 100% square footage to provide adequate heating while warmer temperatures may only require 50%. For example, a 250 sq ft pool in the Southwest of the United States may only require three 4′ × 12′ panels.

The number of panels used may also be limited by the size of the installation area. It may not be possible to have the full 100% area needed on the roof. However, with solar pool heating systems, some heating is better than none. Furthermore, they still save fuel and provide warmer temperatures in the pool.

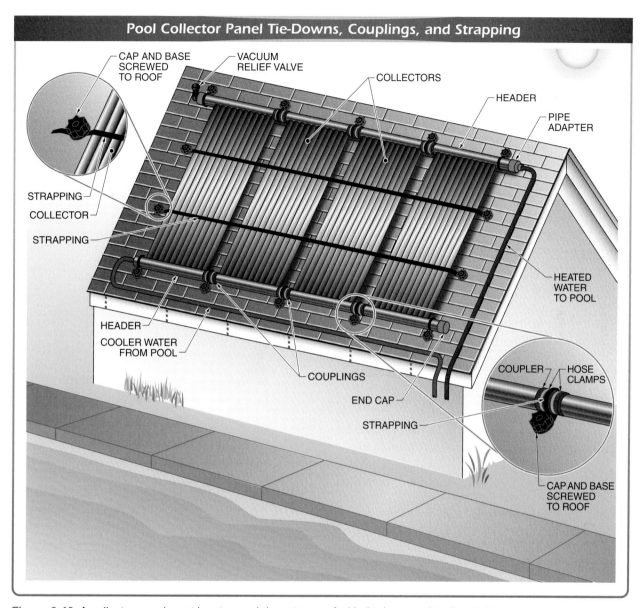

Figure 6-46. A collector panel must be strapped down to a roof with tie-downs rather than bolts and clips.

Solar pool heating systems can be added to existing pool installations or new pool installations. They can also be adapted to pools with or without existing pool heaters. **See Figure 6-47.** In any case, the solar loop piping must be connected to the pool equipment piping at the PVC manifold. The solar collector inlet piping must connect to the manifold through a three-way diverter valve downstream of the pool filter. This ensures that debris does not enter the solar collectors and possibly plug the collector tubing. The collector outlet piping connects to the pool return piping upstream of any existing or future pool heater. This allows the system

to act as a pool-water preheater before the water flows to the heater.

Diverter valves can be either manual or automatic. In a simple system, the diverter valve may be a manual valve. **See Figure 6-48.** A manual valve requires the homeowner to set the diverter valve to allow flow to the solar collectors or back to the pool. For a more efficient system, which is more complex, a motorized or automatic diverter valve operated by an actuator should be used. This requires the installation of a differential controller and temperature sensors before the diverter valve and at the collector outlet piping.

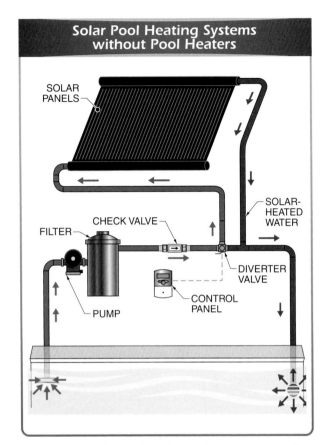

Figure 6-47. Solar pool heating systems can be used on pools with or without existing pool heaters.

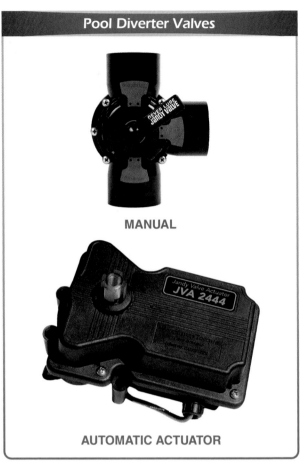

Zodiac Pool Systems, Inc.

Figure 6-48. Diverter valves can be either manual or automatic.

The controller parameters for this system are similar to those of other solar water heating systems. When the proper differential temperatures are sensed, the diverter valve diverts flow to the collector. When the temperatures reach the 4° differential setpoint, the diverter valve directs flow to the pool and closes flow to the collectors.

Instead of a controller, the system may include a pool-pump timer, which activates and deactivates the pump at the proper times. The timer should be set to activate the pump when the weather begins to warm up in the morning and deactivate the pump before it begins to cool in the evening. For example, the timer should be set to turn on the pump at 9 AM and off at 4 PM.

A check valve should always be installed upstream of the diverter valve. This will prevent the backflow of HTF through the pool filter and into the pool, which could carry debris back into the pool. If needed due to high pressures, a check valve can be installed in the collector outlet piping before the connecting tee. This stops any flow into the collector outlet piping when flow is diverted to the pool rather than the collector.

Piping configurations vary depending on installation needs. **See Figure 6-49.** Isolation and drain valves can be installed in the collector piping if the collector needs to be periodically drained and isolated from the system. If the pool pump does not have enough head pressure to lift the HTF to the collector, a booster pump must be installed on the collector inlet piping.

Care should be taken when installing collector inlet and outlet piping. Flow through the unglazed collector must be balanced, as with all solar collectors. HTF enters the collector in the bottom manifold and exits from the opposite end of the top manifold. Since collector piping is not typically run within a building, it may be required to run piping beneath the collector to the inlet to ensure balanced flow. It is recommended that the piping be below the collector for proper collector draining. **See Figure 6-50.** Also, any exposed collector piping, which is typically PVC, must be painted for protection against ultraviolet (UV) ray degradation.

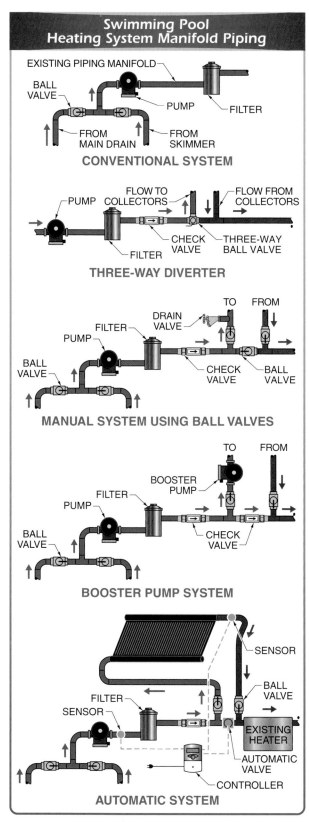

Figure 6-49. Piping configurations vary depending on installation needs.

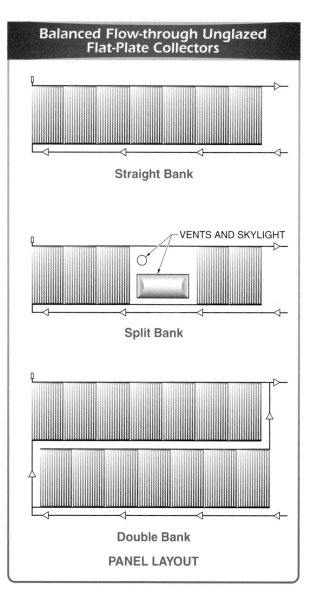

Figure 6-50. Flow through unglazed collectors must be balanced.

Collector HTF protection is provided by draining down the collector. A vacuum relief valve is installed at the top manifold of the collector opposite of the outlet. This allows for the collector to drain by providing a means for air to enter the collector as the HTF flows out from both collector lines. **See Figure 6-51.** It is recommended that isolation and drain valves be added to the system to facilitate the removal of the HTF. The valves are especially needed if the pool circulation system is run year round. Isolation ball valves securely stop flow to the collector piping rather than relying on check valves and the diverter valve.

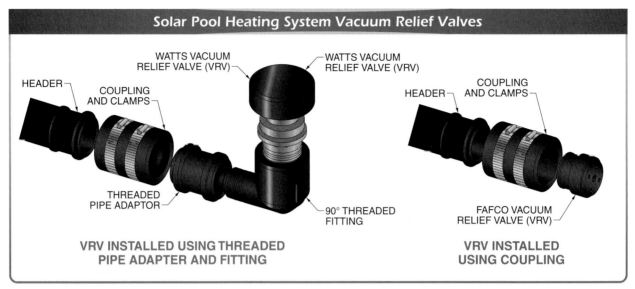

Figure 6-51. A vacuum relief valve is installed at the top manifold of a collector to allow the collector to drain completely.

Under normal operating conditions, the isolation ball valves are opened and the drain valves are closed. The manual diverter valve is adjusted for flow to the collectors. Alternatively, the automatic diverter valve is adjusted by the controller. When the pump is activated, HTF flows to the collector where it is heated by solar radiation. The heated HTF then flows out of the collector and into the pool heater or into the pool. When the pump is deactivated, any HTF remaining in the collectors drains from the collector outlet and into the pool. Unless freezing temperatures are expected, there is no need to drain the collector inlet piping. Having the system full will help the pump prime and create flow.

Solar pool heating systems are very efficient. They are especially efficient when the pool water is protected against evaporation at night, which can cause rapid heat loss. There are several different methods used to prevent the evaporation of pool water. Covering the pool with a pool blanket and then removing it is the best method. A liquid blanket solution is also available. The liquid blanket solution is not as effective as a pool blanket, but the solution requires much less work.

Chapter 6 Review and Resources

Solar Water Heating System Installation Safety

Most of the work performed by solar water heating system installers involves preventing hazardous conditions for other workers and the general public. This responsibility requires them to understand and follow general safety rules established by regulatory agencies and other specific workplace rules. For installers, following general safety rules requires knowing how to perform basic first aid, wearing proper personal protective equipment, safely using tools, practicing proper lifting techniques, working safely with ladders and scaffolds, properly using fall-arrest systems, and working safely with electrical equipment and circuits.

OSHA REGULATIONS

In the past, little emphasis was placed on safety in the workplace. More recently, organizations, such as the Occupational Safety and Health Administration (OSHA) created in 1971, have been established to develop and enforce safety standards for all workers. In addition, organizations such as the National Institute for Occupational Safety and Health (NIOSH) and the National Fire Protection Association (NFPA) have formed codes, standards, and guidelines for safe and healthy work environments.

A *code* is a regulation or minimum requirement. A *standard* is an accepted reference or practice. OSHA Title 29 industrial safety standards were put into effect in the Code of Federal Regulations (CFR) and are identified as OSHA 29 CFR 1910—*Occupational Safety and Health Standards* and 1926—*Safety and Health Regulations for Construction.* Today, the law requires employers to provide their employees with working conditions that are free from unknown dangers. Safety awareness and regulations have dramatically reduced the national industrial fatality occurrences from 6217 deaths in 1992 to 4690 in 2010, according to the U.S. Bureau of Labor Statistics.

On-the-job safety begins with personal safety awareness and understanding of basic first aid, general safety rules, and personal protective equipment. Personal safety awareness is being aware of safety issues and potential hazards to individuals in the workplace. A personal awareness of on-the-job safety is vital in preventing workplace injuries. Installers must have an understanding of certain requirements, health and safety risks, and the steps needed to minimize risks. Common work hazards in the solar water heating industry can be reduced by properly using tools, using fall-protection devices, using proper lifting techniques, and practicing ladder, scaffold, and electrical safety. **See Figure 7-1.**

Personal safety awareness includes keeping a safety mindset while on the job. Activities performed can become risky when a technician's stress level is high or there is an unbalance between work and personal life. Work injuries can occur when individuals either take chances or are unfocused. In addition, taking shortcuts when in a hurry can make individuals make poor decisions regarding their safety and the safety of others. For example, a technician in a hurry may feel that wearing safety glasses is not required for a task. The decision to forgo wearing safety glasses could lead to an eye injury.

CLEAN WORK AREAS

A cluttered and messy work area with excess materials, extension cords, tools, and equipment randomly scattered creates severe fall and trip hazards. Such a work area also represents a workplace in disarray. Work areas must be cleaned and organization must be maintained throughout the workday. An organized work area is not only safer but visually desirable. In addition, to prevent injuries and sprains from slips, trips, and falls, all floors, roofs, and walkways must be dry and clear of obstructions.

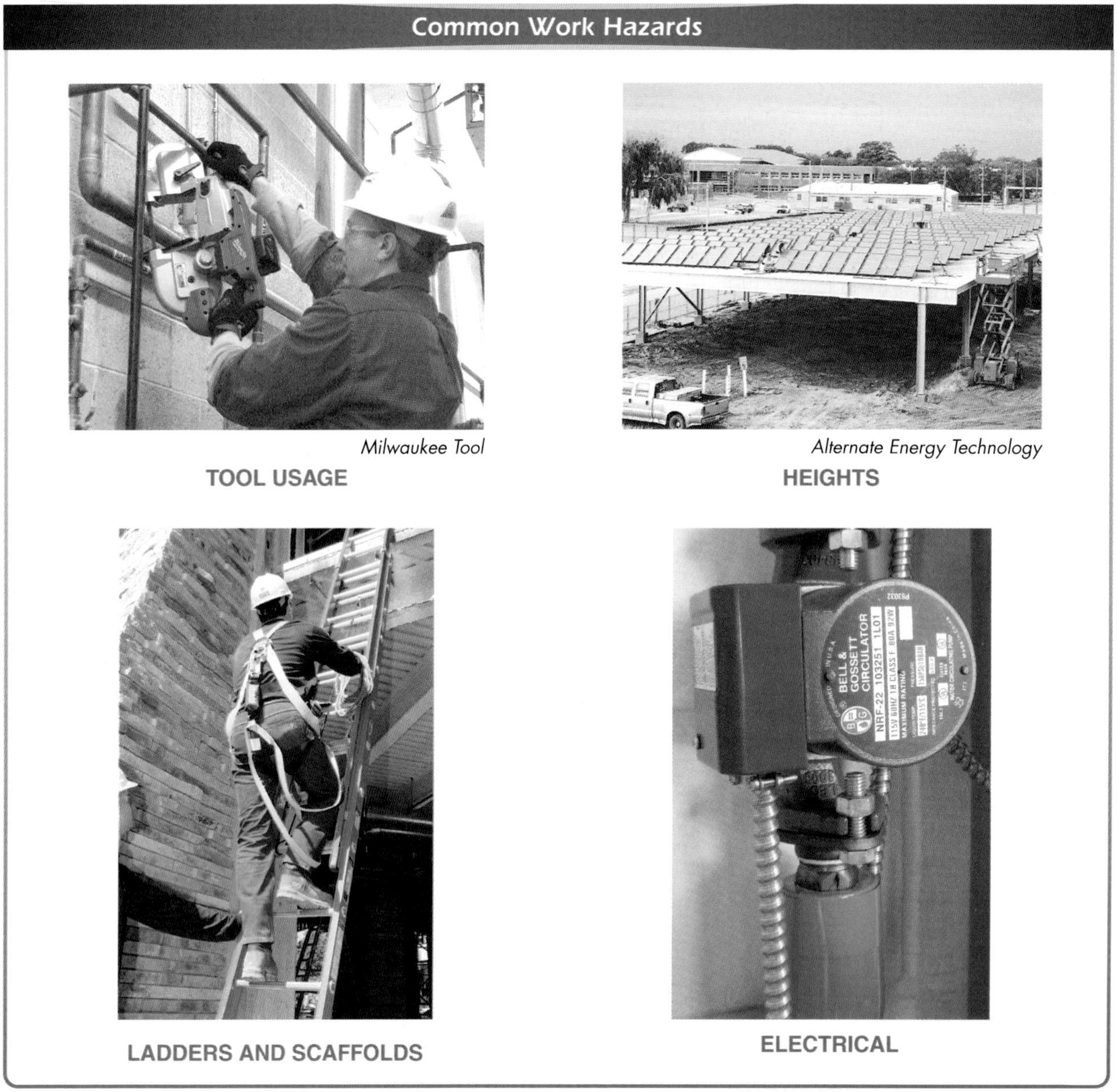

Common Work Hazards

Milwaukee Tool

TOOL USAGE

Alternate Energy Technology

HEIGHTS

LADDERS AND SCAFFOLDS

ELECTRICAL

Figure 7-1. Common work hazards can be reduced by properly using tools, using fall-protection devices, and practicing ladder, scaffold, and electrical safety.

PERSONAL PROTECTIVE EQUIPMENT

Personal protective equipment (PPE) is designed to protect installers and others from serious workplace injuries or illnesses resulting from contact with chemical, radiological, physical, electrical, mechanical, or other workplace hazards. Examples of PPE include safety glasses, hard hats, leather gloves, steel-toe work boots, earplugs, earmuffs, fire-resistant clothing, and dust masks. **See Figure 7-2.**

PPE regulations are listed in OSHA 29 CFR 1910 Subpart G—*Occupational Health and Environmental Control* and OSHA 29 CFR 1910 Subpart I—*Personal Protective Equipment,* which also references various standards for each type. For example, appropriate protective helmets must be worn in areas with overhead hazards, falling objects, or electrical shock, and safety shoes with reinforced steel toes must be worn to provide protection from falling objects.

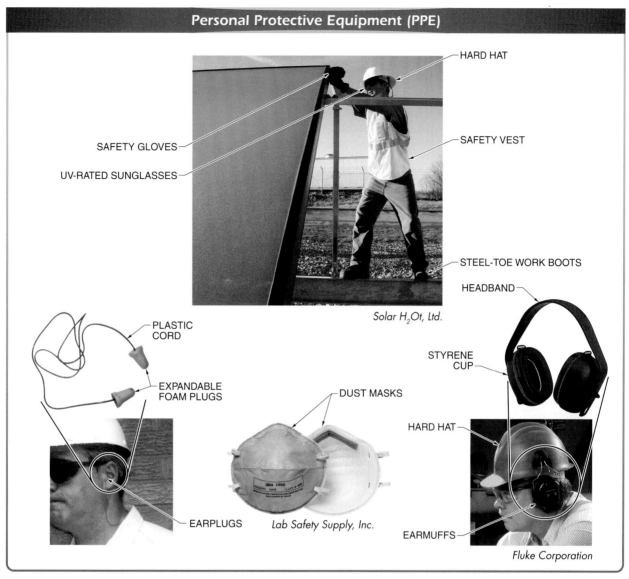

Figure 7-2. Examples of PPE include safety glasses, hard hats, leather gloves, steel-toe work boots, earplugs, earmuffs, fire-resistant clothing, and dust masks.

Facts

Workers must report accidents and safety hazards to the employer or supervisor. Workers must practice job-site safety at all times and wear appropriate personal protective equipment. If injuries occur, medical assistance should be immediately obtained.

Policies and procedures related to PPE should be reviewed if an employee appears to need retraining or when introducing new PPE into the workplace. PPE specific to solar water heating work can include the following:
• eye and face protection (e.g., safety goggles and glasses, face shields, visors)

• head protection (e.g., hard hats, helmets, caps); hard hats are required when there is a risk of objects falling onto a person or hitting the head on an object
• protection of extremities (e.g., steel-toe shoes, safety gloves, kneepads)
• respiratory devices (e.g., respirators, dust masks); used for performing work near structures that may contain lead paint or asbestos; may also be required for performing work in attic spaces around insulation
• hearing protection (e.g., earplugs, earmuffs)
• protective clothing (e.g., long-sleeve shirts, long pants, UV-rated sunglasses, sunblock, and UV protective clothing)

TOOLS AND EQUIPMENT

Installing solar water heating systems requires the use of certain hand and portable power tools as well as torches for soldering and brazing copper tubing. To prevent injuries and damage to property and equipment, proper safety procedures must be followed when using tools and torches. Failure to follow proper safety procedures can lead to accidents and injuries.

Hand Tool Safety

Hand tools must be operated safely. To ensure their safe use and prevent them from sustaining damage, tools must be properly maintained. Injuries can be prevented by using proper techniques, having good work habits, and wearing PPE. Hand tools are typically designed for the efficient and safe completion of a specific task. Hand tools should be used for their intended purpose only.

The safe use of hand tools reduces hand tool accidents. Hand tool safety includes the following:
- Point cutting tools away from the body during use and transportation.
- Fasten or tie off toolboxes on elevated surfaces to prevent tools from falling.
- Organize tools to protect and conceal sharp cutting surfaces.
- Transport sharp tools in sheaths or with the blades pointed down.
- Operate tools according to manufacturer recommendations.
- Keep tools sharp and in proper working order.
- Repair or replace damaged or broken tools.
- Transport hand tools in toolboxes, tool bags, or tool belts. Do not carry tools in pockets.

Portable Power Tool Safety

While some of the work performed by solar water heating installers can be completed with hand tools, certain tasks require the use of portable power tools. Most portable power tools operate using either an alternating current (AC) power source or a direct current (DC) power source (a battery). Portable power tools that operate on DC power are known as cordless tools. Cordless tools typically do not have as much power as power tools that operate on AC power.

The most common portable power tools used by solar water heating system installers include circular saws, reciprocating saws, and drills. **See Figure 7-3.** Most companies perform work using power tools and electric cords and must create clear safety policies for the maintenance and use of this equipment. Guidelines to follow when creating power tool and electric cord safety practices and policies include the following:
- Ensure that extension cords and equipment are protected by a ground-fault circuit interrupter (GFCI) when performing work. Understand OSHA and local municipality rules for GFCI devices.
- Follow testing procedures provided by the manufacturer to ensure that each device is in proper working condition.
- Visually inspect electrical equipment prior to every use for electrical hazards such as missing prongs, frayed cords, and cracked tool cases.
- Remove from service and apply a warning tag to any tools that are damaged. Electric equipment and power cords must always be inspected after an accident or when damage occurs. Power cords with missing ground prongs must never be used.
- Be thoroughly familiar with all manufacturer safety recommendations and operating features of a power tool before using it.
- Ensure that all safety guards are in place and in working order.
- Wear safety goggles or a face shield and a respirator or dust mask when conditions require them.
- Ensure the material to be worked on is free of obstructions and properly secured.
- Ensure that a power tool is turned off before connecting it to a power source.
- Keep attention focused on the work. Never become distracted when using a power tool.
- Immediately investigate any change in power tool sound during operation. It usually indicates a problem.
- Have power tools inspected and serviced by a qualified repair person at regular intervals.
- Inspect electrical cords for fraying or other damage. Correct any problem immediately.
- When work is completed, turn off the power. Wait until the power tool stops before leaving a stationary tool or before laying down a portable tool.
- Do not carry a tool by its cord or yank a cord from a receptacle.
- Locate electric cords so that they do not present a tripping hazard.
- Keep the cord of a power tool safely away from the blade.

- Avoid operating electrical power tools in damp or wet locations.
- Wear a hat or hairnet to keep long hair from becoming tangled in the moving parts of power equipment.
- Never clamp or force open the retractable guard of a power saw.

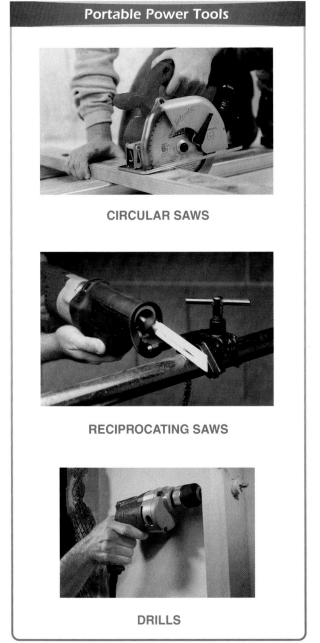

Portable Power Tools

CIRCULAR SAWS

RECIPROCATING SAWS

DRILLS

Milwaukee Electric Tool Corporation

Figure 7-3. The most common portable power tools used by solar thermal technicians and installers include circular saws, reciprocating saws, and drills.

Grounded Power Tools. All power tools must be properly grounded unless they are approved double-insulated tools. A power tool with a three-prong plug should be connected to a grounded outlet (receptacle). It is extremely dangerous to use an adapter to connect a three-prong plug to an ungrounded (two-hole) outlet unless a separate ground wire is connected to an approved ground (common return circuit). The ground ensures that any short circuit will trip the circuit breaker or blow the fuse. **See Figure 7-4.**

WARNING: An ungrounded power tool has the potential to cause a fatal accident. Double-insulated tools have two-prong plugs and are properly identified as such by the manufacturer on the nameplate. The electrical parts of the motor in a double-insulated power tool are provided with extra insulation to prevent electrical shock. Therefore, the tool does not have to be grounded. Both the interior and exterior of the tool should be kept clean and free from grease or dirt that might conduct electricity. Manufacturer recommendations should be followed regarding the service and maintenance of power tools.

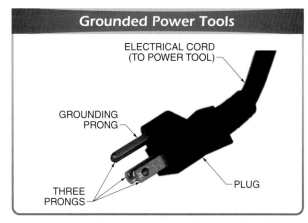

Grounded Power Tools

ELECTRICAL CORD
(TO POWER TOOL)

GROUNDING
PRONG

THREE
PRONGS

PLUG

Figure 7-4. Grounded power tools are identified by a three-prong plug on the end of the power cord.

Fire Safety

Solar water heating system installations often require the use of open flames or tools that produce extreme heat in order to join piping and fittings. These joining operations should be performed in an area that does not contain combustible materials. However, this often is not possible because of the location of existing piping and fixtures. Studs, joists, and other combustible materials must be protected with fire-resistant materials when soldering or brazing operations are conducted in-line, such as when installing water distribution pipes.

Fire extinguishers should be identified and readily available at an installation site in case of a fire. The National Fire Protection Association (NFPA) classifies fires as Class A, B, C, and D based on the combustible material. The appropriate fire extinguisher must be used on a fire to safely and quickly extinguish the fire. Types of fire extinguishers are identified by color and shape. **See Figure 7-5.**

Fire Extinguishers		
Class	**Combustible Material (and Graphic Representation)**	**Extinguishing Chemicals**
A	• Wood • Paper • Rubber • Plastic • Cloth	• Water • Dry chemicals
B	• Flammable liquids • Grease • Gases	• Dry chemicals • Foam • Carbon dioxide
C	• Electrical	Nonconducting agent such as the following: • Dry chemicals • Carbon dioxide
D	• Combustible metals • Magnesium • Titanium • Zirconium • Sodium	• Extinguisher particular to type of metal

Figure 7-5. The appropriate fire extinguisher must be used on a fire to safely and quickly extinguish the fire.

Class A fires use ordinary combustible materials, such as wood, paper, rubber, plastic, and cloth as fuel. Class A fire extinguishers are identified by the color green inside a triangle. Class A fires are extinguished with water or dry chemicals. Carbon dioxide, sodium, and potassium bicarbonate chemicals should not be used on a Class A fire.

Class B fires use flammable liquids, gases, or grease as fuel. Class B fire extinguishers are identified by the color red inside a square. Class B fires can be extinguished with dry chemicals, foam, and carbon dioxide extinguishers.

Class C fires are electrical fires. Class C fire extinguishers are identified by the color blue inside a circle. Electrical fires require a nonconducting agent, such as carbon dioxide or dry chemicals, to extinguish them. Foam extinguishers or water must not be used on electrical fires.

Class D fires are caused by the combustion of metals, such as magnesium, titanium, or sodium. Class D fire extinguishers are identified by the color yellow inside a star. Class D fires cannot be extinguished with Class A, B, or C extinguishers. The chemicals in common extinguishers can intensify fires rather than put them out. Dry-powder extinguishers are available that are made specifically for the types of metal hazards present.

Torch Safety

Soldering and brazing are two common methods used to join copper tube. *Soldering* is a copper tube joining method that uses nonferrous filler metal with a melting temperature of less than 840°F (449°C). Soldered joints are most commonly used on copper tubing. *Brazing* is a copper tube joining method in which a nonferrous filler metal with a melting temperature of 840°F (449°C) or more is used. Brazed joints may be used on copper tubing. However, brazed joints are typically used for specialized piping applications, such as medical gas piping and natural gas supply piping, and in applications in which greater joint strength is required.

The solder, or filler metal, must be heated to its melting point to ensure that the metal flows properly into the fitting socket. Solder used for potable water supply piping requires that the heat produced must adequately heat the tube and fitting. A variety of different types of torches can be used to solder and braze copper tube. **See Figure 7-6.** A torch assembly that uses methyl acetylene propadiene (MAPP) stabilized gas, which is a mixture of acetylene and liquefied petroleum (LP) gases, is commonly used as a heat source when soldering small-diameter copper tube. Large-diameter copper tube is commonly soldered with either an LP gas or acetylene gas torch.

Soldering involves the use of fire and molten metals. Swirl combustion torches emit a loud whistling noise when they are being used. Proper (PPE), including appropriate gloves, eyewear, long-sleeve shirts, and hearing protection, must be worn when soldering. Safe work practices that reduce the risk of injury or damage when using a soldering or brazing torch include the following:

• Use flameproof fire shields or fire-resistant gels to prevent fire.

• Be aware of what is in front of and behind the workpiece to avoid burning through electrical wire or other items that are not visible.

- Have an appropriate fire extinguisher accessible.
- Remove fire hazards by removing flammable materials near the work area.
- When turning off a torch, verify that it is completely off and that the hot torch tip is not near flammable material.
- Wear cotton fabric or flame-retardant (FR) clothing as appropriate. Clothing made of synthetic, non-FR materials can be more susceptible to burning or melting when exposed to heat such as hot solder.

Facts

A serious concern for all workers is the ever-present danger of fire on a job site. Workers must be aware of potential fire hazards, understand what creates fire hazards, and how this danger can be reduced.

COLLECTOR SAFETY

The solar collector is a vital component of a solar water heating system. However, collectors can present several hazards to installers since they can weigh up to 150 lb each and can generate a significant amount heat, even when they are not connected or put into use. The weight and size of some collectors make them hard to lift or handle, especially on a sloped roof. A hot collector, especially the collector inlet and outlet, can create a significant burn hazard for system installers if care is not used during handling. Safe work practices that reduce the risk of injury when handling solar collectors include the following:
- Never work alone when handling large collectors.
- Wear leather work gloves to prevent cuts from sharp edges or burns from a hot collector.
- Never charge a collector that has been recently exposed to the sun. The liquid could flash into steam and cause bodily injury or equipment damage.
- Cover the collector with opaque material to prevent sunlight from heating the collector.
- Wear safety glasses when handling flat-plate and evacuated tube collectors. These collectors are made with glass components that may break when dropped or struck.

Solar collectors should be lifted safely to rooftops. Workers should never be allowed to climb ladders while carrying solar panels. Lifting equipment, such as ladder hoists, swing hoists, or truck-mounted cranes/conveyors, should be used whenever possible.

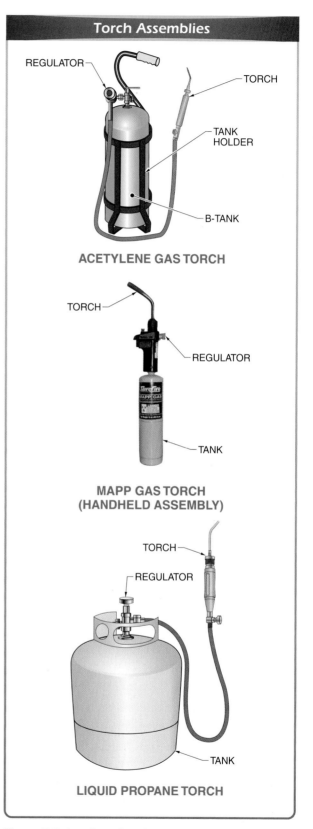

Figure 7-6. A variety of torches can be used to solder and braze copper tube.

LIFTING SAFETY

Back injury is one of the most common injuries resulting in lost work time. Improper lifting and moving techniques can result in back injuries. Back injuries are extremely common but preventable through a combination of safe lifting and moving techniques.

Before lifting, the weight of the object should be determined and the lifting and moving method should be planned based on the object's size, shape, weight, and path of travel. Assistance should be sought when necessary. The entire pathway should be clear of obstacles, obstructions, and other hazards. The knees should be bent and the object grasped firmly. The object is then lifted while straightening the legs and keeping the back as straight as possible. Finally, movement can occur after the entire body is in the vertical position. The load should be kept as close to the body as possible. **See Figure 7-7.**

Carts or hand trucks should be used when objects are excessively heavy or oddly shaped. The load must be balanced to avoid tipping. When moving a load on a cart, the worker's body weight should be used while taking even, short steps. Poor physical condition is a major contributor to all types of back problems. The weight, strength, and flexibility of an individual all play a part in minimizing strain on the muscles and tendons that support the spinal column (core muscles) and contributing to or preventing back problems.

When possible, cranes, hoists, and forklifts should be used to move heavy loads and objects. The use of cranes, hoists, and forklifts can eliminate lifting injuries, especially when lifting large commercial equipment. Lifting large equipment manually, even with several individuals working together, must be avoided as it can be dangerous and lead to serious injuries.

Lifting Loads from Vehicles

During the course of most jobs, heavy loads are often loaded and unloaded from vehicles, such as the bed of a pickup truck or work van. Safe lifting practices must be followed when loading and unloading heavy items from vehicles such as the following:

- Load the heaviest items last and place in an accessible position to avoid placing them too far into the front end of a vehicle bed.
- Always use two or more people to lift and move large and heavy items.
- Store or transport tools and heavy equipment from the shop or installation site on a mobile utility cart for easy unloading and loading of vehicles. Tools and equipment can be transferred to and from a vehicle bed and the upper shelf of a mobile cart without bending.

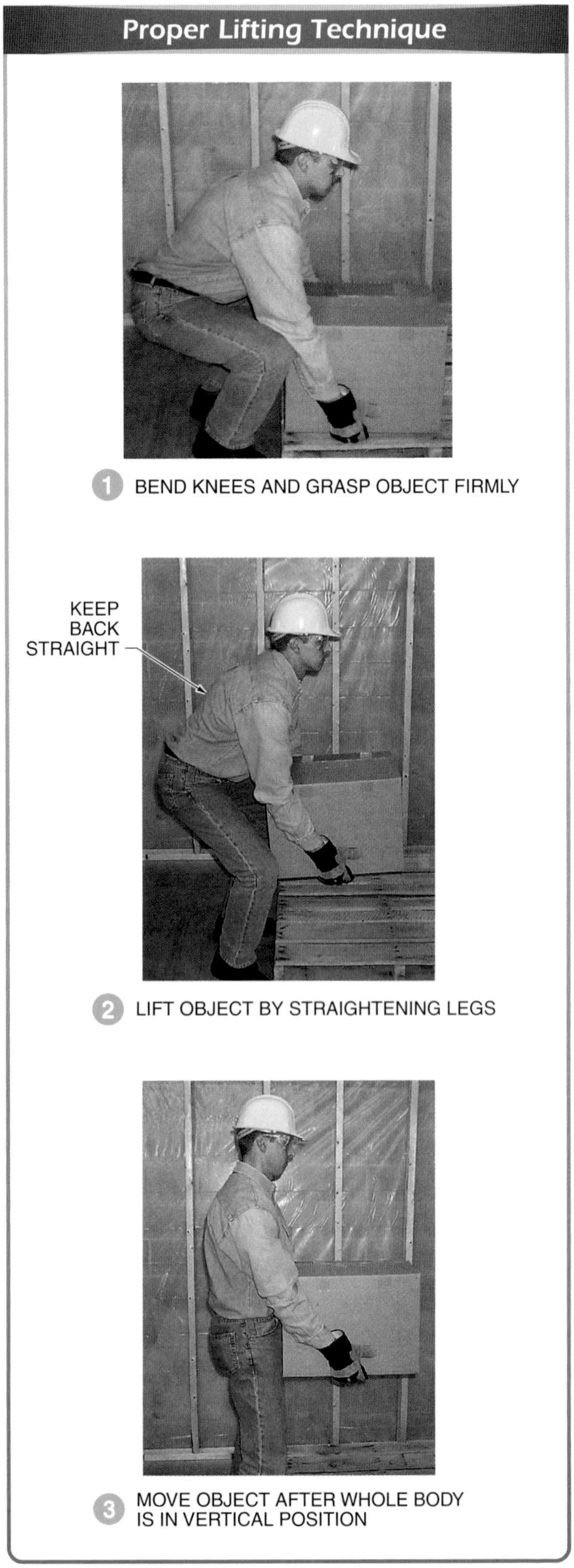

Proper Lifting Technique

1 BEND KNEES AND GRASP OBJECT FIRMLY

KEEP BACK STRAIGHT

2 LIFT OBJECT BY STRAIGHTENING LEGS

3 MOVE OBJECT AFTER WHOLE BODY IS IN VERTICAL POSITION

Figure 7-7. Proper lifting requires that the legs be used to lift an object.

Lifting Loads to Rooftops

Lifting solar collectors onto rooftops is hazardous because collectors can be heavy, cumbersome, and measure up to 4´ × 10´. The most common method used to lift collectors to rooftops is through a ladder-based winch system. Ladder-based winch systems can easily manage the weight of most solar collectors. Whenever using a ladder-based winch system, the original equipment manufacturer (OEM) guidelines for setup, use, and safety precautions must always be followed. Proper fall restraint equipment must be used when hoisting loads onto upper surfaces, such as rooftops, to prevent falls. When ladder winches are not available, a certified crane service should be used. **See Figure 7-8.**

Lifting Collectors

Alternate Energy Technology

Figure 7-8. When possible, cranes, hoists, and forklifts should be used to move heavy loads and objects.

LADDERS AND LADDER SAFETY

A *ladder* is a structure consisting of two side rails joined at intervals by steps or rungs for climbing up and down. Ladders are manufactured in lengths of 3′ to 50′. Ladders are constructed of wood, metal, or fiberglass. All ladders, regardless of the construction material, are manufactured to meet the same standards. Ladder selection is generally based on style, use, and, in some cases, portability. Ladder types include fixed, single, and extension ladders as well as stepladders. **See Figure 7-9.**

All ladders must be used only for the purpose for which they are designed. Ladders must not be used as pry bars or horizontal platforms. All ladders must be equipped with nonslip safety feet such as butt spurs or foot pads. **See Figure 7-10.** A *butt spur* is the notched, pointed, or spiked end of a ladder that helps prevent the ladder butt from slipping. Butt spurs are generally attached to long ladders, such as extension ladders, and may be integrated with a foot pad in the form of a spur plate. A *foot pad* is a metal swivel attachment with a rubber or rubberlike tread that helps prevent a ladder butt from slipping on hard surfaces.

A *fixed ladder* is a ladder that is permanently attached to a structure. Fixed ladders are commonly constructed of steel or aluminum. The fabrication of a fixed ladder, including design, materials, and welding, must be done under the supervision of a qualified licensed structural engineer.

A *single ladder* is a ladder of fixed length having only one section. Typical lengths of single ladders vary from 6′ to 24′. Single ladders offer the convenience of use by one person. However, they are limited in their versatility because a given length ladder may be safely used only within a small height range.

Ladders must only be used on firm and level surfaces free from debris.

Ladders

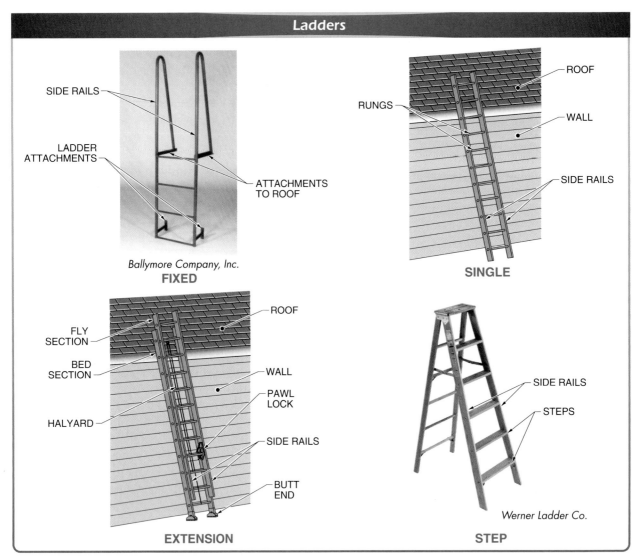

Figure 7-9. Ladder types include fixed, single, and extension ladders as well as stepladders.

Ladder Safety Feet

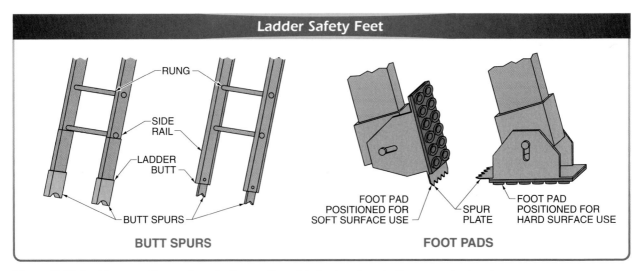

Figure 7-10. Ladders must be equipped with nonslip safety feet such as butt spurs or foot pads.

An *extension ladder* is an adjustable-height ladder with a fixed bed section and sliding, lockable fly sections. The *bed section* is the lower section of an extension ladder. The *fly section* is the upper section of an extension ladder. The fly section may consist of a first fly, second fly, etc. A *pawl lock* is a pivoting hook mechanism attached to the fly sections of an extension ladder. Pawl locks are used to hold the fly sections at the desired height. The three main parts of a pawl lock are the hook, finger, and spring. **See Figure 7-11.**

The rungs nest into the hook of each pawl and prevent downward movement of the fly section. The fly section is lowered by first raising the fly section just enough for a rung to pass below the finger of the pawl lock. The fly section is then lowered. When lowered, the rung forces the finger up, preventing the rungs from nesting into the hook. To hold the fly section in place, the pawl lock must be lowered slightly below a rung to allow the finger to drop. The fly is then raised slightly to allow the spring to force the pawl hook over the rungs.

Extension ladders must have positive stops to prevent overextension of the fly section. The overlap of the fly section must be at least 3′ for extension ladders up to 36′, 4′ for extension ladders over 36′ and up to 48′, and 5′ for extension ladders over 48′ and up to 60′.

Extension ladders are positioned on a 4:1 ratio or approximately a 75° angle of inclination. For every 4′ of working height, 1′ of space is required at the base. Working height is the distance from the ground to the top support of a ladder. The top support is the area of a ladder that makes contact with a structure. For example, a ladder supported 12′ above the ground should be placed with the butt end 3′ from the top support vertical line. **See Figure 7-12.** The tip of a single or extension ladder should be secured at the top to prevent slipping and must be at least 3′ above the roofline, trench wall, or top support. Ladders over 15′ should also be secured at the bottom.

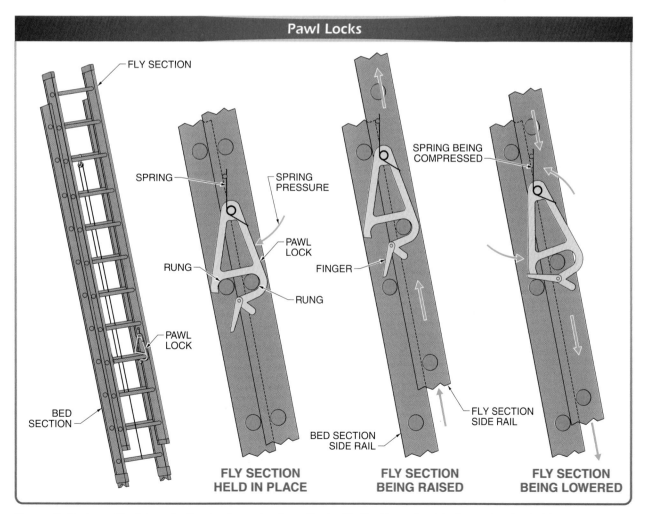

Figure 7-11. A pawl lock is a pivoting hook mechanism attached to the fly section(s) of an extension ladder.

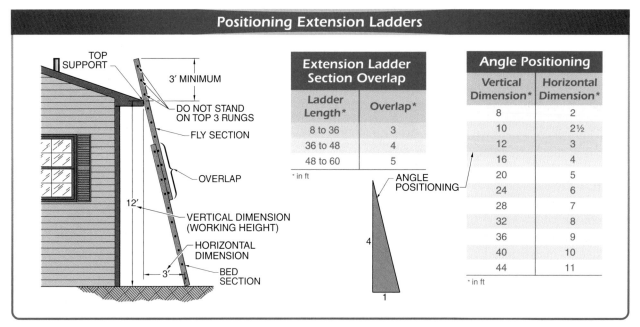

Positioning Extension Ladders

Extension Ladder Section Overlap	
Ladder Length*	Overlap*
8 to 36	3
36 to 48	4
48 to 60	5

* in ft

Angle Positioning	
Vertical Dimension*	Horizontal Dimension*
8	2
10	2½
12	3
16	4
20	5
24	6
28	7
32	8
36	9
40	10
44	11

* in ft

Figure 7-12. Extension ladders are positioned on a 4:1 ratio or approximately a 75° angle of inclination.

A *stepladder* is a folding ladder with flat rungs built into a supporting frame that stands independently of support. Stepladders are available in 2′ to 20′ lengths. Stepladders are easily portable and provide adequate working height for many applications. Individuals must never lean out or reach to one side of a stepladder. Rather, the stepladder must be repositioned so that the work area can be easily and conveniently reached.

Ladder Safety

Proper maintenance of ladders is critical due to the direct relationship of ladders to life safety. The following list of precautions should be observed for proper safety when using ladders:
• Use ladders only for the purpose for which they were designed.
• Inspect ladders carefully when new and before each use.
• Use leg muscles, not back muscles, for lifting and lowering ladders.
• Stand ladders on a firm, level surface.
• Face the ladder when ascending or descending.
• Exercise extreme caution when using ladders near electrical conductors or equipment. All ladders conduct electricity when wet.

• Ladders are intended for use by only one person unless specifically designated otherwise.
• Never use a ladder as a substitute for scaffold planks or for horizontal work.
• Always check for the proper angle of inclination before climbing a ladder.
• Verify that all pawl locks on extension ladders are securely hooked over rungs before climbing.
• Always check for the proper overlap of extension ladder sections before climbing.
• Keep all nuts, bolts, and fasteners tight. Lubricate all moving metal parts as required.
• Ensure that stepladders are fully open with spreaders locked before climbing.
• Do not stand on the top two rails of a stepladder or on the top three rungs of an extension ladder.
• Use the three-point climbing method when ascending or descending a ladder.
• Never place a ladder in front of a door unless appropriate precautions have been taken.
• Never use metal ladders near electrical circuits or lines. Instead, use ladders made of fiberglass or wood.
• Beware of overhead power lines when working with ladders and scaffolding. **WARNING:** An electrical shock received while on a ladder or scaffolding can cause loss of balance and falling, which can result in injury or death.

Ladder Duty Ratings. The ladder duty rating is the weight a ladder is designed to support under normal use. The five ladder duty ratings are the following:

- Type IAA—special-duty, industrial, 375 lb capacity
- Type IA—extra heavy-duty, industrial, 300 lb capacity
- Type I—heavy-duty, industrial, 250 lb capacity
- Type II—medium-duty, commercial, 225 lb capacity
- Type III—light-duty, household, 200 lb capacity

All ladders must be used only for the purpose for which they are designed.

Ladder Climbing Techniques. Climbing may begin only after a ladder is properly secured. Climbing movements should be smooth and rhythmical to prevent ladder bounce and sway. Safe climbing involves the three-point contact method. In the three-point contact method, the body is kept erect, the arms are straight, and the hands and feet make the three points of contact. Two feet and one hand, or two hands and one foot are in contact with the ladder rungs at all times.

In addition, reaching above shoulder level to grasp a rung to maintain balance and unobstructed knee movements should be avoided. Each hand should grasp the rungs with the palms down and the thumb on the underside of the rung. Upward progress should be made by the push of the leg muscles and not the pull of the arm muscles. When climbing, all tools, parts, or equipment must be secured in a pouch or raised and lowered with a rope.

To ascend or descend a ladder safely, the 3-point climbing method is used. In the 3-point climbing method, one hand and two feet or two hands and one foot are in contact with the ladder rungs at all times.

SCAFFOLDS AND SCAFFOLD SAFETY

A *scaffold* is a temporary or movable platform and structure for workers to stand on when working at a height above the floor. Three basic types of scaffolds are pole, sectional metal-framed, and suspension scaffolds. **See Figure 7-13.**

A *pole scaffold* is a wood scaffold with one or two sides firmly resting on the floor or ground. A *single-pole scaffold* is a wood scaffold with one side resting on the floor or ground and the other side structurally anchored to the building. A *double-pole scaffold* is a wood scaffold with both sides resting on the floor or ground and is not structurally anchored to a building or other structure.

A *sectional metal-framed scaffold,* also known as a tube-and-coupler scaffold, is a metal scaffold consisting of preformed tubes and components. Sectional metal-framed scaffolds may be freestanding or mobile. They are easy to use; easily assembled with bolts, pins, or brackets; and may be fitted with locking casters for ease in moving.

A *suspension scaffold,* also known as a swinging scaffold, is a scaffold supported by overhead wire ropes. Suspension scaffolds use either the two-point or multiple-point suspension design. A *two-point suspension scaffold* is a suspension scaffold supported by two overhead wire ropes. The overall width of two-point suspension scaffolds must be greater than 20″ but not more than 36″. A *multiple-point suspension scaffold* is a suspension scaffold supported by four or more ropes. Multiple-point suspension scaffolds must be capable of sustaining a working load of 50 lb/sq ft and are used mainly for repair and maintenance projects. They can be raised or lowered by permanently installed, electrically operated hoisting equipment. Multiple-point suspension scaffolds must not be overloaded.

A scaffold generally consists of wood planks or metal platforms to support workers and their materials. Scaffolds must not be supported by any unstable object, such as boxes, concrete blocks, or loose bricks.

Scaffold footing must be sound and stable and must not settle or displace while carrying the maximum intended load. A *maximum intended load* is the total of all loads, including the working load, the weight of the scaffold, and any other loads that may be anticipated. Scaffolds and their components must be capable of supporting at least four times their maximum intended load.

Scaffolds

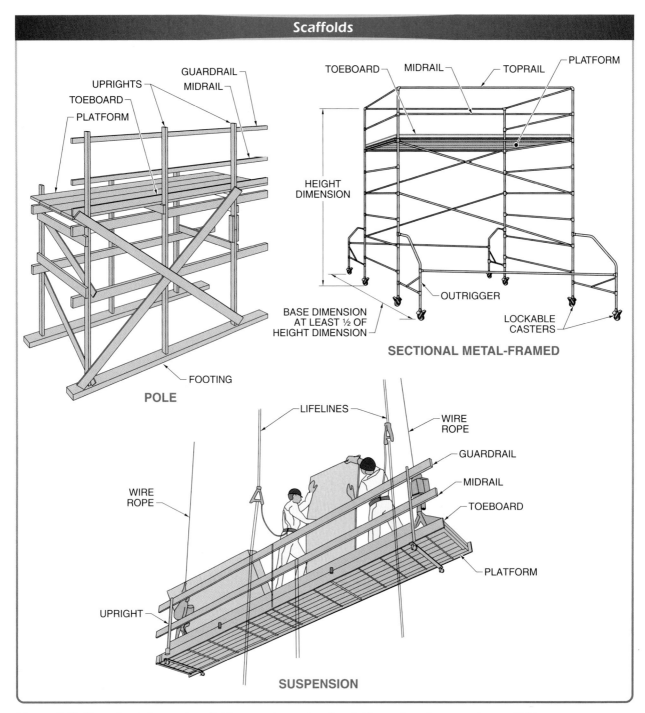

Figure 7-13. Three basic types of scaffolds are pole, sectional metal-framed, and suspension scaffolds.

All scaffolds 10′ or more aboveground must have guardrails, midrails, and toeboards. A *guardrail* is a rail secured to uprights and erected along the exposed sides and ends of a platform. A *midrail* is a rail secured to uprights approximately midway between the guardrail and the platform.

A *toeboard* is a barrier to guard against the falling of tools or other objects. Toeboards are secured along the sides and ends of a platform. Guardrails must be installed no less than 38″ or more than 45″ high, with a midrail. Guardrail and midrail support is to be at intervals of no more than 10′.

Scaffold Safety

OSHA regulations state that scaffolding may be erected, moved, altered, or dismantled only under the supervision of a competent person. The following precautions should be observed:
- Use only 2″ nominal structural scaffold-grade planking that is free of knots for scaffold platforms.
- Platform end extensions must be cleated with a minimum of 6″ extension and a maximum of 18″.
- Always observe working load limits. Scaffolds and their components must be capable of supporting four times the maximum intended load.
- Guardrails, midrails, and toeboards must be installed on all open sides and ends of platforms more than 10′ above the ground.
- Platform planks are to be laid with no openings more than 1″ between adjacent planks.
- Overhead protection must be provided for persons on a scaffold exposed to overhead hazards.
- Work must not be done on a scaffold during high winds or storms.
- Work must not be done on ice-covered or slippery scaffolds.
- Scaffolds with a height-to-base ratio of more than 4:1 must be restrained by the use of guylines.
- Mobile scaffolds must be locked in position when in use.
- All people, tools, and materials must be secured or removed from the platform before a mobile scaffold is moved.
- All personnel in close proximity must be advised and aware of the movement of a mobile scaffold.
- Fall protection must be used on working heights of more than 10′.

FALL PROTECTION

Installers most often work on roofs when installing solar water heating systems. To access these roofs, the installers use ladders or scaffolding. The work required by the installers on the roof exposes them to potential fall hazards. As the solar collectors are installed on the roof, the walking area that may once have been available may no longer be available to the installers. This may force the installers to squeeze by or walk very close to skylights or roof edges.

OSHA has specific requirements for solar construction workers involved in the installation of solar water heating systems exposed to fall distances of 6′ or more. Per OSHA 29 CFR 1926.501—*Duty to Have Fall Protection,* each employee on a walking/working surface (horizontal and vertical surface) with an unprotected side or edge that is 6′ (1.8 m) or more above a lower level shall be protected from falling by the use of guardrail systems, safety net systems, or personal fall-arrest systems.

To protect installers from potential fall hazards through skylights or other holes, employers must ensure that skylights are guarded or that workers near skylights use personal fall protection. Regardless of the roof height, any skylight or hole on a roof that exposes the worker to a fall of over 6′ (1.8 m) must be covered or have a guardrail system erected around it. If the skylight cannot be covered or guarded, the worker must use a personal fall-arrest system.

Certain states may have their own construction safety standards that may contain different or additional requirements. States must set job safety and health standards that are at least as effective as comparable federal standards. States have the option to manage standards covering hazards not addressed by federal standards. The applicable state and federal safety regulations should be known before beginning any work.

Training for solar water heating system installation must include proper instruction on the use of fall protection equipment.

Personal Fall-Arrest Systems

A *personal fall-arrest system* is a safety system used to arrest (stop) a worker's fall. Proper personal fall-arrest systems must be used when other means of fall protection are not provided. Each system consists of an anchorage point, connectors, and a body harness. A system also may include a lanyard, lifeline, deceleration device, or a combination of these devices. **See Figure 7-14.**

An anchorage point provides a secure point of attachment for lifelines, lanyards, or other deceleration devices. Anchorage points for personal fall-arrest systems must be independent of any other anchorage point and cannot be used for more than one person. Normally, they must be capable of supporting a minimum load of 5000 lb. Anchorage points must be designed and installed under the supervision of a qualified person.

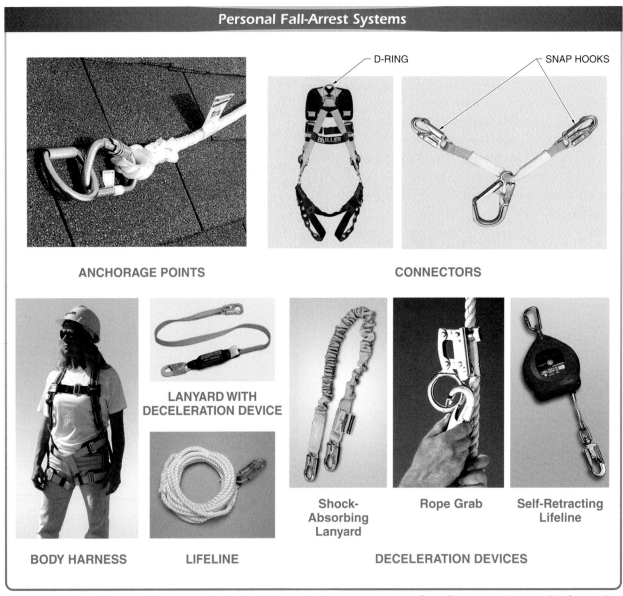

Personal Fall-Arrest Systems

ANCHORAGE POINTS   CONNECTORS

D-RING   SNAP HOOKS

BODY HARNESS   LIFELINE   DECELERATION DEVICES

LANYARD WITH DECELERATION DEVICE

Shock-Absorbing Lanyard   Rope Grab   Self-Retracting Lifeline

Miller Fall Protection/Honeywell Safety Products

Figure 7-14. A personal fall-arrest system is used to stop a worker's fall and must be used when other means of fall protection are not provided.

Connectors, such as D-rings and snap hooks, must have a minimum tensile strength of 5000 lb. They must be proof-tested to a minimum tensile load of 3600 lb. Connectors are typically attached to other personal fall-arrest system components such as anchorage points, body harnesses, lifelines, and lanyards.

Only locking-type snap hooks are permitted to be used for personal fall-arrest systems. One end of the snap hook should be attached directly to the D-ring on the harness. The other end should be attached to an approved tie-off point. Snap hooks should not be attached to each other, to a D-ring to which another snap hook or other connector is attached, to horizontal lifelines unless the snap hooks are designed for such a connection, or directly to webbing, rope, or wire rope. The specific fall protection equipment manufacturer should be consulted concerning restrictions on the use of its products.

A body harness, when worn properly, protects internal body organs, the spine, and other bones in a fall. The attachment point of a body harness should be located in the center of the back near shoulder level, or above the head. It should be inspected before each use to ensure it will provide proper support in case of a fall. Harness webbing should be inspected for wear such as frayed edges, broken fibers, burns, and chemical damage. D-rings and buckles should be inspected for distortion, cracks, breaks, and sharp edges. Grommets should be inspected to ensure they are tight.

Harnesses must fit snugly and be securely attached to a lanyard. A *lanyard* is a flexible line of rope, wire rope, or strap that generally has a connector at each end for connecting a body harness to a deceleration device, lifeline, or anchorage point. A *deceleration device* is a mechanism that dissipates or limits the energy imposed on a worker during a fall arrest. Common deceleration devices include rope grabs, shock-absorbing lanyards, and self-retracting lifelines or lanyards.

A *rope grab* is a deceleration device that travels on a lifeline to automatically engage a vertical or horizontal lifeline by friction to arrest the fall of a worker. Rope grabs protect workers from falls while allowing freedom of movement. A lanyard or lifeline is attached between the body harness and rope grab. A *shock-absorbing lanyard* is a lanyard that has a specially woven, shock-absorbing inner core that reduces the forces of fall arrest. The outer shell of the lanyard serves as the secondary lanyard.

Lifelines are anchored above the work area, offering a free-fall path, and must be strong enough to support the force of a fall. Vertical lifelines are connected to a fixed anchor at the upper end that is independent of a work platform such as a scaffold. Vertical lifelines must never have more than one worker attached per line.

A *self-retracting lifeline* is a type of vertical lifeline that contains a line that can be slowly extracted from or retracted onto its drum under slight tension during normal worker movement. When a fall occurs, the drum automatically locks, arresting the fall.

Horizontal lifelines are connected to fixed anchors at both ends. Workers attach their lanyard to a D-ring on the lifeline, allowing them to freely move horizontally along the lifeline. Horizontal lifelines are subject to greater loads than vertical lifelines and, therefore, must be designed, installed, and used under the supervision of a qualified person.

A lifeline must be properly terminated (anchored) to prevent the safety sleeve or ring from sliding off its end. The path of a fall must be visualized when anchoring a lifeline. Obstructions in the fall path of an anchored system can be deadly.

Personal Fall-Arrest System Requirements. In order to select the proper personal fall-arrest system, the estimated free-fall distance must be determined. **See Figure 7-15.** *Free-fall distance* is the vertical distance of the fall-arrest attachment point on a body harness between the onset of a fall and the point at which a fall-arrest system applies force to arrest the fall. Self-retracting lifelines should be used when the estimated fall distance is less than 18′-6″. Self-retracting lifelines or shock-absorbing lanyards can be used when the fall distance is greater than 18′-6″.

The system must also meet deceleration requirements. The *deceleration distance* is the additional vertical distance a falling worker travels before stopping, excluding lifeline elongation and free-fall distance, from the point at which a deceleration device begins to operate. This measurement of distance begins at the location of a body harness attachment point during a fall the moment the deceleration device activates. It ends at the location of that attachment point after the employee comes to a complete stop.

A personal fall-arrest system must be rigged so that a worker can neither fall more than 6′ nor contact any lower level. When a personal fall-arrest system is used, it must be able to perform the following functions:
- limit the maximum arresting force to 1800 lb when used with a body harness
- withstand twice the potential energy of impact for a free-fall distance of 6′ or the distance permitted by the system, whichever is less
- bring a worker to a complete stop at a maximum deceleration distance of 3′-6″

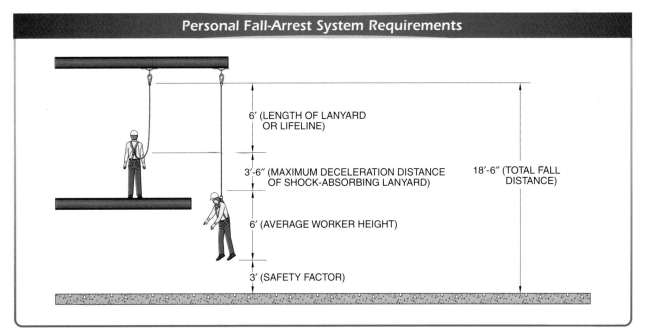

Personal Fall-Arrest System Requirements

6′ (LENGTH OF LANYARD OR LIFELINE)

3′-6″ (MAXIMUM DECELERATION DISTANCE OF SHOCK-ABSORBING LANYARD)

18′-6″ (TOTAL FALL DISTANCE)

6′ (AVERAGE WORKER HEIGHT)

3′ (SAFETY FACTOR)

Figure 7-15. A personal fall-arrest system must be rigged so that a worker can neither fall more than 6′ nor contact any lower level.

Protection from Falling Objects

Uninstalled piping, collectors, insulation, roofing materials, and mounting hardware on a roof can present serious hazards for people on the ground should any of these fall from the roof. These materials must be kept a minimum of 6′ from the edge of a roof. In addition, toeboards can be installed along the edge of the working area to protect employees working below. The toeboards must be capable of withstanding without failure a force of at least 50 lb applied in any downward or outward direction at any point along the toeboard. The toeboards must be at least 3½″ tall and be solid or have openings less than 1″ in any direction. Any person working on the ground where overhead work is being done must wear a hard hat.

HEAT-RELATED WEATHER SAFETY

Working in high-temperature environments, such as on a rooftop or in an enclosed attic, can cause fatigue, fainting, muscle cramps, heat exhaustion, and heatstroke. In addition, working in high-temperature environments increases the risk of injuries due to accidents caused by slippery hands, sweat, fogged glasses, or dizziness.

Muscle cramps usually occur during the first days of hot stressful work, especially if individuals are not accustomed to this type of work. To avoid muscle cramps,

individuals must drink plenty of water to replace the fluids lost from sweating.

A loss of large amounts of fluid from the body can cause heat exhaustion. Individuals with heat exhaustion sweat, but they cannot maintain the correct body temperature because of the amount of heat. During heat exhaustion, the body increases its heart rate and strengthens blood circulation. Individuals suffering from heat exhaustion may feel disoriented, feel dizzy, feel fatigued, have a headache, or have flulike symptoms. The skin has a normal temperature but also has a damp and clammy feeling. Generally, the individual needs to rest, cool down, and drink plenty of liquids.

Heatstroke is the most serious heat-related health problem. Heatstroke occurs when the body stops adjusting to the hot temperature and sweat glands shut down. Symptoms of heatstroke include confusion, collapse, unconsciousness, dry mottled skin, and skin that is warm or hot to the touch. A heatstroke victim can die quickly. Immediate medical attention is required. An individual suffering from heatstroke should be moved to a cool area and cool water should be used to cool the individual.

ELECTRICAL SAFETY

Improper electrical wiring or misuse of electricity causes the destruction of equipment and fire damage to property.

The annual occupational death rate from electrical accidents is 1 per 100,000 workers. Safe working habits are required when troubleshooting an electrical circuit or component because the electric parts that are normally enclosed are exposed. Electrical safety rules should be followed by all personnel working with electricity. **See Figure 7-16.**

The most common OSHA electrical violation is improper grounding of equipment and circuitry. A *ground* is a safety feature designed to prevent shock due to a fault in an electrical system. A ground consists of an added ground wire running from a plug or equipment to the ground. The metal parts of an electrical wiring system that are exposed (switch plates, ceiling light fixtures, conduit, etc.) must be grounded and at 0 V. If the system is not grounded properly, these parts may become energized. Metal parts of motors, appliances, or electronics that are plugged into improperly grounded circuits may be energized. When a circuit is not grounded properly, a hazard exists because unwanted voltage cannot be safely eliminated. If there is no safe path to ground for fault currents, exposed metal parts in damaged appliances or power tools can become energized.

Extension cords with missing or broken ground terminals are unsafe and should never be used. A shock hazard exists if contact is made with a defective electrical device that is not grounded, or is grounded improperly, and a path is completed to ground.

Electric Shock Prevention

Electric shock is a condition that occurs when an individual comes in contact with two conductors of a circuit or when the body of an individual becomes part of an electrical circuit. Electric shock can be prevented by removing power from circuits and equipment before servicing them. Circuits must be deenergized before work is performed on them. To verify that the circuit is deenergized prior to performing work on it, a test instrument such as a clamp-on ammeter or digital multimeter must be used.

Electrical Safety

- Always comply with the NEC®.
- Use UL® approved appliances, components, and equipment.
- Keep electrical grounding circuits in good condition. Ground any conductive component or element that does not have to be energized. The grounding connection must be a low-resistance conductor heavy enough to carry the largest fault current that may occur.
- Turn off, lock out, and tag disconnect switches when working on any electrical circuit or equipment. Test all circuits after they are turned off. Insulators may not insulate, grounding circuits may not ground, and switches may not open the circuit.
- Use double-insulated power tools or power tools that include a third conductor grounding terminal which provides a path for fault current. Never use a power tool that has the third conductor grounding terminal removed.
- Always use protective and safety equipment.
- Know what to do in an emergency.
- Check conductors, cords, components, and equipment for signs of wear or damage. Replace any equipment that is not safe.
- Never throw water on an electrical fire. Turn off the power and use a Class C rated fire extinguisher.
- Work with another individual when working in a dangerous area or with dangerous equipment.
- Learn CPR and first aid.
- Do not work when tired or taking medication that causes drowsiness.
- Do not work in poorly lighted areas.
- Always use nonconductive ladders. Never use a metal ladder when working around electrical equipment.
- Ensure there are no atmospheric hazards, such as flammable dust or vapor, in the area. A live electrical circuit may emit a spark at any time.
- Use one hand when working on a live circuit to reduce the chance of an electrical shock passing through the heart and lungs.
- Never bypass or disable fuses or circuit breakers.
- Extra care must be taken in an electrical fire because burning insulation produces toxic fumes.
- Always fill out accident forms and report any electrical shock.

Figure 7-16. Electrical safety rules must be followed by those working with or near any type of electrical equipment or circuits.

To keep equipment from being accidentally turned on while it is being serviced, circuit lockout and tagout rules should be implemented. *Lockout* is the process of removing the source of electrical power and installing a lock, which prevents the power from being turned on. *Tagout* is the process of attaching a danger tag to the source of power to indicate that the equipment may not be operated until the tag is removed. **See Figure 7-17.**

The power on systems that are capable of being locked out should be locked out. The lockout tag is for the person who is not aware of work being performed on the circuit. All affected persons must be notified of the work in progress. All circuits being worked on must be tagged out at points where that equipment or circuit can be energized. Electrical work must never be performed alone. An individual must work with a qualified partner, and all workers must be trained in cardiopulmonary resuscitation (CPR). More information concerning safe electrical work practices can be found in OSHA 29 CFR 1926.269 — *Electric Power Generation, Transmission, and Distribution.*

Figure 7-17. Electrical equipment must be locked out and tagged out before any type of work is performed.

Overhead Power Lines

Overhead power lines are not insulated and can cause severe injury or death if contact is made with them. More than half of all electrocutions are caused by direct worker contact with energized power lines. When cranes, work platforms, or other conductive materials, such as ladders or aluminum mounting rails, contact overhead wires, the equipment operator or other workers can be electrocuted. Minimum distance requirements must be maintained when working near overhead power lines. The minimum distance requirement for voltages up to 50 kV is 10′. For voltages over 50 kV, the minimum distance is 10′ plus 4″ for every 10 kV over 50 kV. **See Figure 7-18.**

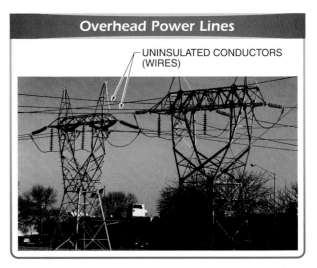

Figure 7-18. To prevent accidental electrocution from contacting uninsulated conductors, minimum distance requirements must be met when working near overhead power lines.

Overloaded Circuits

An *overload* is a small magnitude overcurrent that, over a period of time, leads to an overcurrent, which may turn on an overcurrent protection device (fuse or circuit breaker). Overloaded circuits in an electrical system are hazardous because they can produce heat or arcing. An *arc* is a sustained discharge of electricity across a gap in a circuit or between electrodes. Wires (conductors) and other components in an electrical system or circuit have a maximum amount of current that they can safely carry. If too many devices are plugged into a circuit, if a single device draws too much current, or if incorrect wire size is used, the electrical current can heat the wires to a high temperature.

Excessive heat can melt conductor insulation and cause arcing, which can create ground faults or fires. In order to prevent too much current in a circuit, a circuit breaker or fuse is placed in the circuit. A *circuit breaker* is a device that opens and closes a circuit by nonautomatic means and automatically opens a circuit when a predetermined current overload is reached without damage to itself.

A *fuse* is an electrical overcurrent protective device with a fusible portion that is heated and broken by the passage of excessive current. Circuit breakers and fuses open circuits to shut off any electrical current. If circuit breakers or fuses are too large for the wires they are supposed to protect, an overload in the circuit will not be detected and the current will not be shut off.

Wet Conditions

In addition to exposure to water and rain, wet conditions can include wet/damp clothing, high humidity, and perspiration, which also increase electrocution hazards. Working in wet conditions is hazardous because it increases the chance of completing a path for current flow. For example, if contact is made with any energized (live) wire or electrical component while standing in a wet area, the body will receive a shock. Equipment that is used near wet conditions or a water source should be connected to a ground-fault circuit interrupter receptacle.

A *ground-fault circuit interrupter (GFCI) receptacle* is a fast-acting receptacle that detects low levels of leakage current to ground and opens the circuit in response to the leakage (ground fault). GFCIs detect any difference in electrical current between two circuit conductors. A difference in current can occur when electrical equipment is not working correctly, causing leakage current. If leakage current (a ground fault) is detected in a GFCI-protected circuit, the GFCI switches off the current in the circuit, which removes the shock hazard. Electrical extension cords with portable GFCIs must be used when performing installation work for solar water heating systems. **See Figure 7-19.**

GFCIs are typically set at about 5 mA and are designed to provide protection from electrocution. GFCIs operate differently than other types of circuit breakers because they detect leakage currents rather than circuit overloads. Circuits with missing, damaged, or improperly wired GFCIs or plugs with missing or damaged prongs may not prevent a shock hazard.

Photovoltaic Panels

Some solar water heating systems use a photovoltaic (PV) panel to operate a DC pump or circulator. While these panels may not be large, they still can present an electrical hazard when they are exposed to sunlight. Even low light conditions can create a voltage potential that can lead to a shock or arc-flash.

Portable Ground-Fault Circuit Interrupters (GFCIs)

GFCI-PROTECTED RECEPTACLES

GFCI PLUG

Cooper Wiring Devices

Figure 7-19. Electrical extension cords with portable GFCIs must be used when performing installation work for solar water heating systems.

With solar electric systems, shutting off the main breaker does not prevent a solar electric system from having the ability to produce power. When performing work on a PV panel or associated wiring, safety practices specific to solar electric systems should be followed. These practices include the following:

• Use a current clamp or digital multimeter set to measure voltage or current to check for the presence of hazardous energy prior to working on a PV array.

• Always follow manufacturer instructions and check the equipment for specific operation and safety information.

• The only method of "shutting off" a PV panel is through the removal of the "fuel" source (sunlight). To remove sunlight from a solar panel, cover the array with an opaque cover that blocks sunlight. This prevents the solar panel from generating electricity.

• Prior to working on a solar PV panel, lock out and tag the circuit using standard lockout/tagout procedures. Even small amounts of sunlight can produce a voltage potential and shock or arc-flash hazard. These potential voltages are enough to electrically shock an installer.

• An electric arc-flash hazard exists while adding or removing a PV panel. Never disconnect PV connectors or other associated PV wiring under load.

HAZARDOUS MATERIALS

Solar water heating system installers may encounter hazardous materials at the installation site. A *hazardous material* is any material capable of posing a risk to health, safety, and property. These materials include chemicals used in the installation and startup process or existing materials already on the site. Plumbing chemicals, such as cleaners, primers, and solvent cements, are also classified as hazardous materials. These items are required to have proper labeling and chemical information available. Hazardous materials are also found in such construction materials as insulation, paint, and roofing or may be present due to certain environmental conditions like mold.

Container Labeling

Each hazardous material container must have a label, which should be examined before using the product. Specific hazards, precautions, and first-aid information are listed on the label. Hazardous material containers are labeled, tagged, or marked with appropriate hazard warnings per OSHA 29 CFR 1910.1200(f)—*Labels and Other Forms of Warning*. Material stored in a different container than originally supplied from the manufacturer must also be properly labeled. Unlabeled containers pose a safety hazard since users are not provided with content information and warnings.

Globally Harmonized System

The *globally harmonized system (GHS)* is a method of hazard communication that provides consistent classification of chemical hazards and a standardized approach to labeling elements and presenting information on safety data sheets. The GHS is based on major existing systems around the world, such as the chemical classification and labeling systems of many US agencies. Information about the GHS can be found in OSHA 29 CFR 1926.1200 — *Hazard Communication*.

OSHA regulations are based on a document developed by the United Nations entitled *Globally Harmonized System of Classification and Labeling of Chemicals*, commonly referred to as the Purple Book. This document provides harmonized classification criteria for the health, physical, and environmental hazards created by chemicals. It also includes standardized label elements that are assigned to these hazard classes and categories. It provides the appropriate signal words, pictograms, and hazard and precautionary statements to convey the hazards to users.

Material Safety Data Sheets. A *material safety data sheet (MSDS)*, also known as a safety data sheet (SDS), is a printed document used to relay hazardous material information from a manufacturer, distributor, or importer to a worker. **See Figure 7-20.** The information is listed in English and provides precautionary information regarding proper handling, with emergency and first-aid procedures. MSDSs also provide information regarding the appropriate PPE to use when performing a task. All hazardous materials used in a facility or at an installation site must be inventoried and have an MSDS.

Chemical manufacturers, distributors, and importers must develop an MSDS for each hazardous material. If an MSDS is not provided, the employer must contact the manufacturer, distributor, or importer to obtain the missing MSDS. MSDS files must be kept up to date and readily available to all personnel.

MSDSs may be filed according to product name, manufacturer, or a company-assigned number. If two or more MSDSs on the same hazardous material are found, the latest version is used. Before the GHS, an MSDS had no prescribed format. However, under the GHS, information on the MSDS must be presented using consistent headings in a specified sequence. This format is identical to the format detailed in ANSI Standard Z400.1, *Hazardous Industrial Chemicals Material Safety Data Sheet Preparation*.

Heat Transfer Fluid Hazards

A heat transfer fluid (HTF) may be classified as a hazardous material depending on the type used in the system. Distilled water used in some solar water heating systems is perfectly safe and nonhazardous. Propylene glycol is a safe, nontoxic HTF. Undiluted propylene glycol (PG) does not cause any skin irritation but can cause minor eye irritation. Exposure to PG mist can cause minor upper respiratory irritation, but such exposure is rare. PG is generally recognized as safe by the FDA.

Ethylene glycol is considered a hazardous material. Ethylene glycol exposure can lead to serious health consequences. Brief skin contact is not harmful, but continued or repeated exposure can cause skin redness. Repeated exposure to large amounts of ethylene glycol can lead to absorption of the chemical into the body. Contact with large amounts of heated fluid hot enough to burn the skin could cause absorption of potentially lethal amounts. Also, vapors from heated fluid in unventilated areas could cause respiratory irritation, headaches, or nausea. Ingesting small amounts of ethylene glycol will normally not cause any injury. Ingesting large amounts can lead to intestinal issues, such as nausea or vomiting, some other serious injury, or even death. Continued excessive exposure can have serious effects on the heart, lungs, central nervous system, and kidneys.

Other HTFs that are not as common, such as silicones and hydrocarbon oils, may have health hazards of their own. Regardless of the HTF being used, the MSDS for the particular HTF should be consulted for health concerns and the proper PPE to use.

Construction Material Hazards

Construction materials are not hazardous in their normal use. However, when some materials are altered in any way, such as by drilling, cutting, grinding, or heating, hazardous components or by-products may be released into the environment. Examples of hazardous construction materials include asbestos, lead, concrete, paint,

drywall, and insulation. Also, there is the possibility of exposure to hazardous mold during the installation process. It is important for installers to recognize these hazards and use the appropriate PPE when working with these materials.

Asbestos. *Asbestos* is a mineral that has long, silky fibers in a crystal formation. It was a component of many building materials, such as fireproofing materials, pipe insulation, siding, and tile, installed until the late 1980s. Asbestos is abrasion-resistant and stable at high temperatures. Products manufactured with asbestos release asbestos fibers into the air when the products crumble or are crushed. Airborne asbestos fibers range in size from 0.1μ to 10μ and are odorless and tasteless. Asbestos fibers can stay suspended in the air for long periods. Products containing asbestos may be encountered in remodeling jobs.

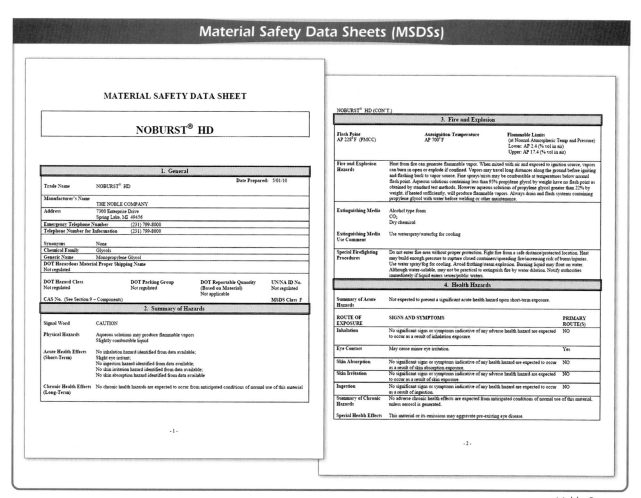

Noble Company

Figure 7-20. An MSDS is a printed document used to relay hazardous material information from a manufacturer, distributor, or importer to a worker.

The inhalation of asbestos fibers may cause asbestosis. *Asbestosis* is a respiratory disease caused by the inhalation of asbestos fibers resulting in the formation of scar tissue (fibrosis) inside the lungs. The severity of asbestosis depends upon the length of exposure to the fibers and the amount of asbestos inhaled. Appropriate PPE when working in areas with airborne asbestos must be used, including a sealed full body suit with respirator. Asbestos must be removed by certified, trained asbestos abatement personnel.

Lead. *Lead* is a heavy and dense material with a low melting point, low strength, and high rate of expansion. Lead can be found in lead sheets used for roof or vent flashing, lead-based paint used in homes built before 1978, and in certain plumbing materials and fixtures. At one time, lead pipe was used for the water service extending from the water main to the building and also to join bell-and-spigot cast iron soil pipe for sanitary waste and vent systems.

Lead-based paint is a common source of lead poisoning and is undetectable to the human eye. Before removing pipe that has been painted, the paint must be tested for lead content. Only workers trained in hazardous material removal should remove lead-based paint.

Lead alloy metal or metal covered with lead-based paint that is cut, sanded, heated, or burned releases lead into the air. Lead is toxic if swallowed or inhaled. Exposure to lead may cause brain and neurological damage. Guidelines to minimize lead exposure include the following:

• Do not eat or drink in the hazardous work area.

• Wear respirators designed to prevent inhaling lead.

• Seal the hazardous area from the rest of the work area and installation site. All rags, boxes, tools, and equipment should be removed from the area and disposed of properly.

• Cover and seal cabinets and surfaces that cannot be removed from the hazardous work area.

• Clean up solid debris using special vacuum cleaners with high-efficiency particle absorption (HEPA) filters. Wet-mop the area after vacuuming.

• Dispose of protective clothing (jump suits) worn in the hazardous work area after completing the cleanup operation. Protective clothing must not be worn in other areas of the installation site.

Lead in Plumbing Materials and Fixtures. Lead is rarely found in source water, but it enters tap water when plumbing materials corrode. Homes built before 1986 are more likely to have lead pipes, fixtures, and solder. However, new homes are also at risk because even plumbing that is legally considered lead free may contain up to 8% lead. Problems most commonly occur with brass and chrome-plated brass faucets and fixtures, which can leach significant amounts of lead into water, especially hot water.

Currently, only three states, California, Vermont, and Maryland, have passed legislation for the reduction of lead contained in the pipes, fittings, and fixtures used to carry drinking water. However, the Reduction of Lead in Drinking Water Act was recently signed into law and will take effect January 1, 2014. This new law reduces the allowable lead content in solder, flux, pipe, and fixtures that carry drinking water to 0.25%.

Lead-Based Paint Renovation, Repair, and Painting Program. The 2008 Lead-Based Paint Renovation, Repair, and Painting Program, developed by the Environmental Protection Agency (EPA), was designed to protect against the hazards of lead-based paint associated with renovation, repair, and painting activities. The program is a federal regulatory program affecting contractors, property managers, and others who disturb painted surfaces. It applies to residential buildings and child-occupied facilities, such as schools and day care centers, built before 1978. It includes prerenovation education, training, certification, and work practice requirements.

General work practice requirements must also be followed. The *Small Entity Compliance Guide to Renovate Right: EPA's Lead-Based Paint Renovation, Repair, and Painting Program* developed by the EPA, provides detailed information on the general work practice requirements.

For example, renovations must be performed by certified companies using certified renovators. Companies must post signs clearly defining the work area and warning occupants and other persons not involved in renovation activities to remain outside of the work area. These signs should be in the language of the occupants. Prior to the renovation, a company must contain the work area so that dust and debris do not leave the work area while the renovation is being performed.

Concrete. Solar water heating installers may have to drill through concrete roofing tiles, walls or floors in order to install the necessary system piping. Drilling through concrete produces dust containing crystalline silica particles. Crystalline silica, also known as quartz, is a natural compound found in the crust of the earth and is a basic component of sand and granite. Silicosis is a lung disease caused by inhaling dust containing crystalline silica particles. As dust containing crystalline silica particles is inhaled, scar tissue forms in the lungs, which reduces the ability of the lungs to extract oxygen from the air. Since there is no cure for silicosis, exposure prevention is the only means of control. The use of an N95 NIOSH-certified respirator is recommended, and in some cases it is required.

Engineering controls, such as power tools or vacuums with HEPA filters, or the use of water-based dust control measures to perform certain drilling or cutting tasks can reduce or eliminate airborne silica particles.

Paint. In certain situations, the installer may have to paint the exterior piping installation to protect the insulation from ultraviolet (UV) rays. Painting operations provide exposure to chemical agents that are ingredients of paint. Solvent-based paints contain aromatic and aliphatic hydrocarbons, ketones, and glycols. Ketones and glycols are harmful if inhaled or absorbed through the skin and are fatal if swallowed. Ketones and glycols cause reactions to skin, eyes, and respiratory tract and also affect the central nervous system.

Water-based paints may contain a biocide (formaldehyde) to prevent fungal growth. Formaldehyde is fatal or causes blindness if swallowed, and its vapor is harmful if inhaled or absorbed through the skin. Formaldehyde causes irritation to skin, eyes, and respiratory tract.

Oil-based paints may contain cadmium or chromium compounds as pigments. Cadmium or chromium compounds cause severe burns and may be fatal if swallowed or inhaled. Their vapors are irritating to the eyes and respiratory tract and cause damage to lungs and kidneys. The risk of cancer depends on the duration and level of exposure. Protective clothing, ventilation, and respiratory protection are required when working at an installation site where painting vapors are in the atmosphere.

Drywall. Installers often must cut, drill through, or remove drywall in the installation process. These actions produce airborne crystalline silica particles. The particles may cause irritation to the skin, eyes, and/or respiratory tract. Prolonged exposure may cause silicosis. The appropriate respiratory protection is required when crystalline silica particles are present in the air.

Insulation. Installers are exposed to insulation when working in attic spaces or insulating piping. **See Figure 7-21.** Cutting and installation of cellulose, foam, or fiberglass insulation may create particles resulting in potential health risks. Solar water heating system installers working where insulation particles exist should wear PPE to avoid health hazards.

Cellulose insulation may cause irritation or abrasion of the mouth and throat, headache, nausea, dizziness, and difficulty in breathing if inhaled. Contact with eyes causes irritation and tearing. Foam insulation may cause irritation to the respiratory tract if inhaled. Repeated exposure may cause lung damage. Contact with eyes causes irritation or corneal injury from abrasion. Fiberglass insulation may cause headache, nausea, dizziness,

and difficulty in breathing if inhaled. Contact with eyes causes irritation and inflammation of the mucous membrane, tearing, and sensitivity to light. Repeated exposure causes inflammation of the eyelids, digestive disturbances, weight loss, and general weakness. Extended exposure may result in respiratory and lung disease, bronchitis, asthma, or pulmonary heart disease.

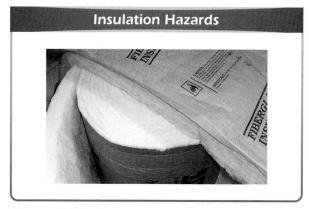

Insulation Hazards

Figure 7-21. Solar water heating system installers are exposed to insulation when working in attic spaces or insulating piping.

Mold. Installers may be exposed to mold during the installation process. Molds are fungi that can be found both indoors and outdoors. The exact number of the existing species of fungi is unknown, but estimates range from the tens of thousands to the hundreds of thousands.

Approximately 10% of the United States population is allergic to one or more types of mold. A certain level of mold must be in the air before a person suffers an allergic reaction. The higher the concentration of mold, the worse the reaction is likely to be.

Leaking pipes or fixtures can create an environment that encourages mold growth. Often, the mold is hidden on the back of drywall or under tile and wallpaper. The easiest solution for removing the mold is to stop the leak and replace the contaminated building materials. The proper PPE should be worn and EPA guidelines should be followed when performing mold remediation of any type.

Water damage is not the only prerequisite for mold formation. A high humidity level is a more likely condition for mold formation than the presence of water. Molds grow best in warm, damp, and humid climates. They spread and reproduce by making spores. Mold spores can survive in harsh, dry environmental conditions, which are not normally considered conducive to mold growth. Mold in buildings can form on any material with cellulose, such as ceiling tiles, paper, and cardboard. Improperly vented attic spaces may also produce mold.

SAFETY PLANS AND JOB HAZARD ANALYSIS

A safety plan is recommended for all work performed in the solar industry. A job hazard analysis is an important tool in the development of a safety plan. Adhering to safe work practices is as important as properly using typical construction tools. Safe installations are more efficient and cost effective. Using equipment properly and working safely results in less downtime due to injuries and resultant delays. Safe work practices also prevent fines from being assessed by agencies responsible for enforcing workplace regulations, such as OSHA. A job hazard analysis is used to develop safety policies that prevent injuries and death and is used as part of a safety plan to ensure a safe working environment.

Safety Plans

A *safety plan* is a comprehensive document that is developed by a company and distributed to its employees that includes the safety regulations and procedures that the company has instituted. This plan should be given to every employee and reviewed annually. A safety plan may include information such as the following:
- company safety policy
- emergency contact information
- emergency procedures
- installation site safety requirements
- employee safety training requirements
- written hazard communication program
- site hazard analysis forms
- site inspection forms

Safety plans should cover all the typical tasks that any member of a company will perform. A job hazard analysis addresses specific safety issues on a particular job location that may or may not be covered in the safety plan.

Evaluating and Identifying Safety Hazards

Evaluation and identification of potential safety hazards is required for eliminating risks and potential injuries. Each workplace can have different occupational hazards. Results of hazard analysis include the following:
- identification of safety training needs
- identify needs such as the need for additional equipment and personal protective gear
- elimination or control of hazards before they can cause injuries

The first step in identifying hazards requires understanding specific conditions concerning the work to be performed, along with when and where the work takes place. A checklist of questions related to employee training and job requirements can be used to identify specific hazards. **See Figure 7-22.**

Evaluating Specific Installation Sites

For each specific installation site, a list of hazards and the potential injuries that could occur must be developed. By understanding the hazards, potential injuries, and possibility of accidents occurring, a suitable safety policy for each specific situation can be developed. Typical situations at a solar water heating system installation site that may create hazards and result in injuries include the following:
- use of hand and power tools
- working with ladders to access equipment and rooftops
- working in extremely hot or cold conditions
- working with solar water heating system collectors
- working with PV panels

After evaluating each specific situation, the hazards and risks associated with those situations must be identified. Then the appropriate corrective action should be selected to address the hazard. **See Figure 7-23.**

Serious injuries, including death, can result from installation site accidents. Potential injuries from installation site hazards include the following:
- broken, fractured, or shattered bones
- bruising
- burns
- death
- electrocution
- eye injuries
- frostbite
- heatstroke
- internal organ injuries
- puncture injuries from falling onto items
- punctured lungs
- serious back or neck injures
- severe cuts or lacerations
- severe head, brain, and/or skull injuries
- spinal injuries
- strains and sprains

Solar Water Heating Installation Jobsite Checklists

- ☐ Are employees properly trained to perform each job with which they are assigned?
- ☐ Are safety policies and procedures implemented and enforced?
- ☐ Are employees trained on company safety policies and procedures?
- ☐ Is a qualified supervisor present at the jobsite? (List supervisor responsibilities)
- ☐ Is required PPE available and in good condition?
- ☐ What preparation and set-up work is required prior to starting work?
- ☐ What are project tasks and conditions?
- ☐ Are employees working from ladders or roof tops? Is fall protection devices available?
- ☐ What tools are being used to complete the project?
- ☐ Are severe outdoor conditions such as wet, windy, hot, or cold weather present?
- ☐ What post-job cleanup and work is required?

Figure 7-22. A checklist of questions related to employee training and job requirements can be used to identify specific hazards.

Installation Site Hazards and Corrective Action

Work Conditions	Hazards/Risks	Corrective Actions
Use of many different power tools and power cords at installation site	Electric shock from worn or frayed power cords and power lines; eye injuries, lacerations, puncture wounds, and bleeding from objects thrown from equipment such as saw blades; and lacerations, puncture wounds, and bleeding from sharp tools	Develop company PPE policy Eliminate extension cord hazards by using battery-operated tools Develop procedures for properly using power tools and extension cords
Carrying and positioning ladders on walls and rooftops as well as climbing and working from stepladders and extension ladders	Lifting injuries from carrying ladders, falls from using ladders, and electric shock from contact with electrical power lines	Use proper lifting and carrying techniques for ladders Develop proper ladder use policies
Working on hot rooftops or in hot attic spaces	Dehydration, fainting, heat exhaustion, heatstroke, or death	Work during cooler hours of day Develop hydration and safe practices for working in hot weather conditions
Lifting, moving, and installing large flat-plate and evacuated tube collectors as well as performing maintenance on collectors	Lifting injuries from lifting heavy and awkward flat-plate collectors and burns from handling collectors that are hot from sitting in sun	Develop policies and procedures for working with collectors Cover collector area with opaque object to prevent collector from heating under sun
Installing and performing maintenance on solar electric PV panels	Electric shock from handling solar electric PV panels in sun	Develop policies and procedures for working with solar electric PV panels

Figure 7-23. After evaluation of each specific situation, the hazards and risks associated with those situations must be identified and the appropriate corrective action determined to address the hazard.

Once the installation site evaluation is complete that identifies where and how work is being performed, safety strategies can be developed to remove or reduce accident risks for that particular job. All employees involved in the installation process should discuss the safety strategies prior to beginning any work.

Chapter 7 Review and Resources

8

Site Assessment

The first and most important step in installing a solar water heating system is to verify that the system will meet the present and future needs of the building owner and that the location is suitable for the installation to harvest solar radiation. A system should not be sold or installed unless these requirements are met. A site visit and assessment, including an interview with the building owner or homeowner, are essential for a successful solar water heating installation.

MEETING BUILDING OWNER NEEDS

Most new construction solar water heating installations have the benefit of inclusion in building design drawings. The site assessment for a new construction installation may only entail a shading analysis. Most of the other parameters for a solar water heating installation can be determined from the drawings and design specifications.

For an existing building, a site visit is necessary for determining system parameters. In many cases, drawings are not available for designing the system. It may be the responsibility of the installation company or the installer to design the system and create the necessary design drawings for the installation.

Typically, a solar water heating installer is not responsible for selling solar water heating systems to customers. Depending on the size of the company, the installer may only be required to properly install systems. In some cases, especially when working for or owning a small company with only a few employees, the installer may be responsible for selling systems and then verifying system suitability. The installer may even be responsible for designing systems. However, only small systems, such as systems for one-family dwellings and light commercial buildings, should be designed by installers, if required. Large installations should be designed by plumbing, mechanical, or solar system design professionals.

In any case, the building owner or homeowner must be interviewed to determine solar water heating needs. This could be accomplished with a phone conversation. However, good business practices dictate that a personal visit to the customer and installation site is preferred. Speaking face to face with a potential customer allows the customer to become familiar and comfortable with the installer and company to whom their future hot water investment is entrusted. An installation site visit also allows for a better understanding of the hot water requirements and for inspection of the site to determine the suitability of a system.

INSTALLATION SITE ASSESSMENT

A site must be assessed to determine the hot water needs of the owner. This allows for the determination of the type and size of solar water heating system to be installed. Equipment and piping installation locations, suitability, and requirements must also be assessed. This allows for the development of an organized installation plan, which will include special tool, material, and equipment needs. A safety assessment must also be made to plan for the protection of all installation personnel, building occupants, and property. The safety assessment should include a plan for complying with all Occupational Safety and Health Administration (OSHA) and building code requirements.

The information acquired from a site assessment should be documented. It is important that an assessment report be developed so that pertinent information can be recorded during the visit. It is also beneficial to document findings by taking photos of the site as the assessment progresses. This will make designing of the system and installation planning much easier as well as save time and money. **See Figure 8-1.**

Some essential information must be received before ever leaving the shop or office to visit the site. The site location address, name of the building owner, and contact information are needed before visiting a site. **See Figure 8-2.**

It is recommended that a computer-generated map also be included with the site assessment report to make it easy for others to find the location. Also, an actual location picture can be reproduced, using Google Earth software, to help in finding the location. The photos of the site can be valuable for other site assessments such as proper collector location and shading analysis. **See Figure 8-3.**

The physical location of the site is critical in determining the type of solar water heating system to be installed. A sun path chart and solar radiation data set for the installation location can be printed out for use in system calculations. Average climate, insolation, and high and low temperatures can be researched before leaving for a site visit. **See Figure 8-4.** This information is available on many government websites, in solar installation software packages, and on the DVD included with this text. This pertinent information should also be added to the site assessment report.

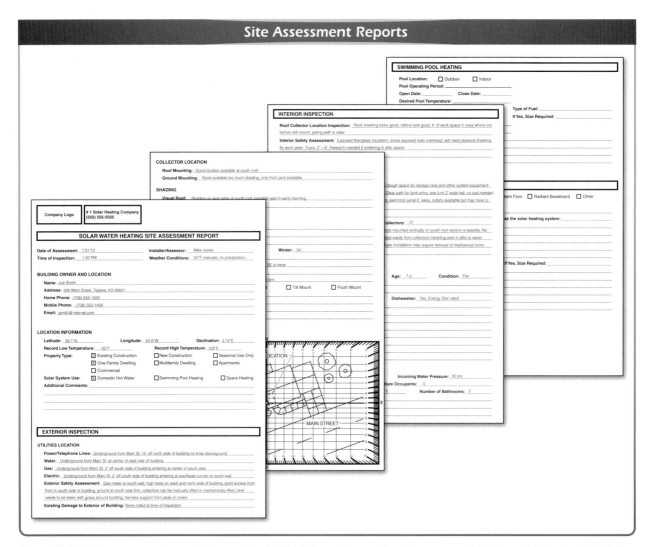

Figure 8-1. A site assessment report documents all of the information required for a successful solar water heating installation.

Owner and Location Information

Date of Assessment: 1/31/12   Installer/Assessor: Mike Jones

Time of Inspection: 1:30 PM   Weather Conditions: 52°F overcast, no precipitation

BUILDING OWNER AND LOCATION

Name: Joe Smith

Address: 000 Main Street, Topeka, KS 66601

Home Phone: (708) 555-1200

Mobile Phone: (708) 555-1400

Email: jsmith@internet.com

LOCATION INFORMATION

Latitude: 39.1°N   Longitude: 95.6°W   Declination: 3.14°E

Record Low Temperature: −32°F   Record High Temperature: 109°F

Property Type:
- [X] Existing Construction
- [ ] New Construction
- [ ] Seasonal Use Only
- [X] One-Family Dwelling
- [ ] Multifamily Dwelling
- [ ] Apartments
- [ ] Commercial

Solar System Use:
- [X] Domestic Hot Water
- [ ] Swimming Pool Heating
- [ ] Space Heating

Additional Comments: _____

Figure 8-2. General information, such as the building type, site location, and owner contact information, should be included in the first part of the site assessment report.

Site Location Software

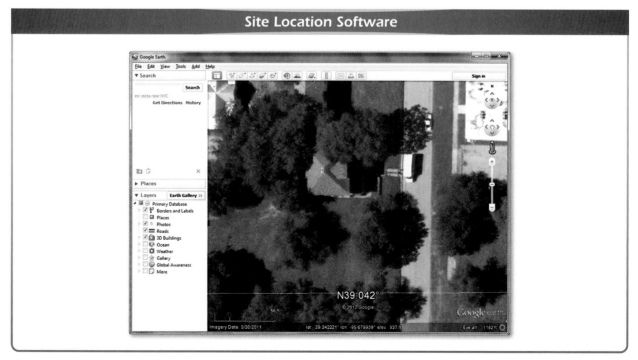

Figure 8-3. Software programs, such as Google Earth, can provide an accurate means of identifying building locations.

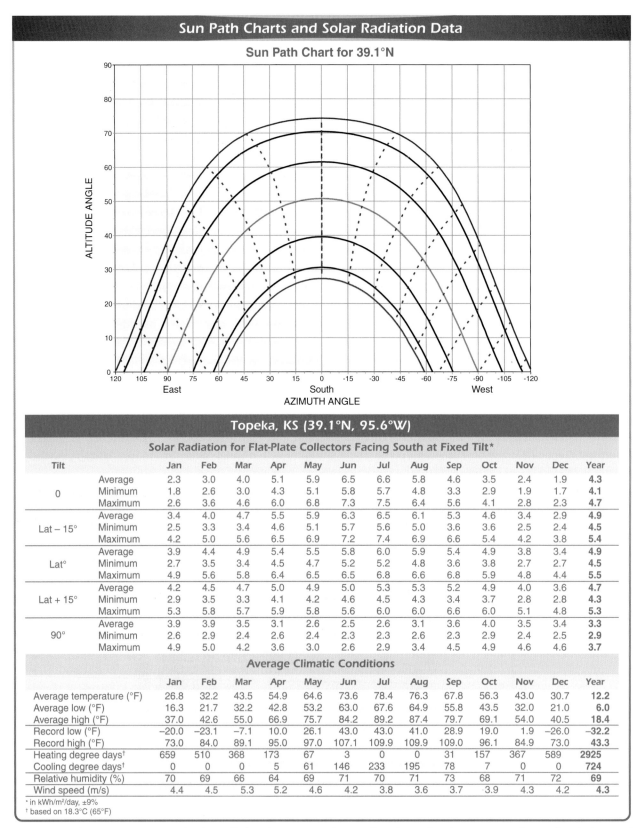

Sun Path Charts and Solar Radiation Data

Sun Path Chart for 39.1°N

Topeka, KS (39.1°N, 95.6°W)

Solar Radiation for Flat-Plate Collectors Facing South at Fixed Tilt*

Tilt		Jan	Feb	Mar	Apr	May	Jun	Jul	Aug	Sep	Oct	Nov	Dec	Year
0	Average	2.3	3.0	4.0	5.1	5.9	6.5	6.6	5.8	4.6	3.5	2.4	1.9	**4.3**
	Minimum	1.8	2.6	3.0	4.3	5.1	5.8	5.7	4.8	3.3	2.9	1.9	1.7	**4.1**
	Maximum	2.6	3.6	4.6	6.0	6.8	7.3	7.5	6.4	5.6	4.1	2.8	2.3	**4.7**
Lat − 15°	Average	3.4	4.0	4.7	5.5	5.9	6.3	6.5	6.1	5.3	4.6	3.4	2.9	**4.9**
	Minimum	2.5	3.3	3.4	4.6	5.1	5.7	5.6	5.0	3.6	3.6	2.5	2.4	**4.5**
	Maximum	4.2	5.0	5.6	6.5	6.9	7.2	7.4	6.9	6.6	5.4	4.2	3.8	**5.4**
Lat°	Average	3.9	4.4	4.9	5.4	5.5	5.8	6.0	5.9	5.4	4.9	3.8	3.4	**4.9**
	Minimum	2.7	3.5	3.4	4.5	4.7	5.2	5.2	4.8	3.6	3.8	2.7	2.7	**4.5**
	Maximum	4.9	5.6	5.8	6.4	6.5	6.5	6.8	6.6	6.8	5.9	4.8	4.4	**5.5**
Lat + 15°	Average	4.2	4.5	4.7	5.0	4.9	5.0	5.3	5.3	5.2	4.9	4.0	3.6	**4.7**
	Minimum	2.9	3.5	3.3	4.1	4.2	4.6	4.5	4.3	3.4	3.7	2.8	2.8	**4.3**
	Maximum	5.3	5.8	5.7	5.9	5.8	5.6	6.0	6.0	6.6	6.0	5.1	4.8	**5.3**
90°	Average	3.9	3.9	3.5	3.1	2.6	2.5	2.6	3.1	3.6	4.0	3.5	3.4	**3.3**
	Minimum	2.6	2.9	2.4	2.6	2.4	2.3	2.3	2.6	2.3	2.9	2.4	2.5	**2.9**
	Maximum	4.9	5.0	4.2	3.6	3.0	2.6	2.9	3.4	4.5	4.9	4.6	4.6	**3.7**

Average Climatic Conditions

	Jan	Feb	Mar	Apr	May	Jun	Jul	Aug	Sep	Oct	Nov	Dec	Year
Average temperature (°F)	26.8	32.2	43.5	54.9	64.6	73.6	78.4	76.3	67.8	56.3	43.0	30.7	**12.2**
Average low (°F)	16.3	21.7	32.2	42.8	53.2	63.0	67.6	64.9	55.8	43.5	32.0	21.0	**6.0**
Average high (°F)	37.0	42.6	55.0	66.9	75.7	84.2	89.2	87.4	79.7	69.1	54.0	40.5	**18.4**
Record low (°F)	−20.0	−23.1	−7.1	10.0	26.1	43.0	43.0	41.0	28.9	19.0	1.9	−26.0	**−32.2**
Record high (°F)	73.0	84.0	89.1	95.0	97.0	107.1	109.9	109.9	109.0	96.1	84.9	73.0	**43.3**
Heating degree days†	659	510	368	173	67	3	0	0	31	157	367	589	**2925**
Cooling degree days†	0	0	0	5	61	146	233	195	78	7	0	0	**724**
Relative humidity (%)	70	69	66	64	69	71	70	71	73	68	71	72	**69**
Wind speed (m/s)	4.4	4.5	5.3	5.2	4.6	4.2	3.8	3.6	3.7	3.9	4.3	4.2	**4.3**

* in kWh/m²/day, ±9%
† based on 18.3°C (65°F)

Figure 8-4. A sun path chart and solar radiation data set can be used to determine the average climate, insolation, and high and low temperatures throughout the year.

Proper site inspections require examining the surrounding property and grounds, the roof, attic spaces, and behind access panels as well as in and around storage, plumbing, and mechanical rooms. The installer should be prepared with the appropriate equipment to conduct these assessments and inspections. **See Figure 8-5.**

Solar System Hot Water Use

The amount of hot water to be used at the site must be known to determine the size of the solar collector array and solar storage tank. The assessment of hot water use entails discussing the current use and future needs of the home or building occupants with the owner. It should also entail a water audit of the building to document existing fixtures and appliances and their water flow or storage characteristics. **See Figure 8-6.**

The interview with the owner should verify the expected solar hot water use within the home or building. It can then be determined if the solar water heating system is intended for domestic hot water heating, space or radiant heating, swimming pool water heating, or a combination of these systems. Each system or combination system has variables that need to be determined such as collector type, heat transfer methods, and solar storage needs. All of these variables should be documented on the site assessment report.

Site Assessment Equipment

Safety and Personal Protective Equipment	Basic Hand Tools	Specialty Assessment Tools	Miscellaneous Equipment
• Safety/sun glasses • Hard hat • Safety/work shoes • Fall protection harness and attachment device • Electrical lockout and tagout devices • Gloves • Vehicle safety cones	• 25' and 100' tape measures • Screwdriver set • Wrench set • Pliers or "channel locks" • Levels: pocket and 2' level • 2' framing square • Hammer • Jab saw • Flashlight	• Solar Pathfinder or other sun path/solar window calculator • Roof angle finder or inclinometer • Compass • Electronic stud finder • Thermometer • GPS device, computer, or tablet computer with wireless capability • Multimeter tester • Water-pressure gauge	• Digital camera • Boot covers • String line • Drop cloths • 6' and 10' ladders

Figure 8-5. Installers should be prepared with the appropriate equipment to conduct site assessments and inspections.

Existing Equipment Information

EXISTING WATER HEATER

Size: 50 gal.   **Fuel Type:** Gas   **Age:** 7 yr   **Condition:** Fair

Retrofit Possible: WH too old and might be too small

Piping Type: Copper

Clothes Washer: Yes, old   **Dishwasher:** Yes, Energy Star rated

Large Volume Fixtures: Whirlpool BT

Gallons per Day Estimate: 80 gal. to 90 gal.

Desired Hot Water Temperature: 125°F

Incoming Cold Water Temperature: 55°F

Estimated Solar Loop Length: 50'

Additional Comments:

Figure 8-6. The size and type of the existing water heater and other fixtures should be included on the site assessment report.

It is also necessary to determine and document the number of occupants in the home or building. It should be determined whether there will be additional occupants in the future or whether there are any plans for home or building expansion.

For example, a homeowner should be asked if additions to the family are expected and if relatives or guests are expected to stay for extended lengths of time. It should also be determined whether there are plans to add rooms to the home. A commercial building owner should be asked if there could be a possible expansion or additional occupants in the building. It should be asked if there could be an increase in hot water use at some point in time. If an increase in building size, the number of occupants, or hot water use is expected, that factor can be added into the calculations used for system sizing or at least measures can be taken to allow for easy expansion of the system. **See Figure 8-7.**

Even if the solar water heating system will be installed in a new construction installation, it will be helpful to speak with the building owner or managing agent to determine the occupant number and whether there are any plans for expansion. The design drawings of new construction projects contain the number of bathrooms and types of fixtures in the plumbing and mechanical drawings. The schedule of fixtures and other equipment that use hot water should also be present within these drawings to allow for the calculation of solar hot water needs.

Site Assessment and Inspection

Once the interview is complete, site inspection and assessment can begin. If design drawings for the building or home exist and are supplied, it will be helpful to have them as a reference and a guide during the inspection and the water audit. It will also be very useful to have a set of drawings on which notations and piping schematics can be made.

This will prevent having to create new drawings by hand or computer aided drawing (CAD) software.

Locations and measurements must be compared with existing drawings and verified to ensure accuracy. The installer does not want to make calculations using incorrect measurements. If drawings do not exist, drawings should be created to assist in the documentation of site conditions, potential safety hazards, and possible collector, storage tank, and solar piping path locations.

The water audit consists of an inspection and documentation of all hot and cold water uses of the site. **See Figure 8-8.** During the inspection, the flow rates of water outlets, fixtures, and appliances are noted as well as their current condition. Any leakage or disrepair should be noted with a recommendation for repairs. Recommendations for flow rate adjustments to faucets and fixtures can also be made at this time. For example, changing fixtures, faucets, or aerators to lower flow rates can be suggested.

The water auditing process can help the owner reduce the initial use of hot and cold water and possibly allow for reduced solar hot water needs. This can save the customer money not only on the system based on size but eventually on water and sewer usage fees. There is no sense in installing a solar water heating system that will save the owner costs in energy use only to see those savings diminish by wasting water.

The exterior of the building and the grounds should be walked and inspected to determine a safe area for equipment storage and ground work. Safety hazards should be noted such as power lines, aboveground and underground utility lines, and utility interconnections. Aboveground work areas where harnessing or lifting equipment is needed should be located, and safe attachment locations and ground areas to place ladders or lifts should be documented. **See Figure 8-9.**

Domestic Hot Water (DHW) Information

DOMESTIC HOT WATER USE

Water Source: ☒ City   ☐ Well   Incoming Water Pressure: _60 psi_

Occupants: _4_   Guests/Future Occupants: _0_

Number of Bedrooms: _4_   Future: _0_   Number of Bathrooms: _3_

Additional Comments: _____

Figure 8-7. Information about the hot water use, including the water source and current and future occupancy, should be included on the site assessment report.

Water Audit Reports

Kitchen water audit

Kitchen Sink — Washing dishes by hand	Once a week	Twice a week	Every other day	Every day
Sink half full or less	☐ 110	☐ 220	☐ 385	☐ 770
Full sink of water	☐ 220	☐ 440	☐ 770	☐ 1,540

Dishwashers	Once a week	Twice a week	Every other day	Every day
Dishwasher cycle - normal	☐ 530	☐ 1,060	☐ 1,855	☐ 3,710
Dishwasher cycle - rinse/hold	☐ 135	☐ 265	☐ 465	☐ 925

Running Faucet — Food Prep/Rinsing/Drinking	One minute per day	Two minutes per day	Five minutes per day	10 minutes per day
One Person	☐ 730	☐ 1,460	☐ 3,650	☐ 7,300
Two People	☐ 1,460	☐ 2,920	☐ 7,300	☐ 14,600
Three People	☐ 2,190	☐ 4,380	☐ 10,950	☐ 21,900
Four People	☐ 2,920	☐ 5,840	☐ 14,600	☐ 29,200
Five People	☐ 3,650	☐ 7,300	☐ 18,250	☐ 36,500

Note: An average is 2 minutes per day per person with a flow rate of one gpm

Total water volume: **Comments:**

    Kitchen sink   _____

    Dishwasher   _____

    Running Faucet   _____

    TOTAL   _____

Recommendations:

Figure 8-8. A water audit should be performed during a site assessment to determine the flow rates and current condition of water outlets, fixtures, and appliances.

Exterior Inspection Information

EXTERIOR INSPECTION

UTILITIES LOCATION

Power/Telephone Lines: Underground from Main St. 10′ off north side of building no lines aboveground

Water: Underground from Main St. at center of east side of building

Gas: Underground from Main St. 3′ off south side of building entering at center of south side

Electric: Underground from Main St. 2′ off south side of building entering at southeast corner on south wall

Exterior Safety Assessment: Gas meter at south wall, high trees on west and north side of building, good access from front to south side of building, ground at south side firm, collectors can be manually lifted or mechanically lifted; care needs to be taken with grass around building, harness support from peak or crown

Existing Damage to Exterior of Building: None noted at time of inspection

Figure 8-9. Information obtained during an exterior inspection of the site, especially related to safety, should be included on a site assessment report.

Any existing damage to equipment or property in work areas should also be noted. Photographs of damage should be taken. This will help to establish responsibility for site or equipment damage and provide protection against claims made in the future.

Proper Collector Location

During exterior site inspection, a preliminary location for collector placement and a determination of roof or ground placement can be made. An initial visual assessment of existing and possible future shading can be used to determine whether collectors should be mounted on the roof or ground. **See Figure 8-10**.

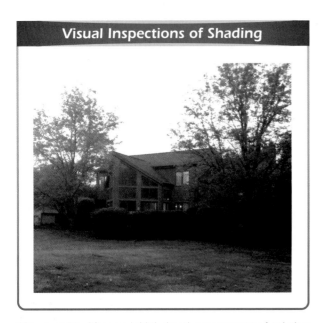

Visual Inspections of Shading

Figure 8-10. After an initial visual assessment of existing and possible future shading, it can be determined whether collectors should be mounted on the roof or ground.

If there is no proper access to the Sun at one location, such as the ground, planning can begin for the other location. For example, if buildings or trees shade most of the ground locations, rooftop mounting of collectors will be necessary. Or conversely, if there is no access on the roof to southern exposure for collector mounting, ground placement may be necessary. **See Figure 8-11.**

A ground-mounted solar collector must be installed in an area where there will be enough space for the collector or collector array. There must also be enough space around the collector or collector array to prevent damage to the system. Soil conditions should also be noted. Concrete footings or pads may be necessary to support the collectors but should not be placed in loose or wet soil. **See Figure 8-12**. Close proximity to the equipment storage room is also necessary.

For roof mounting, a determination of proper roof location and mounting requirements involve getting on the roof. All appropriate safety procedures for roof work must be followed. The roof type, such as shingle, tile, or metal, must be noted to ensure that the proper mounting hardware is ordered.

To save time, money, and effort later, damaged roofing should be repaired or replaced before installing the collector.

An inspection of the roof should be made to determine whether the roof is damaged or weathered. A collector or collector array should not be mounted on a damaged or weathered roof. The roof should be repaired before the collector is mounted. It saves time, money, and effort to incorporate repairs before the system is installed rather than removing the solar system at a later date to repair the roof.

A roof-mounted collector must be installed on a roof where there will be enough space to mount the collector or collector array and that is free of possible shading problems. The location must allow access to the southern sky for the collector to receive the most solar insolation possible. In some instances, it may be difficult to find a roof surface facing directly south. Collectors can be placed on other roof areas, but this will most likely require more elaborate mounting installations and involve the sacrifice of some collector efficiency. **See Figure 8-13**.

Location and Shading Assessment Information

COLLECTOR LOCATION

Roof Mounting: Good location available at south roof

Ground Mounting: None available too much shading, only front yard available

SHADING

Visual Roof: Shading on west edge of south roof, possible east in early morning

Visual Ground: No areas on ground acceptable

Figure 8-11. An initial visual shading assessment can be used to determine the approximate location of the proposed collectors. This information should be included on the site assessment report.

Ground-Mounted Solar Collectors

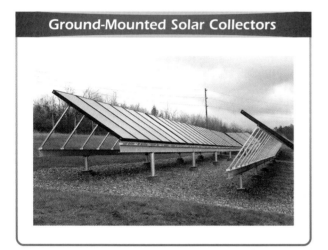

Solar H₂Ot, Ltd.

Figure 8-12. Concrete footings or pads may be necessary to support collectors but should not be placed in loose or wet soil.

The desired location on the roof should be measured and compared with an estimation of area requirements for the future collector or collector array. Two possible collector mounting areas should be found and then assessed to determine the best location for the collector or collector array. It is helpful to have two possible roof areas in case the first area of choice does not meet orientation, collector tilt, or shading parameters.

Facts

The simplest and most commonly used device for measuring sunshine duration is the Campbell-Stokes sunshine recorder. This device consists of a solid glass sphere that generates a focal point on the side that is turned away from the Sun. A correspondingly curved flameproof paper strip is placed around the sphere. The Sun burns a track on the paper strip. When clouds cover the Sun, the burnt track is interrupted.

Other Roof Areas for Collector Mounting

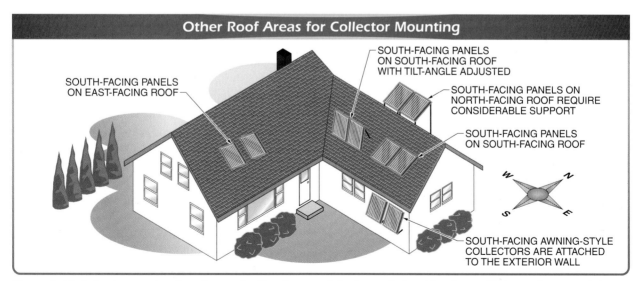

SOUTH-FACING PANELS ON EAST-FACING ROOF

SOUTH-FACING PANELS ON SOUTH-FACING ROOF WITH TILT-ANGLE ADJUSTED

SOUTH-FACING PANELS ON NORTH-FACING ROOF REQUIRE CONSIDERABLE SUPPORT

SOUTH-FACING PANELS ON SOUTH-FACING ROOF

SOUTH-FACING AWNING-STYLE COLLECTORS ARE ATTACHED TO THE EXTERIOR WALL

Figure 8-13. Collectors can be placed on roofs other than south facing roofs, but this will most likely require more elaborate mounting installations and involve the sacrifice of some collector efficiency.

Roof Orientation

The roof orientation, also referred to as roof azimuth, is the compass direction or heading in which the proposed roof areas are situated. For example, the roof area might face true south, 22° east of true south, or some other compass heading. The exact direction the roof faces determines whether a true south, or at least a southern facing, installation is possible. The acceptable range from true south for efficient solar radiation capture is ±45°. At 45° in either direction from true south, the output is reduced by about 5%. **See Figure 8-14.**

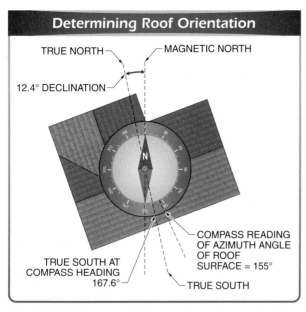

Figure 8-14. The acceptable range from true south for efficient solar radiation capture is ±45°.

Roof orientation can be found using a magnetic compass or digital compass applications such as those on smartphones or tablet computers. Sun path calculators and other solar analysis software can also assist in finding the orientation of the proposed roof area. Many of these software or calculator applications automatically adjust compass headings for true north, making the process quite simple.

The compass needle of a magnetic compass points to magnetic north. However, this is not true, or geographic, north. To find true north, which also designates true south, the compass must be adjusted for the magnetic declination of Earth at the installation location. **See Figure 8-15.** The magnetic declination can be found using maps, digital applications, or the various websites that provide this information.

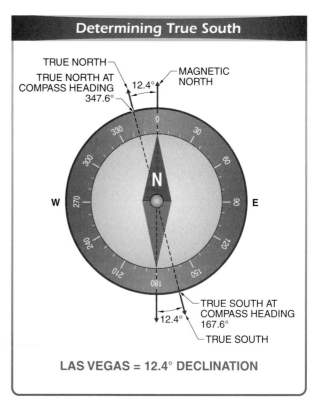

Figure 8-15. To find true north, which also designates true south, a compass must be adjusted for the magnetic declination of Earth at the installation location.

Once the magnetic declination is known, true north can be determined. For example, the magnetic declination for an installation in Las Vegas, Nevada, is 12.4° E. This means that the magnetic compass needle or heading is pointing 12.4° east of true north. True north is actually 12.4° to the west, or left on the magnetic compass dial, at the heading 347.6° (360° − 12.4° = 347.6°).

True south is 180° from true north. In the previous example, true south is at 167.6° on the compass (347.6° − 180° = 167.6°). The true south heading can be compared to the proposed roof orientation (roof angle implies slope) to determine if the roof orientation is within the acceptable range of deviating from true south. For example, a suitable area on the roof of a building in Las Vegas, Nevada, faces 155°. Since true south for that location is 167.6°, the roof section faces 12.6° east of true south (167.6° − 155° = 12.6°). This roof orientation falls within the range of ±45° from true south. The roof orientation is 155° and should meet the parameters for good solar insolation. The roof orientation should be noted in the assessment report. **See Figure 8-16.**

Roof Construction Information

ROOF CONSTRUCTION

Roof Type: Asphalt Shingle

Roof Orientation (Azimuth): 11°E from true south

Roof Angle: 7:12 pitch, 30.26° angle

Optimum Angle at Location: 39°   **Winter:** 54°

Roof Height: 10′ to eave, 18′-6″ to peak

Size (length × width): Triangle shape, from 12″ at peak to 25′ at eave

Obstruction on Roof: None

Condition of Roof: Good, no weathering of shingles, roof firm

Mounting Type Recommended: ☒ Stand-off Mount   ☐ Tilt Mount   ☐ Flush Mount

Additional Comments: _____

Figure 8-16. Information about roof orientation and construction should be included in the site assessment report.

Roof Angle

In addition to the roof orientation, the proposed roof angle or slope should be noted on the assessment report. The roof angle can be easily found by using an angle finder or an inclinometer, also referred to as a clinometer. To find the roof angle, the angle finder or inclinometer is placed on the roof. The angle from level is then read on the angle finder or shown on the inclinometer.

Another method of finding the roof angle is to determine the pitch of the roof and then calculate the angle of the roof. This can be accomplished by measuring the distance from the roof to the bottom of a level or a flat level surface at a point 12″ away from the roof to determine the roof rise. **See Figure 8-17.** For example, if the roof rise is 7″ in 12″, the pitch of the roof is 7:12. The roof angle from horizontal can then be determined by using a roof pitch table. The angle of a 7:12 roof is 30.26°. This angle should be noted and then compared to the proper tilt angle of the collector for the site latitude. This information helps to determine whether the collector or collector array can be flat-mounted on the roof or will require tilt mounting hardware.

The roof also should be examined to determine whether it will support the weight of the collectors.

A determination of weight-bearing capability should be made if heavy storage systems are to be mounted on the roof. The attic also should be examined to determine the suitability of the roof support members. It may be necessary to have a structural engineer make these determinations to avoid any damage liability by the installer.

Shading Analyses

Once a roof or ground location is identified, a shading analysis of the location should be performed. The shading analysis is used to ensure that the proposed collector location is free of obstructions that will prevent solar radiation from striking the collector aperture area. There are various ways to perform a shading analysis. Methods include using a sun path analysis tool, using a solar analysis software package, or performing a simple visual inspection with the use of hand drawings and mathematical calculations.

An installation company or professional installer should use the most-accurate, least-time consuming, and most cost-effective method for conducting and documenting the shading analysis. Conducting the shading analysis using a sun path analysis tool is the method of choice and standard for the industry.

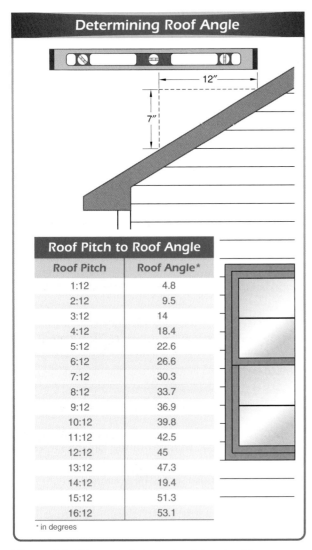

Determining Roof Angle

Roof Pitch to Roof Angle

Roof Pitch	Roof Angle*
1:12	4.8
2:12	9.5
3:12	14
4:12	18.4
5:12	22.6
6:12	26.6
7:12	30.3
8:12	33.7
9:12	36.9
10:12	39.8
11:12	42.5
12:12	45
13:12	47.3
14:12	19.4
15:12	51.3
16:12	53.1

* in degrees

Figure 8-17. The roof pitch can be determined by measuring the distance from the roof to the bottom of a level or a flat level surface at a point 12″ away from the roof.

Several sun path analysis tools are available to the solar water heating system installer. Three of the most commonly used analysis tools used in the solar water heating industry are the Solmetric SunEye, Wiley Electronics ASSET, and Solar Pathfinder™. Each can be used with computer software to assist in developing shading analysis reports. However, the Solar Pathfinder also has a software package to assist in sizing, laying out, and estimating projects. The Solar Pathfinder is a nonelectronic device and does not require special skills or technical knowledge to operate it properly. **See Figure 8-18**.

Solar Pathfinders

Solar Pathfinder

Figure 8-18. The Solar Pathfinder is a nonelectronic device and does not require special skills or technical knowledge to operate it properly.

The Solar Pathfinder uses a highly polished, transparent, and convex plastic dome that provides a panoramic view of the entire site, including possible shading from the north, which also can be possible in southern latitudes. All trees, buildings, and other obstructions to solar radiation or insolation will be plainly visible as reflections on the surface of the dome. **See Figure 8-19**. Since the Solar Pathfinder works on a reflective principle rather than actually showing shadows, it can be used any time of day, any time of year, in cloudy weather, or in clear weather. The actual position of the Sun at the time of the analysis is irrelevant. In fact, the unit is easier to use in the absence of direct sunlight.

A sun path diagram containing information on solar insolation is placed on the base plate of the pathfinder and can be seen through the transparent dome. **See Figure 8-20**. The diagrams are application specific and designated as south-facing for the northern hemisphere, north-facing for the southern hemisphere, vertical for applications of 20° to 90° tilt, and horizontal for applications of 0° to 20° tilt. The diagrams are latitude specific and are designed as graphs.

Solar Pathfinder Domes

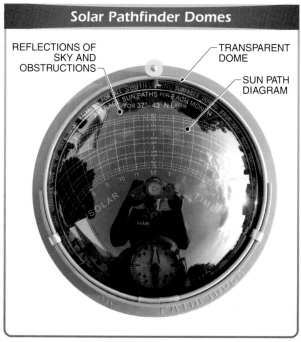

REFLECTIONS OF SKY AND OBSTRUCTIONS

TRANSPARENT DOME

SUN PATH DIAGRAM

Solar Pathfinder

Figure 8-19. Trees, buildings, and other obstructions to sunrays are plainly visible as reflections on the surface of a transparent dome.

The graphs contain ray lines and arcs. The ray lines depict solar time, which is noted below the arcs. The arcs depict the average sun path for a given month. Small numbers appearing in the half-hour increments between the ray lines give the percentage of radiation for that half-hour of the day.

The Solar Pathfinder is designed to be set up at the specific location and height on the roof that the collectors will be mounted. **See Figure 8-21**. The base plate contains a bubble level in the blue triangle that should be adjusted to the center of the black circle, indicating that the device is level. The unit should be adjusted so that the red arm of the compass points north when in the northern hemisphere.

The magnetic declination can be adjusted by pulling out the magnetic declination tab and rotating the diagram to the proper degree marking under the white dot. Negative numbers are left of the "0" for west of north declination, while positive numbers are right of the "0" for east of north declination. **See Figure 8-22**. The dome should be viewed from between 12″ to 18″ above. When taking a digital photograph, the camera should be between 12″ to 18″ above the dome and directly above the vertical centerline on the sun path diagram.

Sun Path Diagrams

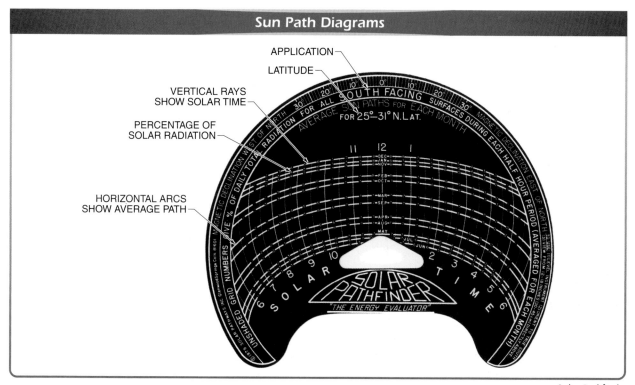

APPLICATION

LATITUDE

VERTICAL RAYS SHOW SOLAR TIME

PERCENTAGE OF SOLAR RADIATION

HORIZONTAL ARCS SHOW AVERAGE PATH

Solar Pathfinder

Figure 8-20. A sun path diagram containing information on solar insolation is placed on the base plate of the pathfinder and can be seen through the transparent dome.

Photographing Solar Pathfinders

Solar Pathfinder

Figure 8-21. The Solar Pathfinder is designed to be set up at the specific location and height on the roof that collectors will be mounted.

Solar Pathfinder Declination Setup

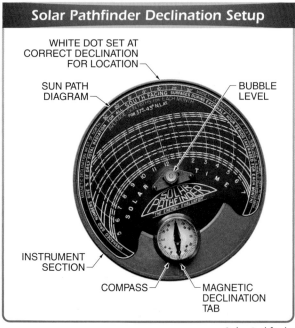

WHITE DOT SET AT CORRECT DECLINATION FOR LOCATION

SUN PATH DIAGRAM

BUBBLE LEVEL

INSTRUMENT SECTION

COMPASS

MAGNETIC DECLINATION TAB

Solar Pathfinder

Figure 8-22. The white dot indicates the amount of magnetic declination for a location. In this case, the Solar Pathfinder is set to 11°E of magnetic north and located within 37°N to 43°N latitude.

The Solar Pathfinder presents two views at the same time. A reflective panoramic view of the site can be seen on the dome, while the sun path diagram can be seen through the dome. The site will be shaded at the time indicated on the diagram where the reflected objects on the dome intersect the sun paths shown through the dome.

The areas of shading can be traced manually on the diagram using a white marking pen. This provides a permanent documented representation of the extent of shading at the proposed collector location. **See Figure 8-23.** A digital photograph should also be taken of the dome to provide further documentation or to be used with the Solar Pathfinder Assistant software.

The Solar Pathfinder instruction manual gives detailed instructions on how to manually calculate the radiation information to generate reports and use the information to size and estimate solar water heating systems. However, the Solar Pathfinder Assistant software can combine local weather data (using NREL/WMO) and the site-specific shade analysis documented in the digital photograph of the reflective dome to create customizable fast, accurate, and professional-looking reports.

The completed shading analysis, roof orientation, and proposed tilt are used to determine if the proposed location for the collector array will allow enough sunlight to strike the collectors for efficient water heating capabilities. A rule of thumb in the industry is that the collector should be located to receive the most intense solar insolation during the day, which is normally from 9 AM to 3 PM.

Interior Building Inspection

Once the shading analysis is complete and the site is selected for collector installation, an inspection of the building interior should be conducted. During an interior inspection, the viability of the collector location in reference to other system requirements should be verified. This involves determining the structural soundness of the roof, solar loop piping path, and installation location for the solar storage tank and other solar water heating equipment.

The interior inspection should also ensure compliance with all pertinent codes. Existing code violations should be documented so that corrections can be made. A safety assessment of the work required within the building should also be completed to prepare for the proper safety measures to be taken to conduct that work.

Shading Analyses

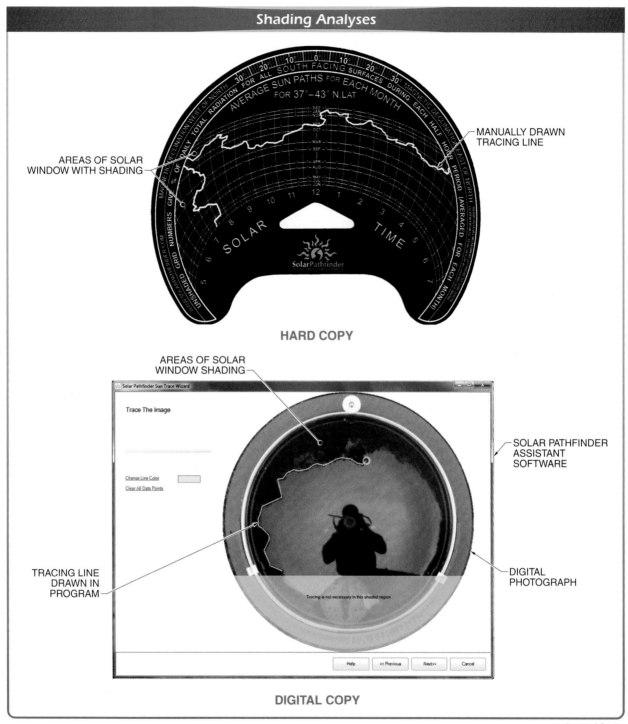

HARD COPY

- AREAS OF SOLAR WINDOW WITH SHADING
- MANUALLY DRAWN TRACING LINE

AVERAGE SUN PATHS FOR EACH MONTH FOR 37°–43° N.LAT

SOLAR TIME

SolarPathfinder

DIGITAL COPY

- AREAS OF SOLAR WINDOW SHADING
- SOLAR PATHFINDER ASSISTANT SOFTWARE
- TRACING LINE DRAWN IN PROGRAM
- DIGITAL PHOTOGRAPH

Solar Pathfinder

Figure 8-23. The areas of shading can be traced manually on the sun path diagram or digitally using software.

The attic underneath the collector mounting area should be inspected to determine whether the roof decking, rafters, and trusses are sound, unweathered, and undamaged and therefore, can be safely used for the attachment of collector mounting hardware. **See Figure 8-24.** The sizes and spacing of the rafter members at the collector location should be noted for determining the type of mounting system to be used.

Attic Inspections

Figure 8-24. The attic underneath the collector mounting area should be inspected to determine whether the roof decking, rafters, and trusses are sound, unweathered, undamaged, and can be safely used for the attachment of collector mounting hardware.

The interior building inspection will reveal how much room there will be to work in the area and the type of safety precautions to be taken. The attic inspection will also help to determine the solar loop piping path. **See Figure 8-25.**

Planned roof penetrations for collector mounting and piping systems should be noted. This will allow the installer to determine how the roof should be sealed as well as plan for sealing other wall penetrations if necessary. It should be noted if fire stopping will be required to ensure that all fire code requirements are met.

During an interior building inspection, any existing water heating equipment must be located. **See Figure 8-26.**

This equipment should be documented on the assessment report. A determination must be made in consultation with the building owner as to the viability and continued use of the existing equipment, specifically the water heater.

Existing Equipment

Figure 8-26. The interior building inspection includes locating the existing water heating equipment.

Interior Inspection Information

INTERIOR INSPECTION

Roof Collector Location Inspection: Roof sheeting looks good, rafters look good, 4′ of work space in area where collectors will mount, piping path is clear

Interior Safety Assessment: Exposed fiberglass insulation, some exposed nails overhead, will need plywood sheeting for work area—3 pcs. 2′ × 6′, firewatch needed if soldering in attic space

Rafter or Truss Size and Spacing: 2 × 6 – 24″ OC

Figure 8-25. An interior building inspection will help determine the size, spacing, and condition of the roof framing below the collector location.

It should also be determined whether the existing mechanical room or water heater space is large enough to house the new solar water heating equipment. **See Figure 8-27.** There should be enough space available to install, maintain, service, and replace water heating equipment. If there is not enough room in the existing space, another area must be located to house the equipment.

Special consideration should be made to determine the proper location of the storage tank. There must be enough space to install and eventually remove and replace the storage tank. Methods to properly support and mount the tank should be noted. Accessibility to all ports on the tank as well as availability of space in close proximity of the tank are necessary for the installation of pumps, valves, controllers, expansion tanks, and the solar loop line set.

It is advantageous to create a drawing of the proposed installation of the solar water heating equipment. This will help to estimate the amount of materials needed to install the new equipment and/or retrofit the existing equipment to meet the requirements of the new system. The drawing should include the solar loop piping path from the collectors to the equipment room. This will allow for an estimation of the needed length of piping and the number and type of piping supports.

Future connections to existing cold and hot water piping should be noted. Availability for power connections for backup water heaters, pumps, and controls should be noted to determine if any electrical work is necessary. Plans should be made for the interconnection of these utilities. The possible shutdown of these systems should be discussed with the building owner and a tentative shutdown schedule made to eliminate any building occupant inconvenience.

JOB PLANNING

Once the site assessment is complete, it is possible to verify the proper system type and locate the collector installation area. It is also possible to estimate the size of the new solar water heating system and create equipment and material lists to prepare a cost proposal for the project. The availability, transport, and delivery of materials should be considered to determine whether costs will be affected, and if so, these costs should be added to the estimate.

As part of the job planning process, a tool list should be created. All tools, including hand tools and site-specific work tools, necessary to complete the project should be listed. All safety equipment, including personal protective equipment (PPE) and site-specific safety equipment, noted in the safety assessment should be listed. Any extra costs incurred from the need for tools, taking safety precautions, and code compliance should be reflected in the job estimate.

Mechanical Room and Utilities Information

MECHANICAL ROOM OR SPACE

Size: 10′ × 8′   **Comments:** Enough space for storage tank and other system equipment

Door Size: 29¼″ opening   **Comments:** Clear path for tank entry, one turn 3′ wide hall, no pad needed

Electric/Gas Availability and Location: Yes from existing, electrical panel 6′ away, outlets available but may have to add for controller or pump or change to four-gang outlet

Piping Access to Roof: Yes   **Distance to Collectors:** 25′

Additional Comments: Installation of two flat-plate collectors mounted vertically on south roof section is feasible. No obstructions in attic areas for piping path. Piping can be installed easily from collectors travelling east in attic to water heater room approximately 12′ from east edge of collectors. Tank installation may require removal of mechanical room door frames if tank is wider than 29″.

Figure 8-27. The interior building inspection will help determine if there is adequate room for the mechanical equipment required for the solar water heating system and the availability and location of required utility connections.

The site assessment provides the information needed to create a project schedule reflecting the time needed to properly install the system. Equipment, tool, and material availability should be taken into consideration when creating the project schedule. This tentative schedule can be discussed with the building owner to ensure suitability. Once terms are agreed to and contracts signed, design drawings can be created and applications made for the proper building permits. Planning for system installation can also begin.

SIZING SYSTEMS

Sizing a solar water heating system can be a daunting task. The collector or collector array needs to be sized as well as the storage system, piping, pump, expansion tank, and heat exchanger if used. In many cases, the installer is not responsible for sizing the system. The site assessment is often sent to a system design department and sized by design professionals. This is the recommended method for large and complicated systems. However, the installer of a small company may be responsible for sizing the system.

There are several ways to size a solar water heating system. The simplest methods include having the manufacturer figure the sizing and using an industry rule of thumb. More complicated methods involve calculating the collector output (in Btu) at the given location and the use of computer-based sizing calculators.

Manufacturer System Sizing Method

A small installation contractor may work with only a few solar collector manufacturers. Many manufacturers require that they themselves design and size the system in order to warranty or guarantee the installation and solar fraction of the system. Since SRCC OG-300 systems are designed and sized by manufacturers, the simplest method of sizing a system is to consult the collector manufacturer. Some manufacturers have developed computer sizing calculators that are designed around the specifications and SRCC OG-100 ratings of their collectors. **See Figure 8-28.** Location parameters, volume of expected domestic hot water (DHW) use, and the desired storage temperatures are entered into the calculator. Then, the collector area and storage tank size are displayed.

The benefit of using the collector manufacturer to design and size the system is twofold. First, the manufacturer is most familiar with the products and will design and size the system to known results. The manufacturer

stands behind these results, especially if there is a relationship with the installation contractor. Second, the manufacturer bears much of the responsibility of system performance. If the installation contractor installs the system correctly and the system does not meet the desired results, the manufacturer typically will provide the remedy to any system problem.

Energy Works US, LLC.

Figure 8-28. Some manufacturers have developed computer sizing calculators that are designed around the specifications and SRCC OG-100 ratings of their collectors.

In most cases, the manufacturer will recommend a packaged system that is SRCC-OG-300-certified. This methodology is simple and relieves the installation contractor of all burdens except for installation. In some cases, the manufacturer may ask for more information to increase the efficiency of the system.

Facts

According to the California Solar Initiative (CSI), as of 2011, there were only 374 contractors in the entire state of California that could install solar thermal heating and cooling solutions and that were eligible to receive a CSI thermal rebate.

Rule-of-Thumb Method

One way to size a solar water heating system is to use rules of thumb. These rules of thumb are guidelines that the industry has followed for some time. The rule-of-thumb method should be used for simple systems in which high efficiency is not necessary. Rules of thumb for sizing collectors and storage tanks are as follows:

- for up to two people, 20 sq ft of collector area per person
- for each person thereafter, an additional 8 sq ft in warm regions (southern United States) or 12 sq ft to 14 sq ft in cool regions (midwestern and northern United States)
- storage capacity of 2 gal./sq ft of collector area in warm regions (southern United States) or 1.5 gal./sq ft of collector area in cool regions (midwestern and northern United States)

For example, a family of four in Topeka, Kansas, requires approximately 68 sq ft of collector area (20 + 20 + 14 + 14 = 68 sq ft) and about 102 gal. of storage (68 gal. × 1.5 = 102 gal.). For hot water storage, a 100 gal. storage tank would be sufficient.

The number of collectors required depends on the aperture area of the collector. The Rheem RS24-BP has an aperture area of 21.88 sq ft. Thus, at least three of these collectors are required (21.88 sq ft × 3 = 65.64 sq ft). However, if using a larger collector, such as the Rheem RS40-BP, fewer collectors may be required. The RS40-BP has an aperture area of 37.12 sq ft. Therefore, only two collectors are needed (37.12 sq ft × 2 = 74.24 sq ft). To determine which collector is the better choice, the costs of the collectors and their installation requirements should be compared while considering the available mounting area.

Calculation Method

The calculation method uses several steps to calculate the demand (in Btu) of the system based on information included on the SRCC Collector Certification and Rating document. This method should be used for larger systems or obtaining higher efficiencies. To size a collector using the calculation method, apply the following procedure:

1. Estimate the solar thermal output required.

2. Determine the average daily solar insolation.

3. Determine the application category, and obtain the appropriate performance rating.

4. Estimate the percentage of collector output by examining the performance rating.

Solar Thermal Output. The first step in sizing a collector system is to estimate the solar thermal output that will be required by the collectors. To find the estimated solar thermal output required, the following formula is applied:

$$OUT_{REQ} = DHW \times T \times 8.33$$

where

OUT_{REQ} = required daily solar thermal output (in Btu)

DHW = estimated daily DHW usage (in gal.)

T = temperature rise (required hot water temperature − incoming cold water temperature) (in degrees)

8.33 = constant (density of water × specific heat)

For example, what is the estimated solar thermal output required for a household that has an estimated daily DHW usage of 80 gal. and a temperature rise of 70°?

$$OUT_{REQ} = DHW \times T \times 8.33$$
$$OUT_{REQ} = 80 \times 70 \times 8.33$$
$$OUT_{REQ} = \textbf{46,648 Btu}$$

Average Daily Solar Insolation. The second step in sizing a collector system is to determine the average daily solar insolation over the course of the year. To accomplish this, the radiation averages are used from the SRCC OG-100 Collector Certification and Rating document. The radiation averages indicate the estimated daily collector output during certain weather conditions, such as the following:

- clear = 2000 Btu/sq ft
- mildly cloudy = 1500 Btu/sq ft
- cloudy = 1000 Btu/sq ft

For greater accuracy, documented averages from sources such as governmental agencies or SRCC Table 1, *Average Daily Total Solar Radiation for U.S. Cities,* should be used when possible. For example, the average daily solar insolation for Topeka, Kansas, according to SRCC Table 1, is 1482 Btu/sq ft (16.83 MJ/sq m) for a 23° tilt and 1489 Btu/sq ft (16.91 MJ/sq m) for a 45° tilt.

By comparing the SRCC radiation averages to the documented averages and using the appropriate weather conditions column on the SRCC Collector and Certification document, expected collector performance under certain weather conditions can be determined. For Topeka, Kansas, since the daily documented average is just under 1500 Btu/sq ft, the Mildly Cloudy column should be used to determine collector performance.

SRCC Application Categories. The next step in sizing a collector is to determine the application category. Once the application category is determined, the performance rating from the SRCC Collector Certification and Rating document can be obtained for the particular collector to be used. The SRCC Collector Certification and Rating document compares the thermal performance of a solar collector under the three solar insolation weather conditions for each of the categories. **See Figure 8-29.** The four SRCC categories applicable to solar water heating systems include the following:

• Category A—pool heating in warm climates
• Category B—pool heating in cool climates
• Category C—water heating in warm climates
• Category D—water heating in cool climates

Facts

The SRCC accreditation process includes submitting an application and arranging to have the collector tested at an SRCC-accredited lab. The certification will be issued if testing is favorable and all requirements and fees are met.

Collector Output. The final step in sizing a collector is to estimate the percentage of collector output for a given collector by using the SRCC Collector Thermal Performance Rating. The percentage is found by dividing the average daily output of the collector by the required

output for the system. To find the percentage of output for a given collector, apply the following formula:

$$PO = \frac{OUT_{COL}}{OUT_{REQ}}$$

where
PO = percentage of collector panel output (in percent)
OUT_{COL} = collector daily average output (in Btu)
OUT_{REQ} = required daily solar thermal output (in Btu)

For example, what is the percentage of collector output for a single Rheem RS24-BP collector used in a household with a required daily solar thermal output of 46.686 Btu? According to the SRCC chart, a single collector produces on average 17,400 Btu/day under mildly cloudy conditions.

$$PO = \frac{OUT_{COL}}{OUT_{REQ}}$$

$$PO = \frac{17,400}{46,648}$$

$$PO = \mathbf{37.30\%}$$

Therefore, a single Rheem RS24-BP collector will provide about 37.30% of the required solar thermal output on a yearly basis. If a second collector is added, the two collectors will provide 74.60% of the required output. An 80 gal. storage tank can be used, but a 100 gal. tank would provide extra storage on cloudy days.

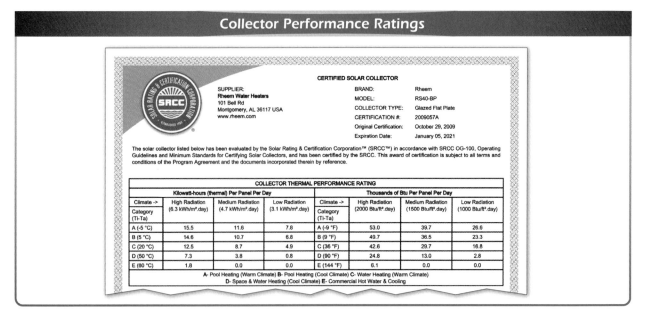

Figure 8-29. The SRCC Collector Certification and Rating document compares the thermal performance of a solar collector under three solar insolation conditions.

Computer Software Method

The computer software method is generally the most accurate, but also the most complicated method of sizing a solar water heating system. The computer software method is designed to get the most efficiency from the solar water heating system. This method should be used for large complicated systems or if high efficiency and strict parameters are desired.

Several component manufacturers and installation companies offer proprietary sizing software that make sizing solar water heating systems quicker and more accurate. Most of these software programs require a license, but they should pay for themselves after several installations of solar water heating systems.

One of the most complicated, yet comprehensive, programs available is RETScreen Clean Energy Project Analysis Software. This program was developed by Natural Resources Canada in collaboration with the National Aeronautics and Space Administration (NASA), Renewable Energy and Energy Efficiency Partnership (REEEP), United Nations Environment Programme (UNEP), and Global Environment Facility (GEF). The software program covers conventional energy technologies and renewable energy technologies, such as solar water heating, photovoltaic (PV), wind, hydro, and geothermal technologies. The software program is widely used around the world and is free to download.

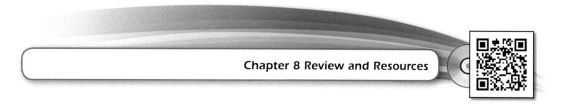

Chapter 8 Review and Resources

9

Solar Collector Installation

The first step in the installation of a solar water heating system is typically to mount the solar collector. The collector, along with the storage tank, has a fixed position and cannot be relocated. The collector is the focal point of the entire system. The piping and wiring run to and from the collector to the storage system. Since installing the collector is the most difficult part of the entire system installation process, it should be installed first.

INSTALLATION CONSIDERATIONS

The solar collector, regardless of the type, is the most critical element in the installation of the solar thermal system for three reasons. First, it is usually the most dangerous part of the system to install due to the work required aboveground at dangerous heights. Therefore, the installer is required to be highly safety-conscious. Second, an improperly mounted collector can impair the efficiency of the solar water heating system and defeat the purpose of installing the system, which is to save the customer money in water heating costs. Third, an improperly installed solar collector mounting system, including improperly placed penetrations and insufficient roof-surface waterproofing, can cause severe damage to the system and possibly to the building. **See Figure 9-1.**

A properly executed site assessment, which involves identifying safety issues and the equipment needed for the installation, is essential for avoiding many of the problems associated with installing solar collectors. The assessment should have made the installer aware of the type of collector attachment assemblies and waterproofing materials necessary for the type of roof involved in the installation. The installer should never rely solely on an assessment completed by someone else and long before work actually begins. It is good practice to check recommendations once at the installation site. Before a solar collector is actually installed, job hazards, code requirements, and tools and materials must be considered.

Damage to Solar Thermal System Due to Improper Installation

Figure 9-1. Improperly installing a solar collector mounting system can result in severe damage to the system and possibly the building.

Job Hazard Analysis

The recommendations in the job hazard analysis, created after conducting a site assessment, should be implemented upon arrival at the installation site. A *job hazard analysis* is an analysis that focuses on the relationship between a worker, job tasks, the tools used, and the work environment in an effort to identify hazards before they occur. Each element of the analysis should be verified on-site to ensure adherence to the federal, state, provincial, and local requirements set forth by the Occupational Health and Safety Administration (OSHA). Before installing solar collectors, the following safety precautions should be taken:

- Installer vehicles should be properly parked and identified. Safety cone markers should be placed around the vehicles to establish safety zones.

- Notices and warnings that work will commence in, on, and around the building should be posted and given to building occupants to prevent accidents and injuries.

- Hazards identified in the site assessment should be identified and marked at the installation site. All workers should be made aware of these hazards at an initial safety meeting.

- All workers should be made aware of the job-specific personal protective equipment (PPE) requirements and ensure all the necessary PPE is in proper condition.

- When working in elevated areas, such as on a roof, there must always be at least two workers. This ensures that if an accident were to occur one of the workers will be able to contact emergency personnel.

- If work is to be performed in an elevated area, fall protection equipment or devices must be installed by the first workers to access the area.

- Any required site-specific fall protection equipment, such as harnesses, safety lanyards, and fall-arrest devices, must be identified and inspected for proper operational condition.

- First aid kits must be available, properly maintained, and their location should be made known to all workers.

- All tools must be inspected for proper operational condition.

- Fire safety equipment should be identified and inspected for proper operational condition.

- Methods of communication with emergency personnel should be made known to all workers.

Additional safety considerations may be required depending on the work to be performed. For example, extra safety precautions must be taken when working on roofs. Fire safety must be considered when using torches to join piping systems. Electrical safety must be considered when working with power tools and to install electrical components, such as the system pump, controller, and sensors. Safety considerations should be taken in all aspects of the installation of solar water heating systems.

Code Compliance

The site assessment includes the identification of applicable codes that must be taken into consideration before installation begins. The installer must be aware of all codes, such as building codes, adopted in the area.

In addition to building codes, codes that require strict adherence when installing solar collectors include plumbing, electrical, and specialty codes. Plumbing codes apply to the installation of piping and water storage systems. Electrical codes apply to the wiring of systems and system components. Specialty codes specific to the solar collector installation may also have been created for various applications. For example, if fuels are used to generate back-up heat, the *National Fuel Gas Code* may need to be referenced. The *NRCA Roofing and Waterproofing Manual,* published by the National Roofing Contractors Association (NRCA), may also need to be consulted when roof penetrations and sealing are involved.

The required permits for work to be performed at the installation site must be obtained and prominently posted before work begins. In addition, any prints or paperwork required to be on-hand should be available for review by the authority having jurisdiction (AHJ). Any code violations that were identified during the site assessment should be marked so they can be corrected during the system installation.

Tools and Materials

All tools and materials needed for the installation should be ordered to arrive or gathered at the installation site before work begins. They should be stored in a safe and secure area at the installation site. If a safe and secure area is not available, enough work materials for two days of work should be transported to the installation site daily. Also, it must be ensured that an installer or other personnel is on time and present to accept all delivered tools and materials or equipment. Such preparation facilitates job productivity and eliminates the downtime associated with waiting for materials.

Once materials and tools are at the installation site, they should be inspected for damage. Materials should also be checked for shortages. Damaged materials should be taken off the installation site and replaced immediately. Tools should be further inspected for safety compliance. Any tool that does not meet safety requirements should be immediately tagged unsafe and either discarded or repaired. Battery-operated tools should be fully charged before work commences and at least one set of spare batteries should be on-hand. **See Figure 9-2.**

Collector Installation Tools

- Angle finder or inclinator
- Battery-operated drill
- Caulking gun
- Chalk snap line
- Channel locks
- Circular saw
- Crescent wrench—12″ adjustable
- Drill bit and hole saw bit sets
- Flashlight
- Framing square
- Hex nut torque chuck set
- Hoisting equipment
- Levels—2′ and 4′
- Marking pen/chalk/crayon
- Open end wrench set
- Power drill
- Ratchet wrench set
- Reciprocating saw
- Screwdriver bits
- Screwdriver set—Phillips and straight
- Stud finder
- Tape measure
- Tie off ropes
- Work gloves
- Work light

Figure 9-2. All tools needed for the installation should be ordered to arrive or gathered at the installation site before work begins.

Equipment and Components

The most important components of the solar water heating system transported to the installation site are the solar collector and the mounting assemblies. The collector should be uncrated and inspected for damage as soon as possible. A damaged collector will need to be replaced, and there may be considerable lead time in obtaining a replacement. If the collector is not damaged, it should be placed in a safe and secure area until ready for installation. The packing material should be saved and used to cover the collector at all times until the collector is filled with heat transfer fluid (HTF) and tested. Even as the collector is physically handled during installation, the aperture area should be covered to avoid the high temperatures that can be generated when the absorber is exposed to the Sun.

Other equipment and system components also should be examined to ensure installation suitability. For example, the collector mounting system should be checked to ensure that it is suitable for the existing roof type and surface material.

INSTALLATION PREPARATION

There are hundreds of different solar collectors manufactured for solar water heating on the market today. However, there are three categories of solar collectors that dominate the market. These are the glazed flat-plate collector, evacuated tube collector, and unglazed flat-plate collector for swimming pool heating. The installation preparation steps are similar for these three types of collectors.

Several actions must be taken in the preparation of installing solar collectors. In addition to taking into account installation considerations, such as performing a job hazard analysis, ensuring the required tools are on-hand, and ordering the proper materials and equipment, it is important to create a project schedule, read the solar collector installation instructions, and review the site analysis report.

A project schedule should be created to ensure the installation is completed in a timely fashion. The schedule should be created in cooperation with other parties involved, such as electrical contractors and suppliers. The schedule should also be shared with the building owner to keep the owner informed of the tasks that are occurring. The schedule should include information on each job task performed by each party. In addition, any times when the building utilities might be shut off should also be noted.

When preparing to install any type of solar collector, it is critical to read the installation instructions provided by the manufacturer. Although collectors may look alike, there are various design characteristics that may be unique to a particular manufacturer. For example, a manufacturer may design the collector attachment system for different collector weights or different wind loadings differently than another manufacturer. In many cases, the collector mounting and attachment methods vary from manufacturer to manufacturer. **See Figure 9-3.** Therefore, the installation instructions provided by the manufacturer must be read thoroughly prior to every installation and then followed.

Before ever reaching the installation site, the site analysis report should be thoroughly reviewed. The report will have identified the proper location for mounting the collector. Solar collectors may be mounted on either the roof or ground. There are a variety of ways the collector can be mounted, depending on the roof structure or ground surface and soil composition. The proper orientation and tilt angle listed on the report are also used to determine how the collector should be mounted and the types of supports necessary for the installation.

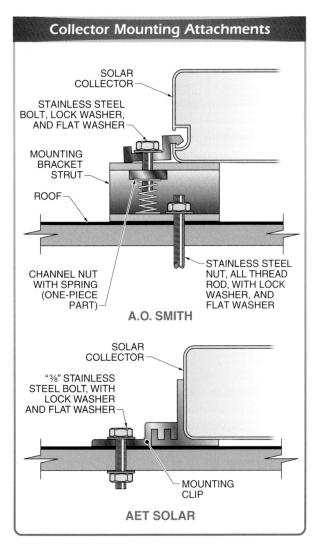

Figure 9-3. Collector mounting and attachment methods may vary from one manufacturer to another. Therefore, the installation instructions provided by the manufacturer must be followed.

SOLAR COLLECTOR MOUNTING METHODS

Solar collectors can be either ground-mounted or roof-mounted. Ground-mounted solar collectors are usually rack mounted on concrete piers. There are four methods of roof mounting solar collectors, which include integral, direct, standoff, and rack mounting methods.

Ground Mounting Methods

Ground-mounted solar collectors have at least one advantage over roof-mounted collectors in that there are less safety hazards for the installer. Work at high elevations should not be involved with this type of installation. This saves time and money since the expenses associated with

preparing and working at high elevations are avoided. In addition, there will be neither roof penetrations and subsequent waterproofing to be concerned about nor much work required for installing pipe in attic spaces of the building.

The solar collector mounted at ground level should be mounted above any ground covering that would encroach upon the collector. The recommended height above the ground is at least 12″. However, it is best to ascertain the average snowpack level and mount the bottom supports well above that level. Also, it should be determined whether any groundwater or stormwater will cause standing or running water around the installation. If so, another area without water issues should be located. If that is not possible, then the installer should ensure that the collector is mounted above any waterline and that the mounting supports are engineered to withstand running or standing water.

The proper supports to use for the solar collector mounts depend upon the ground surface at the installation site. If the collector is to be installed on a concrete surface, the mounts can be attached directly to the concrete. However, supports for the mounts must be placed into the ground in most instances. This is commonly accomplished by forming concrete pillars or bollards in the ground. **See Figure 9-4.**

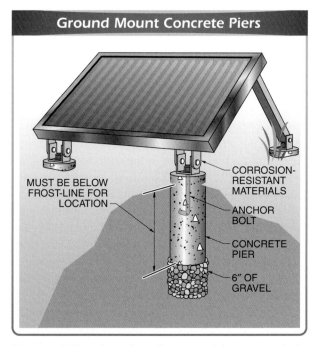

Figure 9-4. Mounting solar collectors on the ground typically involves first placing supports for the mounts into the ground.

The required size and depth of the concrete pillar for each installation will depend on the soil conditions and the weight of the collector array. Local building codes must be followed for proper sizing and compaction of the receiving hole and the size of the pillar or bollard. It is recommended that a structural engineer design the specifications for these systems. Typically, the ground supports are installed and cured before collector installation begins.

An alternative to using pillars or other ground mounting systems is to ballast the rack system. **See Figure 9-5.** The soil for a ground-mounted ballasted system must be very stable for an expected 20 to 30 years. It would be unfortunate to install a system only to have the soil erode and damage the installation.

SunMaxx Solar

Figure 9-5. A ground-mounted ballasted system may be used instead of pillars or other ground mounting systems.

Precautions that should be taken for ground installations include the following:
• The area around the installation should be cleared, roped off, and prepared for work.
• It should be ensured that there are no shading problems from nearby foliage or buildings.
• Corrosion-resistant bolts should be used for the support attachments in the pillars or bollards.
• Threads on the bolts should be protected with plastic coverings or tape before installation.
• Proper antiseize lubricant should be used for attachment of the assembly so that nuts and bolts will not seize up and make disassembly difficult.

• If there are trafficways nearby, collision protection should be provided by installing in-ground traffic bollards around the collector or array.
• Signage should be posted cautioning against hot piping and warning that fragile components of the system are present.

Roof Mounting Methods

The site analysis report includes the information required to determine the proper solar collector roof mounting method to use and the appropriate support assembly to order. The report should include information regarding the best roof location for mounting the solar collector as well as more specific information regarding the location. For example, if it is recommended that the solar collector be installed on the roof, the shape of the roof, such as slightly sloped, steeply sloped, or flat, and the type of roof surface are identified. A determination of the type of method to use to mount the solar collector should also have been made.

Integral Mounting Method

An *integrally mounted collector* is a collector that is recessed into a roof structure and in which the collector surface becomes the roof surface. **See Figure 9-6.** The glazed flat-plate collector is the only solar thermal collector that can be mounted this way. The collector must be designed for this mounting method due to the fact that collector water tightness is crucial to the integrity of the roof. The collector must be supported within and by the roof structure and waterproofed in its entirety. Normally, support assemblies will not be used with this mounting method.

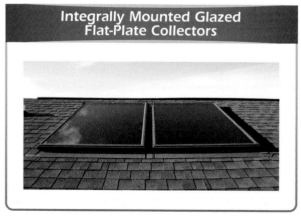

VELUX America Inc.

Figure 9-6. The integral mounting method used for glazed flat-plate collectors involves recessing the collector into a roof structure until the collector becomes part of the roof surface.

Although this method can be adapted to an existing roof, it is recommended that a roofing contractor establish the roof penetration, support, and waterproofing. It is much easier to install an integrally mounted collector on a new roof. However, the roofing contractor should still do the roof work around the collector.

Integrally-mounted collectors should be mounted relatively close to the ridge of the roof. This will ensure that a minimum amount of water running down the roof will encounter the collector. Also, there is no wind loading, and therefore, only the weight of the filled collector and possible snow loading need to be taken into consideration when using this mounting method.

Direct Mounting Method

The *direct mounting method,* also known as the flush mounting method, is a solar collector mounting method that involves attaching the collector directly to the roof surface without any supports or blocking beneath it. **See Figure 9-7.** Like the integral mounting method, collectors should be mounted close to the ridge of the roof, and there is almost no wind loading.

American Solar Living LLC

Figure 9-7. The direct mounting method involves attaching the collector directly to the roof surface without any supports or blocking.

The collector can be mounted before or after roofing material is placed on the roof. If placed before the roofing material is installed, the collector is mounted on the roofing membrane and can be fully waterproofed by the next layers of roofing material. If it is placed on existing roofing material, the collector is usually installed on a waterproof membrane and waterproofed with roofing

material placed around the collector to prevent leaks, mildew, and rotting. Mounting clips attach to the side of the collector. **See Figure 9-8.**

Figure 9-8. Mounting clips are used to attach the collector directly to the roof.

The glazed flat-plate collector and the evacuated tube collector can be mounted directly on the roof. The evacuated tubes will still be above the roof a fraction of an inch, while the manifold and the lower tube support are mounted flush to the roof. The unglazed swimming pool collector is normally mounted directly on the roof. **See Figure 9-9.**

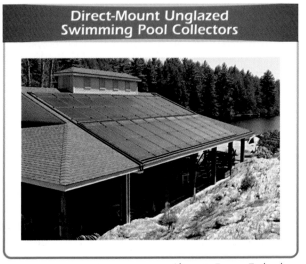

Alternate Energy Technology

Figure 9-9. Unglazed swimming pool collectors are usually mounted directly onto the roof.

Directly mounted collectors should be mounted relatively close to the ridge of the roof. This will ensure that a minimum amount of water running down the roof will encounter the collector. Also, there is almost no wind loading. Therefore, when using this mounting method, only the weight of the filled collector and possible snow loading need to be considered.

Standoff Mounting Method

The *standoff mounting method* is a solar collector mounting method that involves the use of mounting assemblies to support and raise the collector above a roof surface. **See Figure 9-10.** This mounting method allows wind and water to flow under the collector. Waterproofing is only necessary at the support penetrations through the roof.

Schott Solar

Figure 9-10. The standoff mounting method allows wind and water to flow under the collector.

In addition to raising the collector above the roof surface, standoff mounting assemblies can also be used to tilt the collector to a specific angle by using longer upper mounting supports. By using different size standoffs at the top and bottom of the collector, the collector may be tilted as required. **See Figure 9-11.** Standoff sizes vary by hardware manufacturer.

The standoff mounting method is the most widely used method of mounting glazed flat-plate collectors and evacuated tube collectors on sloped roofs. This mounting method cannot be used for unglazed swimming pool collectors.

Quick Mount PV

Figure 9-11. Standoff mounting assemblies can be used to tilt the collector to a specific angle by using longer upper mounting supports.

Rack Mounting Method

The *rack mounting method* is a solar collector mounting method that involves supporting the collector above a roof surface on a frame or rack. This allows the collector to be mounted to precise angles. **See Figure 9-12.** Most frames or racks are adjustable in order to attain the appropriate tilt angle for any installation. The rack mounting method is used when collectors are to be mounted on flat roofs, although it also can be used on sloped roofs to attain a specific angle.

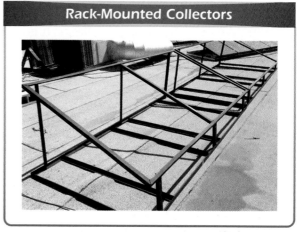

American Solar Living LLC

Figure 9-12. The rack mounting method involves supporting the collector above a roof surface on a frame or rack, allowing the collector to be mounted to precise angles.

The disadvantage of the rack mounting method is increased wind loads on the mounts due to the added pressure of the wind against the raised surface. The raised collector can act as a sail. **See Figure 9-13.** Excessive winds could severely damage the collector and even rip the collector off the roof. In some areas, such as coastal areas frequented by hurricanes, winds can reach up to 146 mph. Such high winds can apply up to 2400 lb of force on the surface of a typical 4′ × 8′ flat-plate collector.

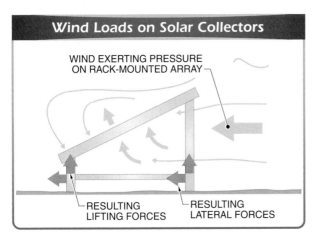

Wind Loads on Solar Collectors

WIND EXERTING PRESSURE ON RACK-MOUNTED ARRAY

RESULTING LIFTING FORCES

RESULTING LATERAL FORCES

Figure 9-13. Wind loads can damage rack-mounted collectors due to the added pressure of the wind against the raised surface.

The added wind load, or force, must be accounted for when selecting mounting assemblies for the collector. Weather and wind information should be given to the solar collector manufacturer when ordering the collector so that the proper mounting assembly for these conditions can be ordered with the collector. It is also recommended that a structural engineer perform wind load calculations, if needed.

The rack mounting method can be used for both glazed flat-plate collectors and evacuated tube collectors. The unglazed swimming pool collector is also mounted in this manner when mounted on the ground. It should be noted that the evacuated tube collector is less prone to wind loading because it does not present a solid surface to the wind. Wind can pass between the tubes, reducing the wind-load effect. When mounting a collector system on a flat roof, the rack may not need to be secured to the roof. Instead ballasts or heavy weights can be used to anchor the rack system to the roof. **See Figure 9-14.** A structural engineer should be consulted for ballast calculations.

Ballasted Roof-Mounted Systems

Solar H₂Ot, Ltd.

Figure 9-14. Heavy weights can be used to anchor the rack system to a flat roof.

ROOF MOUNTING LAYOUT

Once the approach to the building is cleared where the installation will take place, the method for accessing the roof can be readied. After safely gaining access to the roof and all safety precautions are taken, the exact mounting location of the solar collector can be determined. This information, as well as other information needed to install a solar collector, is typically available on the site analysis report. The information provided in the site analysis report is used to perform the tasks required to properly install roof-mounted solar collectors, such as determining the roof orientation, determining the solar collector mounting direction and grade, establishing collector array spacing, and choosing the correct attachment devices to be used. **See Figure 9-15.**

Roof Orientation

The proper orientation, or azimuth angle, of the collector should be determined in order to lay out the collector support assemblies. In most instances, this may be unnecessary if it has already been determined that the roof is oriented within 45° of true south either east or west.

Installing Solar Panels on Roofs

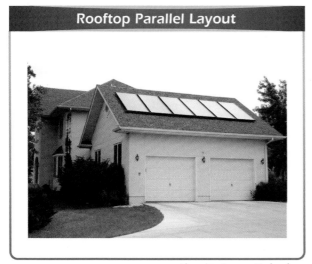

VELUX America Inc.

Figure 9-15. After safely gaining access to the roof and all safety precautions are taken, the exact mounting location of the solar collector can be determined.

If the collector is mounted on a roof surface that faces south within this angle, only a very small percentage of radiant heat capture will be lost. Any lost heat efficiency can be made up by increasing the collector size. If this is the case, the solar collector or collector array can be mounted parallel with the roof ridge line or roof edge line to make an aesthetically pleasing installation. **See Figure 9-16.** If this is not the case, the proper azimuth angle should be laid out and the collector mounts oriented accordingly.

Rooftop Parallel Layout

Alternate Energy Technology

Figure 9-16. A solar collector or collector array can be mounted parallel with the roof ridge line or roof edge line for an aesthetically pleasing installation.

Mounting Direction and Grade

In addition to the roof orientation, it should be determined whether the collector or collector array is to be mounted vertically or horizontally prior to installation. **See Figure 9-17.** There may be restrictions from the manufacturer as to the mounting direction for a particular use. For example, a collector with a serpentine tube may not be able to be installed horizontally in a drainback system.

The installer must verify that the collector can be installed in a particular direction before laying out the collector supports. There is no difference in collector efficiency with either mounting direction. Preference in direction should be based on ease of installation, roof aesthetics, or system type. Customers should always be consulted regarding the aesthetics of the installation, such as the placement of the collector on the roof, since they will have to see it every day.

Another installation parameter that must be determined before physically laying out the supports is whether the collector should be graded. Drainback and steamback system collectors should be graded at ¼″ per foot to the collector inlet. This will allow for the fluid in the collector to completely drain out of the collector. If this is the case, the grade should be included in the layout of the supports. **See Figure 9-18.** *Note:* The grade is much less apparent if the collector is vertically mounted rather than horizontally mounted. If the array is large, the collector can be graded toward the center of the array. This will reduce the noticeable grade and create a chevron effect.

American Solar Living
When installed together on the same roof section, solar collectors and photovoltaic panels are often placed at the same angle and orientation.

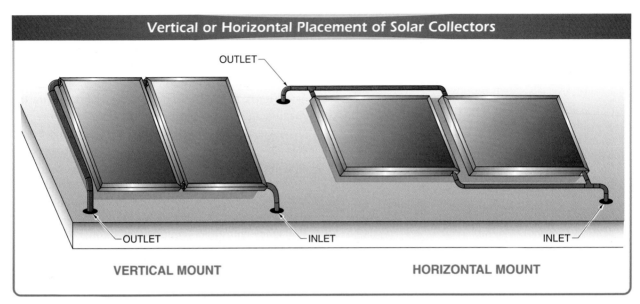

Figure 9-17. Collectors and collector arrays may be mounted vertically or horizontally based on the ease of installation, roof aesthetics, system type, and customer preference.

Figure 9-18. Drainback and steamback system collectors should be graded at ¼″ per foot to the collector inlet to allow for fluid to drain out of the collector. The grade should be included in the layout of the supports.

Collector Array Spacing

The layout for a single collector can begin once the mounting direction and grade have been determined. However, if a multiple collector array is being installed, the distance between collectors in a single row and the distance between multiple rows of collectors must be checked. The required distance between collectors in a single row depends on the piping connection between collectors.

The collector manufacturer will usually recommend the connection method between collectors. The most common method for connecting collectors together is to use copper and brass unions or soldered or brazed copper couplings. Using unions rather than couplings allows for easy connection of the collectors and easy disassembly if a collector in the array must be removed and replaced. However, the layout of the collectors must be very precise for the unions or pipe ends (for couplings) to line up correctly.

The distance between horizontal rows of collectors that are mounted without tilting is usually recommended by the manufacturer, but the installer can use a random practical distance, such as 6″ or 12″. If the collector array is tilted, the possible shading of the rear row of collectors by the front row of collectors must be addressed and the array rows staggered appropriately. **See Figure 9-19.** In most cases, installation instructions provided by the manufacturer include the staggered spacing measurements.

The stagger distance depends on the lowest angle of the Sun. Therefore, the angle of the Sun at winter solstice on December 21 should be used to calculate the distance. The latitude, roof angle, and tilt angle are required to calculate the spacing. These calculations can be done by the installer. However, it is recommended that the manufacturer be consulted if a collector spacing table is not included in the installation instructions.

Roof-Mount Attachment Devices

Once the collector array spacing is known, the roof penetration points for the mounting assemblies can be physically marked on the roof. The collector manufacturer usually provides marking guides. **See Figure 9-20.** The guides contain the proper measurements for each type of installation. All that remains to do at this point is to physically mark them on the roof. If there is no guide, the installer may need to physically place the mounting assemblies on the collector and measure for the roof penetration. It is easier and safer to do this on the ground.

The precise placement of the roof penetrations for securing the mounting assemblies to the roof depends on whether there is access to the attic space under the roof. If there is no attic space available, the mount supports will need to be secured by lag screws, which screw directly into the roof truss. If attic space is available, it will not be necessary to penetrate the roof truss. In this case, the mounting assembly can be secured to the roof using spanners or J bolts.

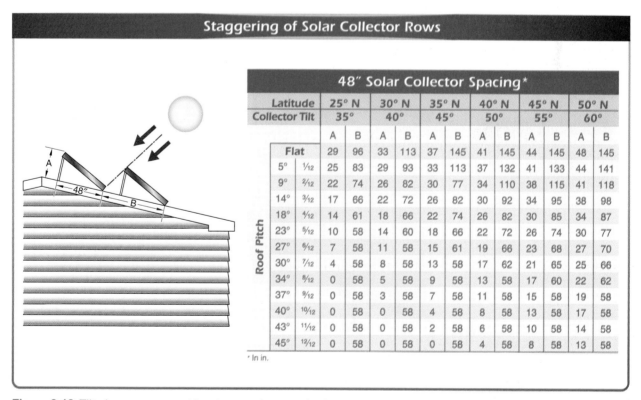

Figure 9-19. Tilted array rows must be staggered appropriately to prevent shading of the rear row of collectors by the front row. Usually, the manufacturer provides the measurements for stagger spacing in a table.

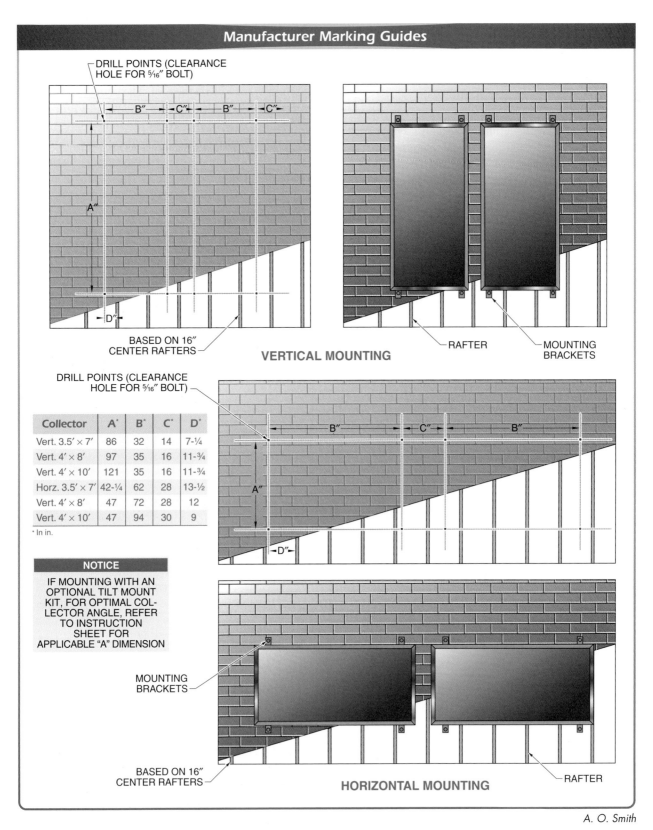

Manufacturer Marking Guides

Collector	A*	B*	C*	D*
Vert. 3.5′ × 7′	86	32	14	7-¼
Vert. 4′ × 8′	97	35	16	11-¾
Vert. 4′ × 10′	121	35	16	11-¾
Horz. 3.5′ × 7′	42-¼	62	28	13-½
Vert. 4′ × 8′	47	72	28	12
Vert. 4′ × 10′	47	94	30	9

* In in.

NOTICE

IF MOUNTING WITH AN OPTIONAL TILT MOUNT KIT, FOR OPTIMAL COLLECTOR ANGLE, REFER TO INSTRUCTION SHEET FOR APPLICABLE "A" DIMENSION

DRILL POINTS (CLEARANCE HOLE FOR ⁵⁄₁₆″ BOLT)

BASED ON 16″ CENTER RAFTERS

VERTICAL MOUNTING

RAFTER

MOUNTING BRACKETS

HORIZONTAL MOUNTING

MOUNTING BRACKETS

RAFTER

A. O. Smith

Figure 9-20. Manufacturer-provided marking guides can be used to mark the roof penetration points for the mounting assemblies once the collector array spacing is known.

Lag Screws. A *lag screw* is a screw used to penetrate a roof truss and secure a collector mount to the roof. **See Figure 9-21.** The lag screw must be corrosion-resistant, as it is intended to last the life of the installation. It must be able to support the weight of a filled collector and piping. It must also withstand the pullout force of wind against the collector if it is installed using the standoff or rack-mounted method. The load calculations will have already been accounted for if the mount is supplied by the manufacturer and the installation parameters are given with the order.

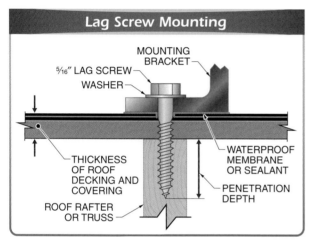

Figure 9-21. A lag screw penetrates a roof truss and must be able to support the weight of a filled collector and piping.

Because the majority of roof rafters or trusses are comprised of 2 × 4 or 2 × 6 lumber, the most common size of lag screw used in solar water heating installations is a ⁵⁄₁₆″ lag screw. A smaller size lag screw may not have the pullout resistance required for the installation. A larger lag screw cannot be used because there will not be enough wood surrounding the screw when fastened into the edge of the rafter or truss. **See Figure 9-22.**

To install the lag screw, a pilot hole must be drilled approximately 60% to 70% of the width of the screw. This pilot hole should be drilled to a depth within three-quarters of the penetration depth of the screw. The screw must be long enough to account for the width of a washer, the mount support, the flashing or sealing membrane, and the roof thickness. It must also penetrate the truss 1″ to 2″ depending on the pullout strength needed. The lag screw is screwed into the pilot hole using a wrench or power drill, taking care not to strip the screw threads in the wood or break the hex nut off the screw.

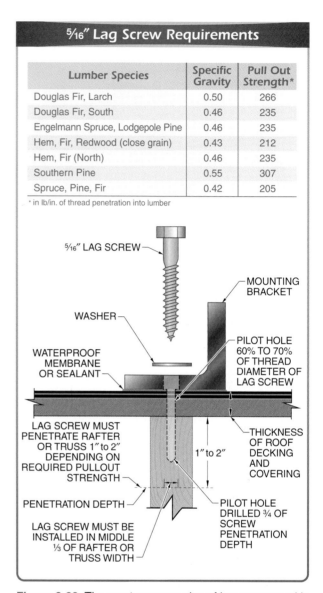

Lumber Species	Specific Gravity	Pull Out Strength*
Douglas Fir, Larch	0.50	266
Douglas Fir, South	0.46	235
Engelmann Spruce, Lodgepole Pine	0.46	235
Hem, Fir, Redwood (close grain)	0.43	212
Hem, Fir (North)	0.46	235
Southern Pine	0.55	307
Spruce, Pine, Fir	0.42	205

* in lb/in. of thread penetration into lumber

Figure 9-22. The most common size of lag screw used in solar water heating installations is ⁵⁄₁₆″.

Locating Roof Rafters and Trusses. To lay out lag screw mounts, roof rafters or trusses are located and the centers of the rafters or trusses are marked for penetration. This is sometimes difficult to do depending on the roof surface. The best way to find the rafter/truss is to first locate its position by finding its nailing point at the fascia of the roof and transferring that line onto the roof. The roof-penetration points are then marked on the roof using a chalk snap line and indelible marker, chalk, or crayon. **See Figure 9-23.**

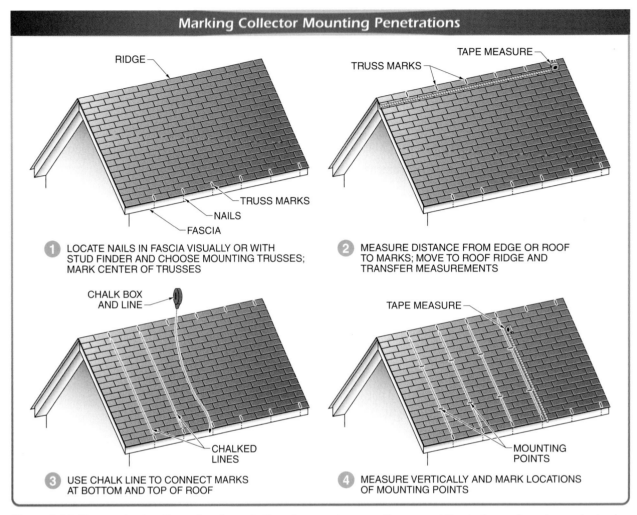

Figure 9-23. The precise locations of roof penetrations for securing mounting assemblies to the roof can be marked using a simple procedure.

The rafter/truss center should be verified by using a digital stud finder with the appropriate depth capacity. The width of the rafter/truss should be marked. If a stud finder is not available, the rafter/truss can be verified by using a hammer. This is done by gently hammering along a line perpendicular to the rafter/truss while listening for a change in tone and feel of the hammer stroke, which normally indicates the framing member beneath.

In most cases, the attachment points on the collector are adjustable so that the mounting assembly can be placed on the rafter or truss line. The layout for the collector may need to shift or the assemblies shift along the collector to match the rafter or truss line. Once the penetration points are marked and the locations double-checked, the pilot holes for the lag screw can be drilled.

Spanners. A *spanner* is a piece of 2 × 4 or larger lumber or angle iron that spans across at least two trusses and is used to secure a collector mount to a roof using a long threaded rod or bolt, nuts, and washers. **See Figure 9-24.** The rod or bolt must also penetrate a spacing board or two blocks placed alongside the rod to create a solid mass from the roof decking to the backing board. If this is not included in the spanner assembly, the installer could overtighten the rod and deform the roof, which might cause it to leak or create an area of standing water. The spanner rod should be of a corrosive-resistant metal.

Mounting solar collectors using spanners requires two installers: one on the roof and one in the attic space. The bottom nut must be held by the installer in the attic space while the rod is tightened by the installer on the roof. The rod must be tightened enough to secure the mount. The spanner board will also need to be screwed or nailed to the truss and the spacer or blocks to ensure it will stay in place.

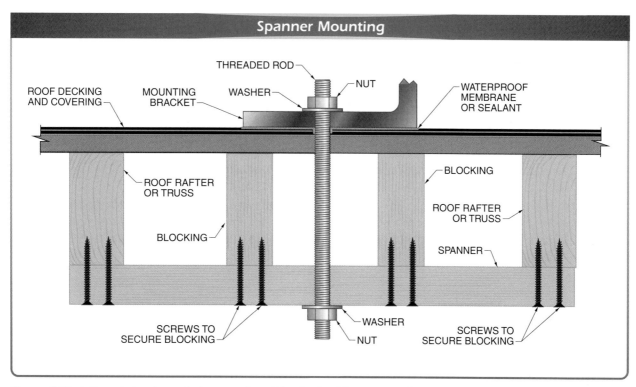

Spanner Mounting

THREADED ROD

ROOF DECKING AND COVERING

MOUNTING BRACKET

WASHER

NUT

WATERPROOF MEMBRANE OR SEALANT

ROOF RAFTER OR TRUSS

BLOCKING

BLOCKING

ROOF RAFTER OR TRUSS

SPANNER

SCREWS TO SECURE BLOCKING

WASHER

NUT

SCREWS TO SECURE BLOCKING

Figure 9-24. A threaded rod penetrates a roof past the depth of the truss and spanner underneath to secure a mount to a roof using nuts and washers.

When a spanner is used, there is no need to find the center of the rafter or truss. However, the rafter or truss should be located so that the penetration hole is not drilled into it. The mounting points can be laid out and marked relative to the rafter or truss layout. The penetration hole should be large enough for the rod, usually a ½″ threaded rod, to slide through it.

J Bolts. A *J bolt* is a fastener that hooks around a secure framing member and has a threaded end that is used with a nut and washer to secure a collector mounting assembly. **See Figure 9-25.** The J bolt length must be measured correctly or be long enough for the excess thread to be cut above the nut. J bolts should be made of a corrosive-resistant metal.

The J bolt is laid out in the same manner as the spanner, except the penetration must be next to the roof truss. If there is room in the attic space, the threaded end of the J bolt may be pushed through a drilled hole in the roof. However, the roof penetration may need to be more than a drilled hole if there is not enough attic space. A slit wide enough for the J section of the bolt to fit through the roof must be cut into the roof. In any case, there must be enough spacing under the truss to maneuver the J bolt onto the truss.

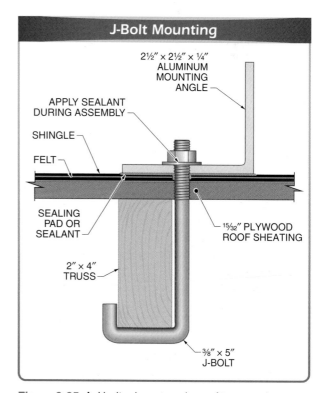

J-Bolt Mounting

2½″ × 2½″ × ¼″ ALUMINUM MOUNTING ANGLE

APPLY SEALANT DURING ASSEMBLY

SHINGLE

FELT

SEALING PAD OR SEALANT

2″ × 4″ TRUSS

$^{15}\!/_{32}$″ PLYWOOD ROOF SHEATING

⅜″ × 5″ J-BOLT

Figure 9-25. A J bolt grips around a roof truss and secures a mount using nuts and washers.

WATERPROOFING ROOF PENETRATIONS

Roof penetrations must be waterproofed in an acceptable manner and tested to ensure a watertight installation before mounting collectors. Waterproofing typically requires roofing membrane material, roofing sealant, and possibly roof flashings. Water should be run over the penetration area for 15 to 20 min to determine whether the installation is watertight or leaks. A water hose with enough flow to simulate the amount of water running down the roof during a normal rainfall should be sufficient.

The proper method for waterproofing the penetrations depends on the type and condition of the roof surface to be penetrated. Slanted roofs are made of asphalt or wood

shingles, clay or concrete tile, or corrugated or standing seam metal. Flat roofs are made of built-up layers of tar or asphalt or a membrane material, such as polyvinyl chloride (PVC), thermoplastic olefin (TPO), or ethylene propylene dimonomer (EPDM). **See Figure 9-26.**

Slanted-Roof Penetrations

Roof penetrations are a requirement on most solar water heating system installations. The biggest consideration when making roof penetrations for these systems is waterproofing. Slanted-roof penetrations are easier to waterproof than flat-roof penetrations because there is considerably less risk of standing water on a slanted roof than on a flat roof.

Roofing Materials

Solar Service Inc.

ASPHALT SHINGLES

Alternate Energy Technology

METAL

American Solar Living LLC

CLAY OR CONCRETE TILE

Alternate Energy Technology

MEMBRANE

VELUX America Inc.

WOOD SHINGLES

Lochinvar, LLC

BUILT-UP

Figure 9-26. Slanted roofs are made of asphalt or wood shingles, clay or concrete tile, or corrugated or standing seam metal. Flat roofs are made of built-up layers of tar or asphalt or a membrane material.

Shingle-Roof Penetrations. The penetration of an asphalt shingle roof in good condition can be made right through the shingle. **See Figure 9-27.** For a lag screw installation, the screw should be coated with roof sealant and roofing membrane placed under the mounting bracket. Roof sealant should be placed on the underside of the membrane. The screw can then be tightened into the roof. This same waterproofing method can be used with spanners if the drill hole is snug around the spanner rod.

Shingle-Roof Penetrations

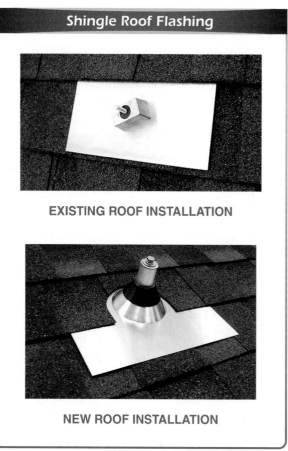

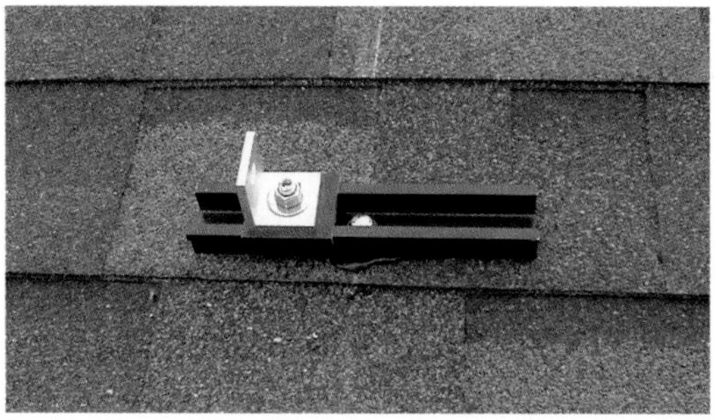

SolarWorld Industries America

Figure 9-27. The penetration of a fiberglass or asphalt shingle roof in good condition can be made right through the shingle.

Facts

The **NRCA Waterproofing Manual** *provides the latest information on proper design, quality of materials and workmanship for waterstops, material storage and handling, weather considerations, proper slope and drainage, and flashings for properly sealing roofs.*

The procedures for installing flashing on an asphalt shingle roof vary depending on the flashing manufacturer and whether it is a new or existing roof. The flashing manufacturer instructions should be carefully read and followed to ensure a waterproof installation. **See Figure 9-28.**

If standoff mounting assemblies are used, the mounting bracket can be attached to the roof using membrane material and sealant. Flashings can be placed over the bracket and under the top shingles. The collector attachment bracket must extend through and above the flashing. **See Figure 9-29.**

Shingle Roof Flashing

EXISTING ROOF INSTALLATION

NEW ROOF INSTALLATION

Quick Mount PV

Figure 9-28. The procedures for installing flashing on an asphalt shingle roof vary depending on the flashing manufacturer and whether it is a new or existing roof.

Collector Mounting Assemblies above Roof Flashing

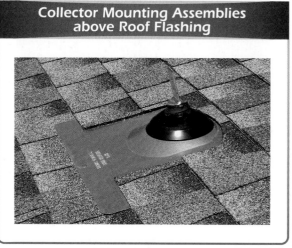

Schott Solar

Figure 9-29. The collector attachment bracket of a standoff mounting assembly must extend through and above the flashing.

Tile-Roof Penetrations. If the roof tile is in good condition and can remain on the roof, the area of penetration should be cut out of the tile and the mounting bracket secured using any of the three roof-mount attachment devices. In some cases, the tile can be repaired. If not a flashing should be placed over the tile and sealed in place with waterproofing sealant appropriate for the tile composition. **See Figure 9-30.**

Tile Roof Flashing

BARREL TILE

FLAT TILE

Quick Mount PV

Figure 9-30. Flat and barrel-shaped tiles may require custom flashing, which should be sealed in place using the appropriate sealing material for the tile composition.

Barrel-shaped tiles may require a custom flashing made to fit the tile shape to complete the waterproofing. The flashing should be sealed in place using the appropriate sealing material for the tile composition.

A solar collector mounted on a tile roof must be mounted using the standoff mounting method. **See Figure 9-31.** Because the standoff method requires drilling a hole in the roof tile, the appropriate waterproofing measures must be taken where the standoff penetrates the roof tile. A standoff must be used on barrel tile because the tile will not support the collector.

Care must be taken when walking on tile roofs. Most tile materials are fragile and may break when stepped upon. In some cases, the tile may need to be removed to install a collector on the roof. This will require much more work and should be taken into consideration during the site analysis.

Metal-Roof Penetrations. Solar collectors are mounted to metal roofs according to the type of metal roof. A corrugated metal roof requires penetrations similar to an asphalt tile roof. However, installing flashing is more difficult since there are no shingles into which to tuck the flashing. The flashing must be properly waterproofed using the appropriate roofing sealant.

In many cases, standing seam roofs do not require penetrations for the mounting of the collectors. Manufactured seam clamps can be used to secure the collectors to the seams on the roof. The clamps are simply tightened onto the roof seams and the mounting hardware is then bolted to the clamps. **See Figure 9-32.**

Alternate Energy Technology
Seam clamps allow collectors to be attached to standing seam roofs without penetrating the roofs. The only penetrations required are for the solar supply and return piping.

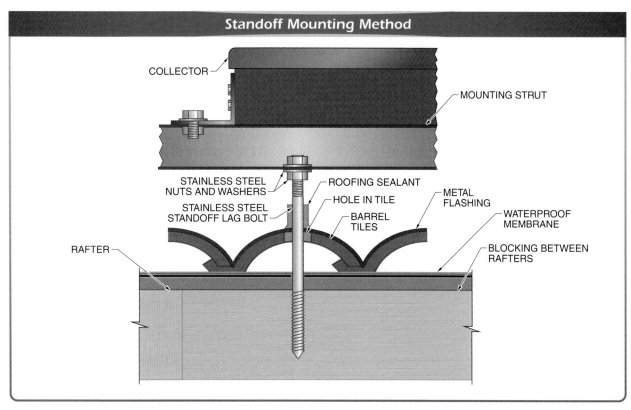

Standoff Mounting Method

COLLECTOR

MOUNTING STRUT

STAINLESS STEEL
NUTS AND WASHERS

STAINLESS STEEL
STANDOFF LAG BOLT

ROOFING SEALANT

HOLE IN TILE

BARREL
TILES

METAL
FLASHING

WATERPROOF
MEMBRANE

BLOCKING BETWEEN
RAFTERS

RAFTER

Figure 9-31. The standoff method of attaching solar collectors to a tile roof involves drilling a hole through the tile and attaching a standoff lag bolt to the framing below.

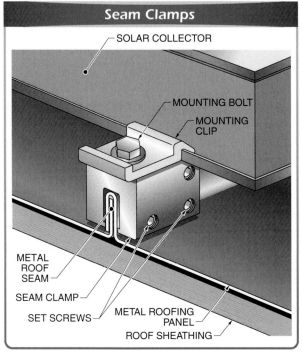

Seam Clamps

SOLAR COLLECTOR

MOUNTING BOLT

MOUNTING
CLIP

METAL
ROOF
SEAM

SEAM CLAMP

SET SCREWS

METAL ROOFING
PANEL

ROOF SHEATHING

Figure 9-32. Seam clamps are used to attach solar collector mounting hardware to metal standing seam roofs.

Flat-Roof Penetrations

Flat-roof penetrations can be made in a manner similar to slanted-roof penetrations, but because of the susceptibility to damage from standing water, flat-roof penetrations may require pitch pockets. A *pitch pocket,* also called a pitch pan, is a small reservoir made around a roof penetration to keep standing water away from the penetration. A pitch pocket is especially required when the mounting bracket is bolted to a metal roof. **See Figure 9-33.**

The pitch pocket can be fabricated of copper or galvanized sheet metal that will cover an area of about 4″ × 4″ around the penetration. The pitch pocket must have a raised dam that will keep standing or running water away from the penetration. On a built-up roof, the area under the pitch pocket should be cleared of gravel and sealed under and around the pocket. The interior of the pitch pocket also should be sealed. The loose gravel should then be replaced around the pocket. On a membrane roof, the pan edges should be sealed with the appropriate roofing material. In some cases, a professional roofing contractor may be needed to perform the waterproofing in order to protect the integrity of the roof and maintain a warranty against leaks.

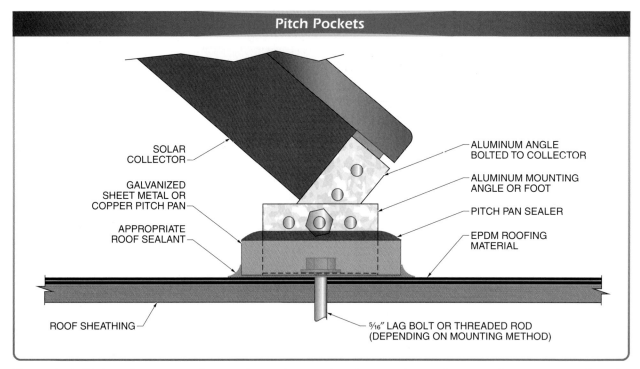

Pitch Pockets

SOLAR COLLECTOR

GALVANIZED SHEET METAL OR COPPER PITCH PAN

APPROPRIATE ROOF SEALANT

ROOF SHEATHING

ALUMINUM ANGLE BOLTED TO COLLECTOR

ALUMINUM MOUNTING ANGLE OR FOOT

PITCH PAN SEALER

EPDM ROOFING MATERIAL

⁵⁄₁₆" LAG BOLT OR THREADED ROD (DEPENDING ON MOUNTING METHOD)

Figure 9-33. Pitch pockets are small reservoirs made around roof penetrations to keep standing water away from the penetrations.

MOUNTING SOLAR COLLECTORS

Once the solar collector mounting assemblies are secured and waterproofed, the collector can be attached to the mounting brackets. This will entail raising the solar collector onto a roof. The average $4' \times 8'$ flat-plate solar collector can weigh up to 160 lb. Due to the weight of the collector, it is highly recommended that safe lifting methods be used to mount the collector onto the roof.

The proper method for transferring equipment to the roof depends on the roof elevation, shape of the roof, and type of roof material. Hoisting or lifting equipment is required if the collector is being mounted on a steeply sloped roof. A crane may be required to lift the collector if a high elevation is involved. **See Figure 9-34.** This is especially true if a multiple-collector array is to be mounted. *Note:* All federal, state, provincial, and local safety requirements must be followed while working on a roof.

Once the solar collector is lifted to the roof, it can be attached to the mounting assemblies. There are many different assemblies and attachment methods for the various solar collectors on the market. The safest and most appropriate method of mounting a solar collector involves using the mounting assembly and attachment

system provided by the manufacturer. In many cases, the manufacturer warranty will not be honored if the collector is not mounted per the mounting instructions and the assembly designed by that particular manufacturer is not used.

Using Cranes to Lift Collectors

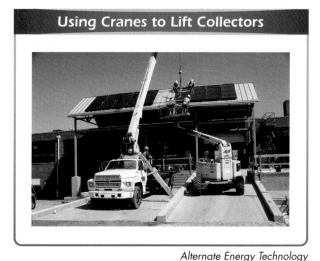

Alternate Energy Technology

Figure 9-34. A crane may be required to lift a collector to a high elevation.

Regardless of the mounting method and type of solar collector used, the mounting assembly connections must be checked. All attachment screws or bolts must be tightened securely. In addition, the angle of the mounting assembly should be checked with an angle finder or inclinometer to ensure it is correct. **See Figure 9-35.** Once the mounting assembly is installed, the collector can be mounted.

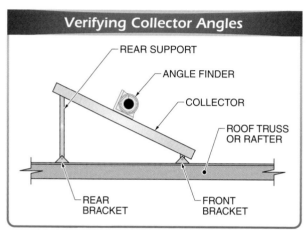

Figure 9-35. The angle of the mounting assembly should be checked to ensure it is correct.

Facts

All workers who participate in signaling or flagging a crane and giving direction as to where and what should be lifted must be qualified when the following circumstances exist: the point of operation is not in full view of the operator, the view of direction of travel is obstructed, or site-specific safety conditions exist. Qualification must be obtained through a third-party organization or by an employer's qualified evaluator.

The evacuated tube solar collector is often mounted on a roof using the rack mounting method. **See Figure 9-36.** Once the rack system is completed and checked for the appropriate angle and tightness, the evacuated tubes can be installed. The tubes may also need to be adjusted for the proper orientation if the absorber is adjustable.

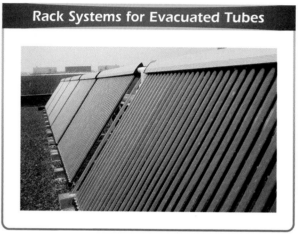

SunMaxx Solar

Figure 9-36. Evacuated tube solar collectors are often mounted on a roof using the rack mounting method.

Once the solar collector is mounted, the entire assembly should be checked again to verify that all the attachment screws or bolts are tight and the collector is at the proper angle. The collector aperture area should continue to be covered to protect it from generated heat. (This is not necessary for an unglazed flat-panel swimming pool collector.) After the solar collector has been mounted, the piping can be installed.

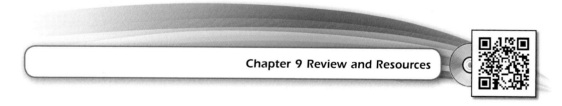

Chapter 9 Review and Resources

10

Installation of Solar Storage Tanks, Piping, and System Components

The installation of solar collectors and a solar storage system provides a solar water heating installer with the two focal points of the solar water heating system. Once these two primary components of the system are installed, the piping and other system components can be fitted thereby linking the two primary components together and creating the solar water heating system.

INSTALLATION CONSIDERATIONS

Once the installation of a solar collector is complete, the remaining majority of the installation process takes place within a building or home. At this point, care must be taken to protect the property. This is especially true when working in a residential home. Homeowners are more conscious of what takes place within their home. Therefore, the installer should be considerate of the location while working and practice cleanliness, professionalism, and customer care.

Other important installation considerations concerning the installation of a solar storage system, piping, and components should be addressed. These considerations include safety and knowledge of local and federal codes. Safety is an especially important consideration. For example, the roof can be the most dangerous area for the installer, and working within a building or home entails another set of concerns.

Furthermore, a solar water heating system, in most cases, uses potable cold water and provides potable hot water. Therefore, thorough knowledge of plumbing and mechanical code regulations is required. Improper installation of storage tanks, system piping, and components can not only lead to a poorly working solar water heating system, but it can cause damage to property or harm the occupants.

Safety

A job hazard plan created during the preplanning stage should account for the three most prevalent safety concerns while working on the storage tank, piping, and system components. These safety concerns include transporting heavy tanks or crates through a building and possibly up stairs, using torch equipment within a building, and working on ladders, ceilings, or rafters.

Back injuries that result from carrying heavy objects are one of the most common injury types in the construction industry. Solar storage tanks are heavy. Even if a large storage tank can be assembled on site, it still consists of several crated pieces that can be heavy. The average storage tank is either an 80 gal. or 120 gal. tank, and the inclusion of an internal heat exchanger increases its weight. For example, an empty Rheem Solaraide HE 80 gal. tank is 222 lb. The 120 gal. unit is 380 lb. **See Figure 10-1.**

The first precaution that must be taken is to ensure that someone assists the installer in handling the tank. Moving equipment, such as a heavy-duty powered hand truck, should be used. **See Figure 10-2.** One worker may be able to cart the tank using the hand truck, but another worker should be assisting and guiding its progress through the building. A powered hand truck that can climb stairs is helpful if stairs are present. The route through the building must be carefully chosen based on width and height restrictions. Turning corners with a large item that is tilted requires extra room.

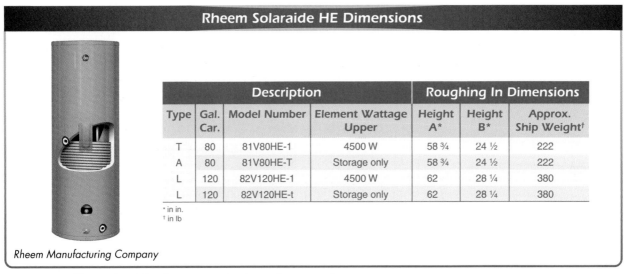

Rheem Solaraide HE Dimensions

		Description			Roughing In Dimensions	
Type	Gal. Car.	Model Number	Element Wattage Upper	Height A*	Height B*	Approx. Ship Weight†
T	80	81V80HE-1	4500 W	58 ¾	24 ½	222
A	80	81V80HE-T	Storage only	58 ¾	24 ½	222
L	120	82V120HE-1	4500 W	62	28 ¼	380
L	120	82V120HE-t	Storage only	62	28 ¼	380

* in in.
† in lb

Rheem Manufacturing Company

Figure 10-1. Size and weight of storage tanks depend on capacity.

Powered Hand Trucks

L P International Inc.

Figure 10-2. Powered hand trucks allow a single person to easily move and maneuver large solar water heating tanks.

Once the tank is situated in the equipment room, it may have to be placed on a drip pan. This will also take more than one individual and some effort to prevent damaging the pan or tank. While working with heavy equipment or packages, the proper lifting techniques should be used to reduce the risk of back injuries.

A back brace may be worn, but back braces are neither discouraged nor recommended by the Occupational Safety and Health Administration (OSHA).

While installing the piping for the solar water heating system, it may be necessary to solder copper piping on or within the building. Copper piping is soldered using a torch flame, although there are electric-resistance soldering tools. **See Figure 10-3.** If a flame torch is used, the appropriate fire extinguisher must be on hand. If working in attic spaces or within walls, another worker should be on firewatch in case insulation or wall materials are ignited. When working within wall spaces, it is a good practice to wet the interior wood or drywall with a heat-resistant spray to keep the surface cool.

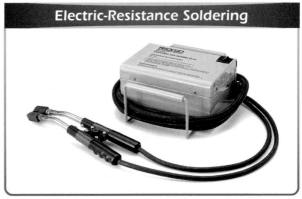

Electric-Resistance Soldering

RIDGID®

Figure 10-3. Electric-resistance soldering connects pipe without using a flame.

Although most of the remaining work is inside the building once the installation of a solar collector is complete, the danger of falling, such as off a ladder, still exists. Ladders are used for out-of-reach tasks such as installing piping runs and sensor wiring for the system controls. Twenty percent of the 123 fatal falls in 2009 were falls from ladders. Most of these falls were due to inattention. Installers should be aware of the various safety precautions for using ladders.

Another safety precaution that some installers may not take seriously is maintaining a clean work area. A clean work area is a productive and safe work area. In a properly maintained work area, the ground is free of scraps that can create a fire hazard and debris, tools, or pipe that can create a tripping hazard. Also, a good impression is made when a customer notices a neat and respectful installer.

Codes

Requirements for the installation of water heaters or solar-heated water storage tanks can be found in two of the most adopted national plumbing codes: the *Uniform Plumbing Code®* (UPC®), published by the International Association of Plumbing and Mechanical Officials (IAPMO), and the *International Plumbing Code®* (IPC®), published by the International Code Council (ICC). Both codes reference or use language from the *National Fuel Gas Code,* published by the National Fire Protection Association (NFPA), which covers the installation of gas water heaters and their venting systems. For electric water heaters or electric solar storage tanks, *NFPA 70®: The National Electrical Code®*, published by the NFPA, can also be used for reference. In some instances, one of the following green codes may be adopted:

- *Uniform Solar Energy Code* (USEC) published by IAPMO
- *Green Plumbing and Mechanical Code Supplement* published by IAPMO
- *National Green Building Standard* published by the ICC
- *International Green Construction Code* published by the ICC

There may be regulations adopted by individual state or provincial codes that affect installation. The installer must know which codes are adopted in the local area. If no codes are adopted, any of these codes are a good reference to guide the installation.

The local authority having jurisdiction (AHJ) should be consulted to confirm which codes are adopted in the local area and the provisions of the codes that may apply to the installation. Also, all manufacturer installation instructions should be followed when installing solar water heating system components. The following are just some of the code requirements that may apply to solar water heater/storage tank installation, multiple storage tank installation, and solar system piping installation:

- All water heaters and storage-type water heaters must have a primary temperature control, an overtemperature limiting device, and a temperature-and-pressure (T&P) relief valve per UPC 506.2.
- Clearances from walls and ventilation openings for listed water heaters must be based on the manufacturer installation instructions. For unlisted water heaters, there must be a 12″ clearance on all sides. Floors containing combustible materials must be protected from tanks installed on the floor per UPC 505.3.
- Water heaters or tanks with an internal heat exchanger must have heat transfer fluid of either potable water or a nontoxic fluid with a toxicity rating of Class 1. The tank may also be limited to a maximum pressure of 30 psig per UPC 506.4.
- If gas fired, water heaters or storage tanks must comply with combustion and ventilation requirements per UPC 507.0.
- In seismic zones C, D, E, and F, water heaters must be strapped to resist horizontal displacement with two straps located in the upper and lower third of the tank per UPC 508.2. **See Figure 10-4.**
- If a water heater or storage tank is located on bare ground, it must be supported by a concrete pad a minimum of 3″ thick and 3″ above the ground per UPC 508.3.
- If a water heater or storage tank is located in an area above a ceiling, it must have a watertight pan and drain placed underneath for protection from leaking per UPC 508.4.
- If a water heater or storage tank is not a flammable or vapor resistant model and is located in a garage, it must be installed with the ignition point 18″ above the garage floor. If a water heater or storage tank is in the possible path of a vehicle, it must be protected from physical damage per UPC 508.14.
- When testing solar system piping, an open system must be tested at 50 psi with air, and a closed system must be tested at 1½ times working pressure per USEC 316.2.

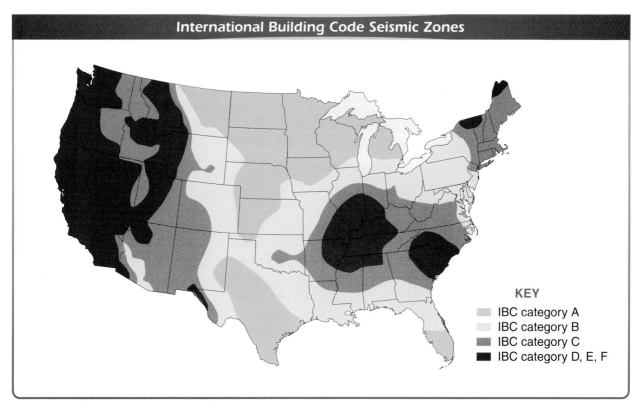

Figure 10-4. IBC seismic zone installation requirements determine how equipment must be installed.

- Piping hanger and support spacing must be installed to adequately support the pipe per USEC Table 3-2.
- Thermal expansion tanks and pressure-relief valves must be sized and installed as required per USEC 408.0 and 602.0.
- Underground solar piping installation must be installed as required per USEC 411.1.
- Thermal insulation sizing must be based on requirements of USEC 802.0 and Table 8-1.
- Potable water piping materials used must meet the requirements of UPC 604.0.
- PRV valves and drains used must meet the requirements of UPC 608.4 and 608.5.

SOLAR STORAGE SYSTEM INSTALLATION

Site analysis and project preplanning will help identify the type of solar storage system to be installed. Site analysis and project preplanning will also detail the area within the building or residence in which the solar storage system is to be located. There are several different types of solar storage systems, but the most prevalent storage systems and tanks include nonseparable thermosiphon storage systems, solar storage tanks, and water heaters.

The initial step in the installation process for a solar storage system is to receive and inspect the solar storage tank or solar water heater. Upon delivery, either at the shop or the installation site, the system should be uncrated and inspected for defects.

It is most important with piping installations to check the tank ports for damage to piping threads. If either male or female threads are galled, it will be difficult to make a watertight connection. If the tank or system is damaged in any way, the manufacturer should be notified. The installer should never try to repair the defect because this might nullify the warranty. The manufacturer will make recommendations as to what should be done. It is also important to remember that each system or tank should be installed per the manufacturer's installation instructions and applicable code requirements.

Nonseparable Thermosiphon Storage Systems

A nonseparable thermosiphon system, which incorporates an attached solar storage tank, is commonly installed in areas of North America with the mildest temperatures. **See Figure 10-5.** This type of solar storage system is normally installed on the roof at the same time as the collector.

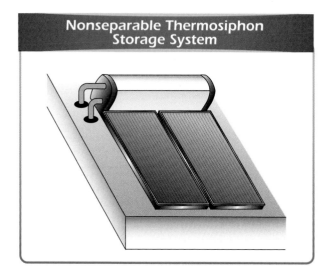

Figure 10-5. Nonseparable thermosiphon storage systems are installed as a unit.

The most important installation parameter for this type of storage system is to ensure that the roof structure can withstand the weight of the tank. If not, the roof structure must be reinforced per the local building codes or structural engineer. **See Figure 10-6.**

Separable thermosiphon system tanks that are not attached to the solar collector should be mounted at least 12″ above the collector. This will allow for more efficient convection between collector and storage tank. Proper freeze protection measures should be taken if freezing temperatures are likely.

Another unique feature of a separable or nonseparable thermosiphon system is that it may require electrical power on the roof to energize an element in the tank, which acts as a back-up heating system. **See Figure 10-7.** As with all outside electrical connections of power systems, a licensed electrician is normally required to perform electrical work. In most cases, the average solar water heating installer does not have the required electrical license to perform this task.

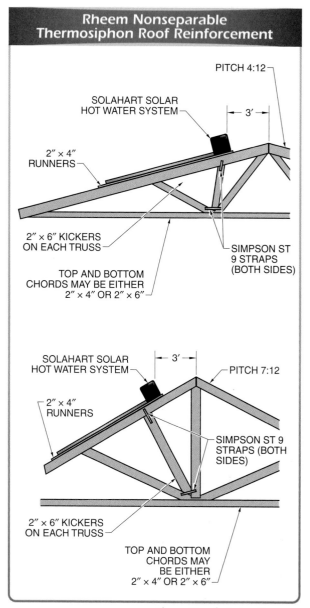

Rheem Manufacturing Company

Figure 10-6. Roof structure must be reinforced per the local building code or structural engineer requirements.

Facts

A roof truss is an engineered combination of structural members arranged and fastened in triangular units to form a rigid framework for supporting loads over a long span. Ends of trusses bear directly on the opposing exterior walls. Today over 75% of new homes in the United States and 90% of the new homes in Canada are constructed with roof trusses.

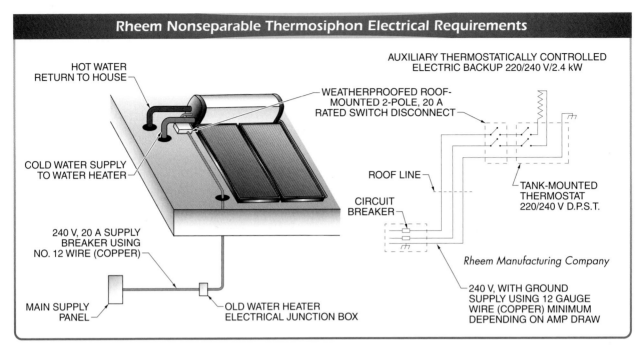

Rheem Nonseparable Thermosiphon Electrical Requirements

HOT WATER
RETURN TO HOUSE

AUXILIARY THERMOSTATICALLY CONTROLLED
ELECTRIC BACKUP 220/240 V/2.4 kW

WEATHERPROOFED ROOF-
MOUNTED 2-POLE, 20 A
RATED SWITCH DISCONNECT

COLD WATER SUPPLY
TO WATER HEATER

ROOF LINE

CIRCUIT
BREAKER

TANK-MOUNTED
THERMOSTAT
220/240 V D.P.S.T.

240 V, 20 A SUPPLY
BREAKER USING
NO. 12 WIRE (COPPER)

Rheem Manufacturing Company

MAIN SUPPLY
PANEL

OLD WATER HEATER
ELECTRICAL JUNCTION BOX

240 V, WITH GROUND
SUPPLY USING 12 GAUGE
WIRE (COPPER) MINIMUM
DEPENDING ON AMP DRAW

Figure 10-7. All electrical work must be installed according to code.

Existing Water Heater Retrofit or Use

A common solar storage system used in a residential solar water heating system is an existing water heater. In many cases, the existing water heater can be retrofitted to use the solar thermal collector as a preheater. This can be accomplished whether the system is passive or active and whether it is direct or indirect.

First, the installer must assess whether the existing water heater is in good working condition. According to the U.S. Department of Energy, most residential water heaters have a life span of 10 to 15 years. Realistically, if the water heater is 7 to 10 years old, a new water heater should be used. For more efficient use of the system, a true solar storage tank should be installed. If the existing water heater is used, the anode should be checked visually to determine whether it has deteriorated. If so, the anode should be replaced.

An existing water heater for a solar water heating system is most commonly used in a direct system. The cold water supply to the water heater should be connected to a three-way valve that connects the cold water inlet and the solar collector inlet. The collector outlet also connects to the cold water inlet. The hot water outlet piping should be piped with a tempering valve connected to the cold water to temper the domestic hot water (DHW) to 120°F (49°C). **See Figure 10-8.**

There are several ways to use an existing electric water heater, depending on how much the installer wants to repipe the water heater. In some cases, the water heater drain at the bottom of the tank is used as the solar collector supply or inlet connection and the cold water inlet. The cold water port is used for the solar collector outlet or return piping. Two things should be done in this installation. The cold water (solar return) dip tube should be cut to just below the top element, and the lower element thermostat should be set to the lowest setting or disconnected. This allows for the coolest water to be sent to the collector and for the hot water from the collector to stay at the top of the tank. The top element will be used as the back-up water heater rather than both elements being used to heat the entire tank.

The most effective use of the existing water heater is as a second storage tank, which also acts as the back-up water heater. This also allows for a smaller solar storage tank to be used since both tanks provide solar storage rather than one large tank. In this installation, the tanks are interconnected with the hot outlet from the solar tank to the cold water inlet of the existing water heater. Three ball valves are used to provide a bypass between the two tanks for the repair or removal of either tank. **See Figure 10-9.**

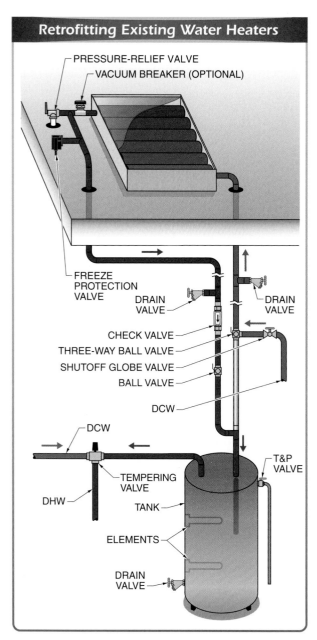

Figure 10-8. Existing electric water heater systems can be retrofitted as part of a solar water heating system.

hottest water from the solar collector to be used first. It also has a flue at the center of the tank, which can lead to standby heat loss.

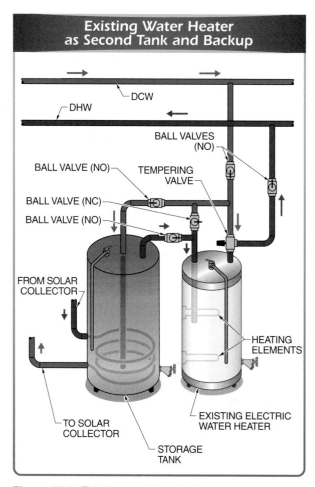

Figure 10-9. Existing electric water heaters can serve as a back-up tank and heater for a new solar heating system.

An existing or new gas water heater can also be retrofitted or used in a solar water heating installation. The gas water heater can be an efficient second tank or auxiliary back-up water heater. **See Figure 10-10.** It is not good practice to use a gas water heater as a single-tank system unless the solar system is used as a preheater only. This is because the gas water heater heats from the bottom up to heat the entire tank and does not allow the

One advantage of using an existing water heater in a solar water heating installation is that there is no need to bring in a tank or find a location to install the tank. However, the existing water heater and its new piping installation must meet current code requirements once the installation is complete. Also, because the DHW heater usually does not have adequate insulation, an insulation blanket or foam insulation made for this purpose should be placed around the tank. Piping in and around the tank should also be insulated from at least 2′ to 5′ from the tank. If possible, an electric water heater should be insulated underneath the tank as well.

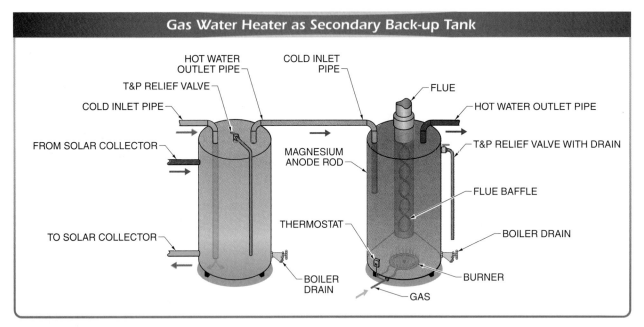

Gas Water Heater as Secondary Back-up Tank

Figure 10-10. Existing gas water heaters can serve as a back-up tank and heater to a new solar heating system.

Solar Storage Tanks

Using a solar storage tank or water heater designed for solar water heating installations has several advantages over using an existing or new water heater. One advantage is that, since the tank is in new condition, it will be able to last much longer than an existing water heater. Many solar storage systems are designed to last about 20 years. Another advantage is that they contain much more insulation than ordinary water heaters, which makes them far more efficient. Most importantly, the piping connections or ports on the solar storage system are specifically designed for use in a solar water heating system. Furthermore, the majority of dedicated solar storage systems incorporate an integral heat exchanger.

There are several different kinds of solar storage systems available for use in a solar water heating system. In some cases, the solar storage tank or solar water heater is similar to the average water heater. The tank, ports, insulation, and shell are preassembled and come in one carton. **See Figure 10-11.** A one-piece system is relatively easy to install since all that is necessary is to get it to the proper location, stand it up, and provide seismic strapping if necessary. Piping can be installed and electrical connections can be established almost immediately.

Large solar storage tanks may come in parts. **See Figure 10-12.** For example, the Vaughn Hotstow storage tank is delivered as an uninsulated tank. This particular tank has a ½″ interior stone lining that is impervious to adverse water conditions. The tank has mounting brackets at the bottom to mount the tank to the floor. This should be done as soon as the tank is set in the proper location. The insulation is preformed, 3″ thick, and placed around the tank. A plastic outer shell is provided to wrap around, secure, and protect the insulation. The insulation and plastic jacket must be installed before piping can begin.

Water heaters include sensors, controls, and safety devices for operating their heat sources safely and efficiently.

Preassembled Storage Tanks

Alternate Energy Technology

Figure 10-11. Many solar storage tanks are preassembled and ready for installation and connections.

Vaughn Hotstow solar storage tanks are designed to be pressurized. However, there are nonpressurized, or open, tanks that are designed for solar water heating. These tanks are normally large tanks with capacities of hundreds of gallons. For example, the Hydroflex flexible water storage tank is a pressure-free tank with internal heat exchangers made from 1″ copper tubing. **See Figure 10-13.** The tank has an EPDM line pinned to the top of the insulated wall. The wall of the tank is prefabricated using two layers of 1″ isocyanurate foil-faced insulation covered with an embossed aluminum structural outer skin. The tank lid is comprised of 2″ isocyanurate insulation. The tank bottom is comprised of a layer of 1″ isocyanurate and a layer of 1″ foam. Additional insulation may be added around the tank if needed. The tank is designed to be built on-site, comes in pieces small enough to fit through a common door, and is available in capacities ranging from 100 gal. to 5000 gal. This type of tank may require a lot of work to install, and some manufacturers require installation by their own trained personnel.

Vaughn Solar Storage Systems

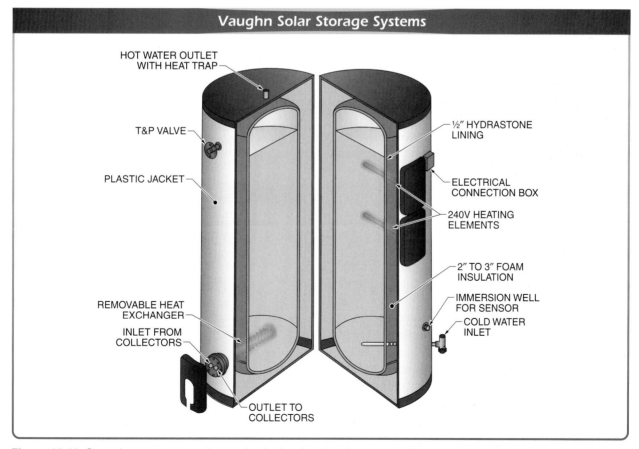

Figure 10-12. Some large storage tanks need to be insulated and protected after installation and before connections.

Hydroflex Storage Tanks

Hydroflex Systems, Inc.

Figure 10-13. Some storage tanks have multiple heat exchangers to increase efficiency.

Some of these large tanks require a large installation area for the carting in, uncrating, and setting up of the tank. These tanks are normally used in commercial installations. **See Figure 10-14.** Many manufacturers can custom build tanks for specific capacities and installations. Regardless of the type of solar storage system used, it is necessary to follow the manufacturer recommendations and installation instructions and meet local code requirements.

Dip Tubes

The use of new or existing water heaters in a solar water heating system requires modification of the dip tubes within the tank for system efficiency. In most hot water storage tanks or water heaters, there is a natural stratification of water temperatures. For example, the top of the tank will contain the hottest water at about 120°F (49°C) or higher depending on the hot water needs. The hottest water should be used by the demand fixtures or appliances. In the center of the tank, there may be a layer of 90°F (32°C) water that is in the process of being heated by a heat exchanger or heating element. Toward the bottom of the tank, there is a layer of water

at about 70°F (21°C) that is ready to be sent back to the solar collector or being heated moderately by the lower temperatures at the bottom of the heat exchanger coil. This is ideal thermostratification, or hot water layering, within the tank. **See Figure 10-15.**

Commercial Storage Tanks

SunMaxx Solar

Figure 10-14. Commercial storage tanks provide high-capacity storage.

Thermostratification

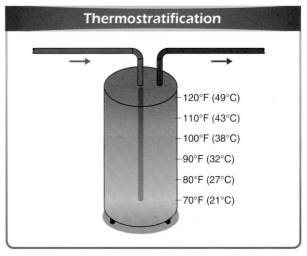

Figure 10-15. Water temperatures vary in storage tanks with the hottest water at the top.

Normally, a dip tube is placed in the cold inlet port of the water heater. The dip tube can extend to within 6″ of the bottom of the tank. This is so that entering cold water will be introduced at the bottom of the tank and does not cool the hottest water at the top or middle of the tank. In a retrofit installation, the solar-heated water is supplied to the DHW heater through the cold water port. The temperature of the water entering from the collectors may not always be at the temperature of the uppermost water in the tank, especially on low-insolation days. Therefore, the incoming solar-heated water should be introduced at a lower level of the tank. In most cases, it will not be as cool as incoming cold water, thus it should not enter at the bottom of the tank where it would be cooled by entering cold water.

In order to take advantage of the thermostratification within a water heater, the dip tube must be modified. In a retrofitted electric water heater, the dip tube should be cut a few inches below the top element. This will allow for the incoming water to enter the water heater without disturbing the hotter water at the top of the tank or being cooled by the cooler water at the bottom of the tank. The lower element should be adjusted to its lowest temperature or disconnected. This can also be done with a gas water heater. However, the water heater will not be as efficient because the water within the tank is heated from the bottom only and thermostratification is less evident. Some manufacturers offer solar storage tanks and DHW heaters with multiple ports at the top of the tank. **See Figure 10-16.**

The tank should be supplied with the proper dip tube placement or strategy to take advantage of thermostratification. If not, the dip tubes should be modified or installed before piping can begin. **See Figure 10-17.** The proper dip tube strategy for this arrangement is as follows:

- The hot water outlet should have no dip tube, allowing only the hottest water to leave the tank when called for.
- The cold water inlet dip tube should extend to within 6″ of the bottom of the tank, allowing for the coolest water to be introduced at the bottom of the tank.
- The solar out port that conveys water to the collector inlet should have a dip tube that extends to within 3″ of the bottom of the tank to convey the coolest water possible to the collector for heating. This allows for the most efficient use of the solar collector.
- The solar in port, which introduces heated water into the tank from the collector, should have a dip tube extending roughly halfway down the tank. This will allow for the solar-heated water to enter the tank without being cooled by the cooler water at the bottom and will not disturb the hottest water at the top of the tank.

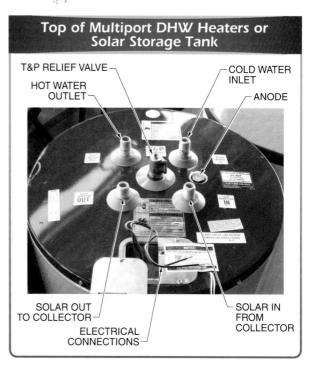

Figure 10-16. Multiple ports on storage tanks are connected to pipes ending at various levels to take advantage of the various temperature levels within the tank.

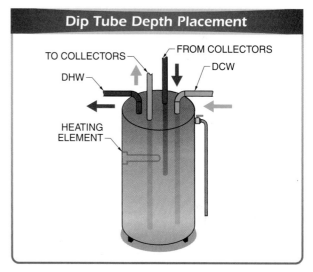

Figure 10-17. Dip tube pipe lengths can be adjusted to end at various temperature levels within the tank.

Integral Heat Exchangers

Many solar water heaters or storage tanks have one or more integral or immersed heat exchangers. These systems provide for very efficient heat transfer from the HTF to the DHW. Another benefit for the installer is that the integral heat exchanger makes the solar water heating installation easier because there is no external heat exchanger to install unless the integral heat exchanger is removable and needs to be installed (rather than coming installed from the factory). The heat exchanger will normally be a finned tube design that will slide and thread into a port on a tank designed for this purpose.

The installer should be able to recognize which solar system piping lines should be connected to which ports on the tank. These ports may be identified or marked on the tank itself, but in some cases they are only identified in the manufacturer installation instructions. The installer should always consult the instructions to ensure the piping is connected at the correct location.

A single integral heat exchanger storage tank has a heat exchanger installed in the lower portion of the tank. **See Figure 10-18.** The upper port of the heat exchanger is connected to the solar supply from the collector outlet. This allows for the solar-heated water to enter the heat exchanger coil at the temperature midpoint in the tank. This corresponds to the dip tube strategy and allows for heat transfer to the DHW to occur in the temperature midpoint of the tank without disturbing the hottest DHW in the upper portion of the tank. The bottom port of the heat exchanger coil is piped to the solar return collector inlet. This allows for the cooled HTF to travel back to the collector after transferring its heat to the DHW in the bottom portion of the tank.

Solar storage tanks or heaters with multiple heat exchangers allow for greater use of the solar-heated HTF. Heat exchangers can be used for not only heating DHW but for any HTF use, such as radiant space heating. Solar water heating systems utilizing multiple heat exchangers are normally designed by design professionals. Heat exchanger use is determined by the temperature needs of the piping systems connected to them and will correspond to the thermostratification within the tank.

A relatively simple solar water heating design will incorporate a solar storage tank with two heat exchangers. **See Figure 10-19.** The DHW is contained in the body of the tank in this system. The lower heat exchanger is connected to the solar collectors, while the upper heat exchanger is connected to a back-up boiler for auxiliary heating on low insolation days. The cold water input is at the top and utilizes a dip tube to introduce the cold water at the bottom of the tank. DHW is taken from the top of the tank.

Another design utilizes two integral heat exchangers in a combi system. **See Figure 10-20.** The DHW is contained and heated in the tank by the solar HTF in the lower heat exchanger coil. The upper heat exchanger coil is utilized by a radiant space heating system. In this system, the HTF in the radiant heat exchanger acts as a preheater with primary heating accomplished by a boiler system. In some cases multiple pumps are needed to transport HTF throughout the system while the DHW flows by line pressure.

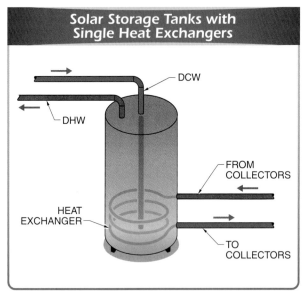

Figure 10-18. Single heat exchangers are placed near the bottom of solar storage tanks.

Solar H₂Ot, Ltd.
Solar storage tanks provide preheated water to adjacent conventional water heaters.

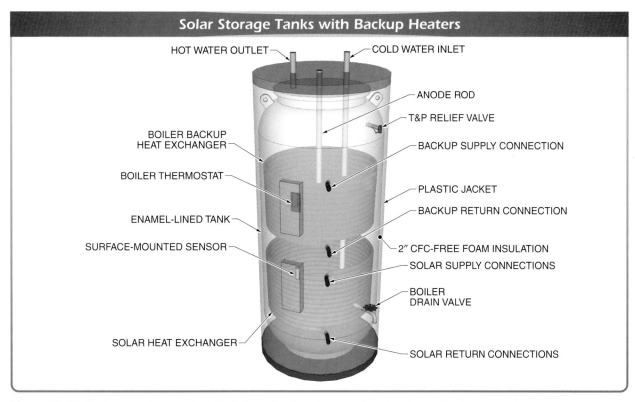

Figure 10-19. Back-up heaters are installed above heat exchangers in solar storage tanks.

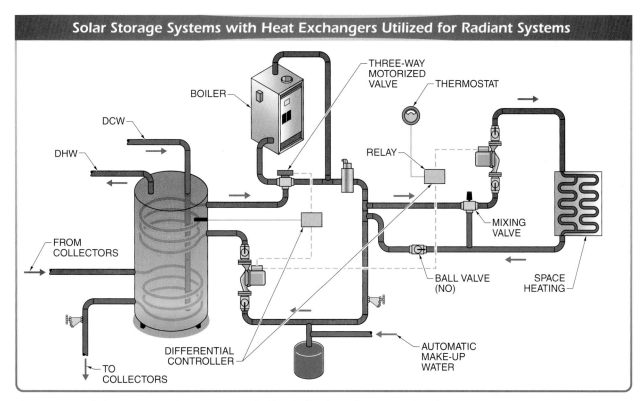

Figure 10-20. Solar storage tanks can be used with a boiler for radiant space heating systems.

This combi system may also use a solar storage tank that contains three integral heat exchangers. **See Figure 10-21.** These heat exchangers are connected to piping systems according to temperature needs or system efficiency. In this case the top center heat exchanger coil is connected to the DCW and the DHW.

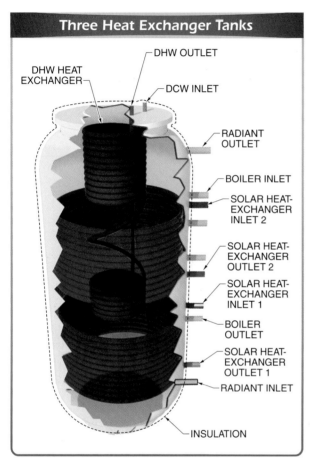

Three Heat Exchanger Tanks

DHW OUTLET

DHW HEAT EXCHANGER

DCW INLET

RADIANT OUTLET

BOILER INLET

SOLAR HEAT-EXCHANGER INLET 2

SOLAR HEAT-EXCHANGER OUTLET 2

SOLAR HEAT-EXCHANGER INLET 1

BOILER OUTLET

SOLAR HEAT-EXCHANGER OUTLET 1

RADIANT INLET

INSULATION

Figure 10-21. Three heat exchanger solar storage tanks are connected to piping systems according to their temperature needs.

Cold water enters and then travels through the coil and is heated by the HTF within the tank. It then exits the coil at the top of the tank as DHW. The HTF within the tank is heated by the lower two heat exchanger coils connected to the solar collector array. These are piped together through a motorized three-way valve so that on heavy insolation days more of the solar-heated HTF can be utilized to heat the HTF within the tank. The HTF within the tank is used to supply the space heating needs of the building and transfer heat to the DHW. The system heat is backed up by a boiler. **See Figure 10-22.**

Combi systems can be designed in many different ways and are only limited by the designer's imagination. It is then up to the installer to turn these designs into reality.

Multiple Tank Installations

Many solar water heating installations use multiple storage tank/water heater arrangements. There are several reasons for this. For instance, the system may need more solar storage than a single tank can provide, or there may not be enough space for a single large storage tank but two smaller tanks will fit in the best location for the storage system. In some cases the designer may want to use a dedicated solar storage tank and then use a solar water heater as the back-up heater rather than a nonstorage boiler. The design can also use tanks with integral heat exchangers.

The use of an existing water heater with a solar storage tank, whether natural gas or electric, is a popular multiple tank installation. The solar system and tank act as a preheater for the DHW entering the supply inlet to the existing water heater. There are typically two ways to interconnect the two storage tanks. One way to connect the tanks is to simply connect the solar storage tank outlet to the water heater inlet using piping. There are no valves between tanks in this installation. One drawback of this system is that the tanks cannot be isolated if one of the tanks or heating systems malfunctions.

Another way to connect the tanks is to install a bypass system that can isolate either tank if necessary. Although this system might be more expensive to install, in the long run it will save the customer time and will be more convenient when changing out the tanks. Large tanks can also be interconnected in this manner.

In some cases, especially where DHW use is in low demand, a smaller electric water heater can act as a back-up heat source and as extra solar storage. **See Figure 10-23.** Two common types of installations of this kind include installing a smaller electric water heater lower than the taller storage tank and elevating the smaller electric water heater above the storage tank. The second type will be somewhat more efficient because the hottest water will rise to the smaller tank by thermosiphoning, thus lowering the demand for back-up heat. In either installation, the DHW outlet of the storage tank is connected to the inlet of the electric water heater. These systems can also be installed with bypasses to facilitate repair or change out.

Combi Systems

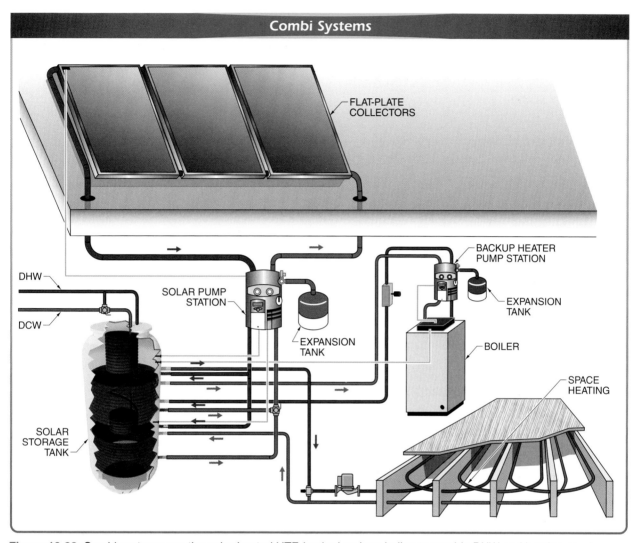

Figure 10-22. Combi systems use the solar-heated HTF, backed up by a boiler, to provide DHW and heating needs.

Small Electric Heaters as Back-Up Heaters and Extra Storage

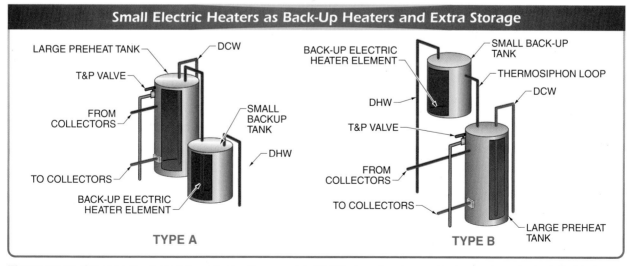

TYPE A

TYPE B

Figure 10-23. Small back-up electric water heaters provide DHW when the demand is low and can be installed below or above the storage tank.

Large commercial or industrial solar water heating systems may use multiple large tanks. These systems should be designed by a solar system design professional. In this situation the solar water heating installer concentrates on installing the system per the design professional's schematics and not on system design.

PIPING INSTALLATION

Once the solar collectors and the solar storage system are in place, the connecting piping system can be installed. The solar water heating installer should be experienced in the pipe joining methods that will be used for the installation. The possibility of leaks in piping, which can cause damage to the solar system and the customer's property, are too great to have inexperienced installers perform the piping installation. An inexperienced installer should help, watch, and learn before attempting to preform critical tasks such as soldering or brazing copper tube. In many areas the installer may need to be a licensed journeyman plumber to do these tasks. This is due to the fact that many jurisdictions require a journeyman plumbing license to modify potable DCW and DHW lines.

Piping Materials

The type of piping materials (pipe and fittings) used to convey solar-heated HTF must be suitable for the application. In the case of a direct solar water heating system or any modifications made to the potable water piping, the piping materials must be suitable for the transport of potable water (DCW and DHW). Plumbing codes allow a number of piping materials to convey potable water within a building. However, there is one critical requirement that solar loop piping materials must meet that severely limits the type of piping materials that can be used for the system. The piping materials suitable for use in the solar loop (the piping to and from the collector) must also be suitable for temperatures that may reach above 300°F (150°C).

These requirements limit the suitable piping materials for use in the majority of the solar water heating systems to malleable iron (commonly called black iron pipe), copper tube, and stainless steel tube. Black iron pipe is not normally used for potable water and is susceptible to rusting. It can be used in an indirect solar loop. However, rust inhibitors or HTF that has rust inhibiting qualities should be used. Cutting and threading black iron pipe can also be costly and time consuming, which

makes its use for small or moderate residential systems impractical. Black iron pipe can be sensible for use in large indirect systems installed in commercial or industrial buildings. Residential building systems usually use copper and corrugated stainless steel tubing.

Copper Tube

Copper tube is by far the most commonly used piping material in solar water heating systems. **See Figure 10-24.** Type L copper tube is the most commonly used piping material in solar loops. There are several methods of joining copper tubing. However, the methods that meet the temperature requirements of the solar loop are soldering using a torch flame or electric resistance tool and brazing.

Copper Tubing and Fittings Used In Solar Water Heating Systems

HARD-DRAWN OR HARD-TEMPERED COPPER TUBE

SOFT-COILED OR ANNEALED COPPER TUBE

VALVES

FITTINGS

Copper Development Association Inc.

Figure 10-24. Copper tubing, valves, and fittings are available in many standard sizes and lengths.

Other joining methods, such as press-connect (commonly called press-fit) and push-connect joints, are suitable for temperatures of up to 250°F (121°C). However, some manufacturers have press-connect fittings that use a sealing element by FKM industries rather than the typical ethylene propylene diene monomer (EPDM) sealing element. Fittings that use the FKM sealing element are rated for temperatures of up to 284°F (140°C) with spikes up to 356°F (180°C). These fittings can be used in the solar loop piping, however they are not yet approved for potable water use.

Soldered Joints. A *soldered joint* is a method for joining copper tube and fittings using capillary action through the heating of metals and filling the space between the tube and fitting with a metal solder alloy that melts at temperatures below 840°F (449°C). **See Figure 10-25.** This is normally performed using a flame torch to heat the tube and fitting metal, which in turn will melt the filler metal and fill the spaces between the tube and the fitting. This can also be performed using an electric resistance tool to heat the metal. Electric resistance soldering uses a tool to deliver a high amperage electrical current to heat the copper and melt the filler metal. This is accomplished without a flame and thus can be safer than using a torch. When using either method, care must be taken to heat the fitting cup and tube end fully to achieve full penetration of the filler metal throughout the fitting cup.

Brazed Joints. A *brazed joint* is a method for joining copper tube and fittings using capillary action by heating the metals and filling the space between the tube and

fittings with a filler metal at temperatures above 840°F (449°C). The filler metal used for brazing is a much harder metal than the filler metal for soldering and has a melting range between 1100°F (593°C) and 1500°F (816°C). Brazing is rarely done for solar water heating piping due to the high temperature flame needed to melt the filler metal. However, if brazing is used, a more permanent and durable joint is made. Care should be taken when brazing small-sized tube. The high heat flame can easily melt through the copper tube and fittings if left in one position for too long. Also the high heat needed to braze will tend to anneal and soften the smaller copper tube making it easy to bend or crimp.

While soldering or brazing copper tube, cotton work gloves must be worn and kept on at all times. There are two reasons for this. First, the hands must always be protected while working with sharp or hot materials such as tubing ends or heated pipe. Hands must also be protected from abrasive or caustic materials such as sand cloth and fluxes. The second reason is that the body's skin contains oils that can be transferred to items that are touched. These oils create finger prints. If the clean end of the copper tube or fitting cup is touched with a bare hand, the oils will deposit on the tube or fitting and if not removed could create a void in the soldered joint. This void could lead to a leak in the joint.

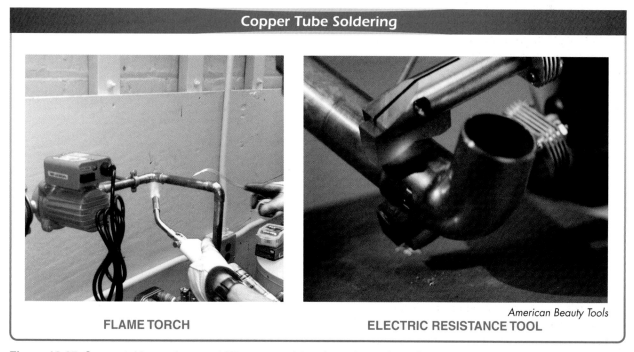

Copper Tube Soldering

FLAME TORCH

ELECTRIC RESISTANCE TOOL

American Beauty Tools

Figure 10-25. Copper tubing, valves, and fittings are soldered together using a flame torch or electric resistance tool.

There are eight steps to creating a properly soldered or brazed joint, and each step is equally important to perform a water tight joint. The eight steps are as follows:

1. Measure the pipe lengths correctly so that the tube end fits snugly to the end of the cup. Do not try to make up for a bad measurement by slipping the tube end from the end of the cup. This will reduce the strength of the joint and could cause a leak. All tubing cuts should be squarely cut with proper cutting tools.

2. All tubing ends should be reamed or deburred to prevent turbulence within the tube. Turbulence is caused by fluids rushing by a raised surface within the tube. It can cause erosion of the tube, especially in small-sized tubes at high temperatures. **See Figure 10-26.**

3. Lightly abrading the tube end the full length that will be fit into the fitting will remove oxides and impurities that, if left on the tube, can cause voids and leaks. Fitting cups must be cleaned also. Although fittings may look new and clean, they may contain light oil coatings that need to be removed.

4. A light coating of flux should be applied with the proper brush. Too much flux can cause voids in the joint, and the flux may deposit on the inside of the tubing, which later can cause erosion of the tube. For soldering, always use a water soluble flux meeting the requirements of ASTM B813-10, *Standard Specification for Liquid and Paste Fluxes for Soldering of Copper and Copper Alloy Tube*. A brazing flux is necessary for brazing dissimilar metals such as copper and brass.

5. Fit the tube securely in the cup and support the joint so that the tube ends are square and plumb. A cocked tubing end can also cause voids and leaks in a joint.

6. Apply heat evenly around the tube and then the fitting cup. Never heat just one side of the tube or fitting even with small sizes. The solder will tend to flow to the hot side and may not fully fill the opposite side of the cup. Do not overheat the metal as this will burn the flux and cause voids leading to possible leaks.

7. Apply the solder or filler metal evenly around the fitting. Do not just fill from one side of the cup as this may also cause uneven filling of the cup and cause a leak. Do not use too much solder or filler metal as this may create a buildup or ridge in the fitting and cause turbulence.

8. Allow the tube and fitting to cool before moving or jostling the tube. If the solder has not hardened fully, it could crack when jostled and cause a leak. Clean all flux off tubing and fittings. Flux is caustic and may cause erosion of the tube if not removed.

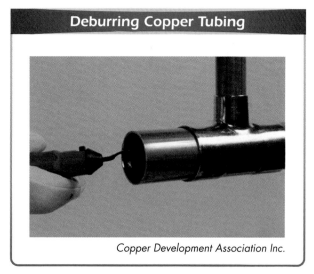

Copper Development Association Inc.

Figure 10-26. Copper tubing is deburred after cutting to provide a smooth end.

Corrugated Stainless Steel Tubing

Corrugated stainless steel tubing (CSST) is becoming a popular choice in solar water heating installations when running piping to and from the solar collector. The flexibility and length of CSST allows it to be run up through attic spaces more easily and with fewer joints than rigid copper tube.

CSST connections are made by a proprietary compression or flared joint. **See Figure 10-27.** The compression joint seal is made using the supplied nut, washer, and adapter fitting, which should be tightened with adjustable or flat-jawed wrenches. The tubing can be cut with a CSST cutter, which is similar to a copper tubing cutter. A copper tubing cutter may be used, but a hardened CSST cutting wheel should be used rather than the normal copper tube cutting wheel. Each CSST manufacturer incorporates a proprietary joining method for their tubing. Therefore, it is prohibited to use one manufacturer's joining method with another manufacturer's tubing. Each manufacturer's installation instructions for joining these materials must be followed.

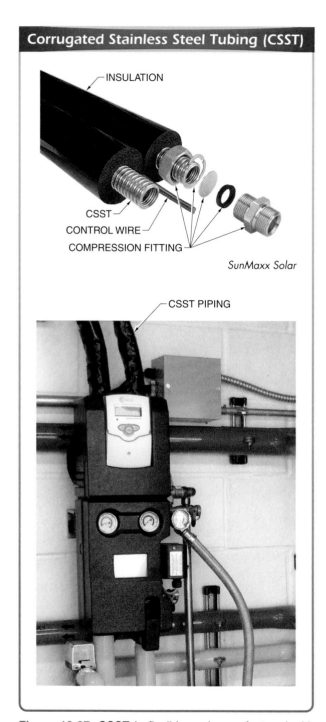

Figure 10-27. CSST is flexible and manufactured with insulation and attached control wire in standard lengths.

DCW and DHW Piping

Piping materials used for modifying or interconnecting the DCW and DHW piping should be the same material used for the potable DCW and DHW piping already in place. Common materials that are used for this piping are copper tubing, PEX, PEX-AL-PEX, PVC, or

CPVC. Many of these piping materials, such as PEX, have proprietary joining methods that may be different from manufacturer to manufacturer, and they cannot be substituted by another manufacturer's joining device. Each manufacturer's installation instructions for joining these materials must be followed. **See Figure 10-28.**

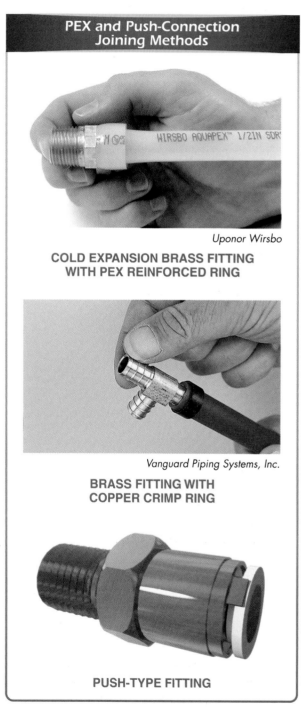

Figure 10-28. Manufacturers have developed several methods for connecting PEX piping to fittings.

Plastic type piping materials should not be connected directly to the water heater or storage tank. Some plumbing and mechanical codes require at least 18″ of metallic piping before connecting plastic piping to these appliances. This requirement is due to the fact that if temperatures in the water heating systems reach their automatic shut off point, that heat may be conducted to the plastic piping and possibly melt or loosen the piping connections.

PVC Piping

The installation of a swimming pool solar water heating system normally consists of installing only the solar loop. The swimming pool system primarily uses PVC schedule 40 pipe and fittings for the solar loop. The temperatures involved in this system will be in the 90s at the very high end. PVC schedule 40 is suitable for temperatures up to 140°F (100°C). Schedule 80 PVC is a thicker pipe and is used primarily in higher pressure commercial installations. **See Figure 10-29.**

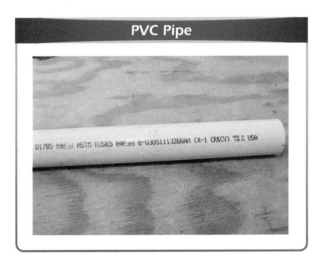

Figure 10-29. PVC pipe is commonly used for swimming pool solar water heating systems.

PVC should be cut with plastic pipe saws or cutters and joined using the proper method for joining PVC pipe for pressurized water lines. **See Figure 10-30.** PVC pipe and fittings must be protected from ultraviolet (UV) radiation. This can be accomplished by painting the pipe and fittings with an appropriate paint, as swimming pool piping is not normally insulated. For more information on PVC piping, the *PVC Piping Systems for Commercial and Industrial Applications, Design Guide,* published by the Plastic Pipe and Fittings Association, can be consulted. Care must be taken when joining PVC pipe and fittings with glue products. Gluing pipe and fittings should always be done in an open or ventilated area. Glue and primer can be harmful if inhaled.

Solar H₂Ot, Ltd.
Solar water heating components can be preassembled for connections at the installation site.

Piping and Fittings Insulation

Solar water heating piping and fittings must be insulated to achieve the efficiencies expected of solar radiant systems. Pipe and fittings must be completely encased in insulation and sealed to contain heat within the piping system. Any voids or spaces in the insulation will cause a steady loss of heat, which can lead to an inefficient system.

In order to gauge the efficiency of an insulated piping system, the insulation is rated by its R-value. The R-value is a measure of the ability of a material, such as insulation, to impede heat flow. The higher the R-value, the better the ability of the insulation to impede heat flow. R-values will change with the thickness of the insulation. The most commonly used insulation for solar water heating piping is fibrous glass and flexible tubing insulation. **See Figure 10-31.** The 1″ thick Micro-Lok® fiberglass insulation has an R-value of 4.3. The R-value of 1″ thick flexible tubing, such as AP/Armaflex®, is approximately 7.

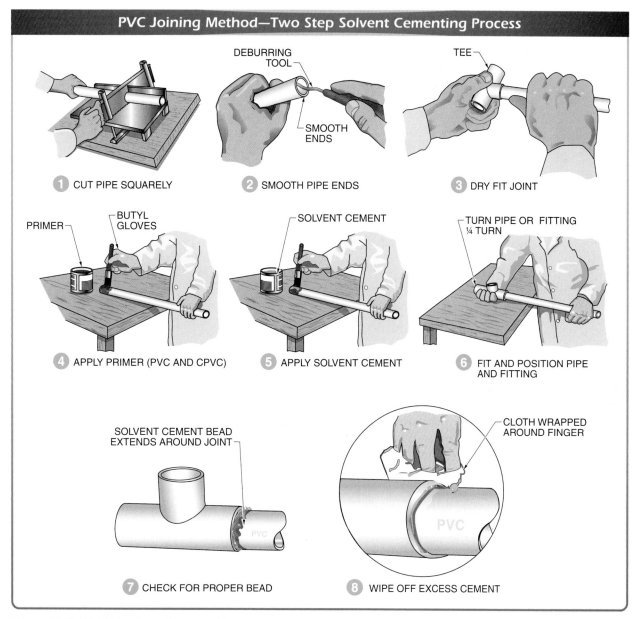

PVC Joining Method—Two Step Solvent Cementing Process

1 CUT PIPE SQUARELY

DEBURRING TOOL
SMOOTH ENDS
2 SMOOTH PIPE ENDS

TEE
3 DRY FIT JOINT

PRIMER
BUTYL GLOVES
4 APPLY PRIMER (PVC AND CPVC)

SOLVENT CEMENT
5 APPLY SOLVENT CEMENT

TURN PIPE OR FITTING ¼ TURN
6 FIT AND POSITION PIPE AND FITTING

SOLVENT CEMENT BEAD EXTENDS AROUND JOINT
PVC
7 CHECK FOR PROPER BEAD

CLOTH WRAPPED AROUND FINGER
PVC
8 WIPE OFF EXCESS CEMENT

Figure 10-30. PVC pipe is easily cut and joined with solvent cement.

Insulation is ordered by pipe size outside diameter (OD) and insulation thickness. For example, for ordering AP/Armaflex® for ¾″ copper tube one would order a ⅞″ diameter AP/Armaflex® tube to fit the outside diameter. The proper thickness of insulation and the installation of insulation are regulated by various codes and standards. For example, the *Uniform Solar Energy Code* by IAPMO states the following:

802.1 Required. Pipe and fittings, other than unions, flanges, or valves, shall be insulated. Insulation material shall be suitable for continuous operating temperatures of not less than 220°F (104.4°C). See Table 8.1. **See Figure 10-32.**

802.2 Fittings. Fittings shall be insulated with mitered sections, molded fittings, insulating cement, or flexible insulation.

802.3 Installation. Insulation shall be finished with a jacket or facing with the laps sealed with adhesives or staples so as to secure the insulation on the pipe. Insulation exposed to the weather shall be weatherproofed in accordance with standard practices acceptable to the Authority Having Jurisdiction. In lieu of jackets, molded insulation shall be permitted to be secured with sixteen (16) gauge galvanized wire ties not exceeding nine (9) inches (229 mm) on center.

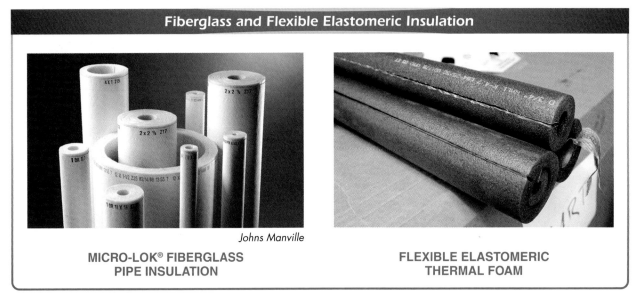

Fiberglass and Flexible Elastomeric Insulation

Johns Manville

MICRO-LOK® FIBERGLASS PIPE INSULATION

FLEXIBLE ELASTOMERIC THERMAL FOAM

Figure 10-31. Fiberglass and flexible elastomeric thermal foam are commonly used to insulate copper tubing.

The proper thickness of insulation, therefore, is determined by using Table 8-1 in the *Uniform Solar Energy Code*. This table is based on R-values from 4.0 to 4.6 and meets the R-value of Micro-Lok®. Using this table and Section 802.1, insulation for solar water heating should be sized at the 201°F to 250°F range. For pipe sized 1″ and less, 1½″ inch insulation should be used. This is also the recommendation of the manufacturer of Micro-Lok®.

In order to size insulation thickness for AP/Armaflex®, the second formula in Note 1 of Table 8 should be used. The R-values of 1″ thick AP/Armaflex® for pipe sizes 1″ and less average to be 7.0. **See Figure 10-33**.

Insulation installed outdoors must be protected from the weather. This may entail a jacketing around the insulation. Jacketing can be metal or PVC material that is placed around the insulation. **See Figure 10-34.** The manufacturer recommendations for jacketing of pipe insulation must be consulted and followed.

> **Facts**
>
> *Insulation on piping reduces heat loss in the fluid temperature by several degrees Fahrenheit when compared to uninsulated piping. Hotter HTF increases the heat transferred to the storage tank.*

Fluid Temperature Range in °F	Pipe Diameter in Inches				
	1 and Less	1.25 to 2	2.5 to 4	5 to 6	8 and Larger
	Insulation Thickness in Inches*				
306 to 460	2.5	2.5	3.0	3.5	3.5
251 to 305	2.0	2.5	2.5	3.0	3.0
201 to 250	1.5	1.5	2.0	2.0	2.0
105 to 200	0.5	1.0	1.5	1.5	1.5

USEC Table 8-1 Minimum Pipe Insulation

* Insulation thickness is based on materials having thermal resistance in the range of R = 4.0 to 4.6 per inch. For materials with thermal resistance less than R = 4.0 per inch, the minimum insulation thickness shall be determined as follows:

$$\frac{4.0 \times (\text{Table thickness})}{\text{Actual R}} = \text{New minimum thickness}$$

For materials with thermal resistance more than R = 4.6 per inch, the minimum insulation thickness shall be permitted to be reduced as follows:

$$\frac{4.6 \times (\text{Table thickness})}{\text{Actual R}} = \text{New minimum thickness}$$

IAPMO

Figure 10-32. Minimum recommended pipe insulation is based on the HTF temperature and the size of the pipe.

R-value for AP/Armaflex				
Pipe Insulation ID Size*	Wall Thickness of Insulation*			
	Nom.⅜*	Nom.½*	Nom.¾*	Nom.1*
⅜	2.9	3.4	5.7	7.4
½	2.7	3.3	5.5	7.2
⅝	2.5	3.3	5.5	7.1
¾	2.4	3.3	5.4	6.9
⅞	2.3	3.3	5.4	6.9
1⅛	2.2	3.2	5.3	7.2
1⅜	2.1	3.1	5.1	7.3
1⅝	2.4	3.1	4.9	7.2
1½ IPS	2.3	3.1	4.8	6.9
2⅛	2.3	3.1	4.7	6.7
2 IPS	2.2	3.1	4.6	6.6

* in in.

Figure 10-33. R-value depends on wall thickness of the insulation and the pipe size.

Metal Insulation Jacketing

Lochnivar, LLC.

Figure 10-34. Metal jacket is installed over insulation for protection.

Piping Supports

All piping must be supported at certain intervals to prevent sagging and deforming the fluid filled pipe. There are several different methods to support piping both horizontally and vertically. The methods used will vary with the type of structures or framing members used to mount or hang supports.

Much of the solar water heating piping will be run in attic spaces. It is possible to allow the piping, which is encased in insulation, to rest on the ceiling trusses. However, this will not allow for proper grading of piping. At best, the piping supported on trusses will be level

rather than graded to the storage system. In order to provide the proper ¼″ per foot grade for the solar loop piping, it should be supported above the roof trusses. The piping support or hanger should go around the piping insulation rather than trying to cut the insulation for the pipe to rest on the hanger. **See Figure 10-35.**

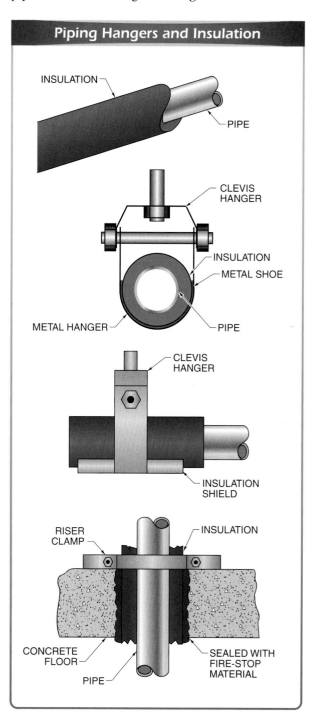

Piping Hangers and Insulation

Figure 10-35. Piping hangers are installed outside of the pipe insulation.

Reinsulating around the piping support is tedious work and can lead to voids in the insulation. The best solution is to use a lightweight hanger supported from an above truss or rafter. This can be done using a side-beam threaded rod hanger nailed or screwed to the side of the truss with a ⅜″ threaded rod carrying a clevis or swivel loop hanger. **See Figure 10-36.** These hangers can be hung before installing the piping and will help support the pipe during the joining process. This style of supports is adjustable so that grading the pipe can be easily accomplished.

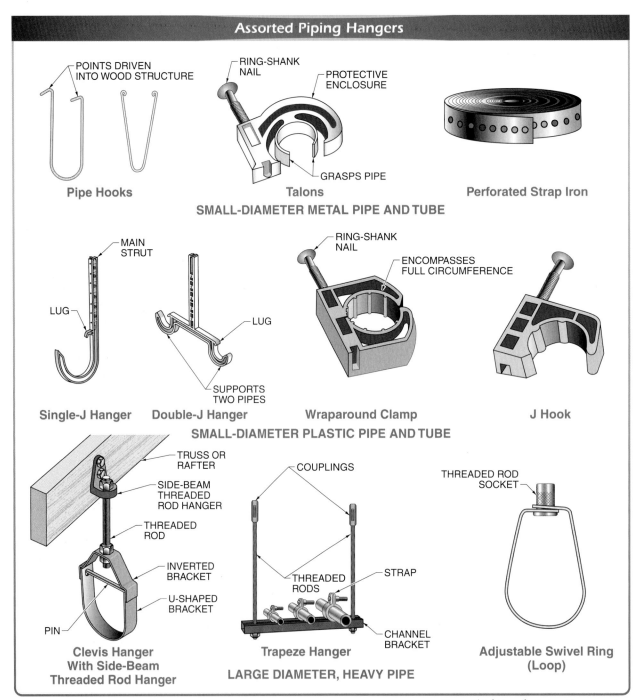

Assorted Piping Hangers

Pipe Hooks — POINTS DRIVEN INTO WOOD STRUCTURE

Talons — RING-SHANK NAIL, PROTECTIVE ENCLOSURE, GRASPS PIPE

Perforated Strap Iron

SMALL-DIAMETER METAL PIPE AND TUBE

Single-J Hanger — MAIN STRUT, LUG

Double-J Hanger — LUG, SUPPORTS TWO PIPES

Wraparound Clamp — RING-SHANK NAIL, ENCOMPASSES FULL CIRCUMFERENCE

J Hook

SMALL-DIAMETER PLASTIC PIPE AND TUBE

Clevis Hanger With Side-Beam Threaded Rod Hanger — TRUSS OR RAFTER, SIDE-BEAM THREADED ROD HANGER, THREADED ROD, INVERTED BRACKET, U-SHAPED BRACKET, PIN

Trapeze Hanger — COUPLINGS, THREADED RODS, STRAP, CHANNEL BRACKET

Adjustable Swivel Ring (Loop) — THREADED ROD SOCKET

LARGE DIAMETER, HEAVY PIPE

Sioux Chief Manufacturing Company, Inc.

Figure 10-36. Many types of hangers are available for supporting piping.

Surface wall horizontal and vertical piping support is normally accomplished using metal channels, such as Unistrut®. The metal channel is secured to the wall, and channel clamps are used to secure the piping. The clamps should go around the insulation, or the insulation must be cut and fitted around the clamp. **See Figure 10-37.** The clamps should not crush the insulation. These two support methods are just two of the many methods for supporting the piping. There are hundreds of different piping supports that can be used. Any method used, however, must meet local and national code standards.

Figure 10-37. Clamps connected to Unistrut® channels are used to support piping and equipment.

Metallic supports should never come in contact with copper tubing. This can cause galvanic action, erode the tubing, and lead to leaks. Copper tubing should be protected and isolated from any dissimilar metallic surface by heat-resistant tape or another isolating material before allowing it against a metallic support.

Piping supports should be spaced apart by the appropriate distance recommended to prevent sags in the fluid filled piping. The distances between supports are regulated by local and national codes. For example in the *Uniform Solar Energy Code,* these requirements are contained in Table 3-2, Hangers and Supports. Both horizontal and vertical support spacing are included in the table. **See Figure 10-38.** Copper tubing 1½″ and less are to be supported at 6′ intervals. Schedule 40 PVC should be supported at 4′ intervals. CSST is not included in this table and as of yet is not in any support schedule for fluid piping. Therefore, the manufacturer's recommendations for support of CSST must be consulted and followed.

PIPE AND FITTINGS INSTALLATION

The solar water heating piping system consists of piping in two possible circuits or loops. These circuits or loops are the primary and secondary circuits or loops. The primary circuit or loop is the solar loop. The solar loop consists of the piping to and from the solar collector, which connects to the solar storage system. The secondary circuit or loop consists of the DCW supply piping to the storage system and the DHW piping from the storage system. In direct solar water heating systems, there is no secondary loop. The DCW flows to the solar collector, then to the solar storage system, and finally to the demand fixtures. In combination systems, there can be a third or even a fourth loop supplying HTF to other hydronic systems. **See Figure 10-39.**

Solar Loop Piping

Installation of solar loop piping begins at the solar thermal collector. In most cases, the piping will be installed immediately after the collector is mounted and secured. This is because the roof access and safety equipment are already in place, and the tools necessary for roof penetrations will also be present and used for penetrating the roof for the piping and sensor wiring.

A solar water heating system with a single solar collector is a typical system design used in residential installations. This system design will entail connecting piping to the collector inlet and outlet, penetrating the roof, and then routing and installing the remaining piping to the storage system inlet and outlets. This piping will also contain the majority of the remaining system components to be installed, such as pumps, expansion tanks, drain valves, air vents, etc.

USEC Table 3-2 Hangers and Supports

Materials	Types of Joints	Horizontal	Vertical
Cast	Lead and oakum	5' (1524 mm), except 10' (3048 mm) where 10' lengths (3048 mm) are installed[*,†,‡]	Base and each floor a maximum of 15' (4572 mm)
	Compression gasket	Every other joint, unless over 4' (1219 mm), then support each joint[*,†,‡]	Base and each floor a maximum of 15' (4572 mm)
Cast iron hubless	Shield coupling	Every other joint, unless over 4' (1219 mm), then support each joint[*,†,‡,§]	Base and each floor a maximum of 15' (4572 mm)
Copper tube and pipe	Soldered or brazed	1½" (40 mm) and smaller, 6' (1829 mm), 2" (50 mm) and larger, 10' (3048 mm)[*,†,‡,§]	Each floor a maximum of 10' (3048 mm)[‖]
Steel and brass pipe for water or DWV	Threaded or welded	¾" (20 mm) and smaller, 10' (3048 mm), 1" (25 mm) and larger, 12' (3658 mm)	Every other floor, a maximum of 25' (7620 mm)[‖]
Steel, brass, and tinned copper pipe for gas	Threaded or welded	½" (15 mm), 6' (1829 mm) ¾" (20 mm) and 1" (25 mm), 8' (2438 mm), 1¼" (32 mm) and larger, 10' (3048 mm)	½" (15 mm), 6' (1829 mm) ¾" (20 mm) and 1" (25 mm), 8' (2438 mm), 1¼" (32 mm) every floor level
Schedule 40 PVC and ABS DWV	Solvent cemented	All sizes, 4' (1219 mm); allow for expansion every 30' (9144 mm)[‡,#]	Base and each floor; provide mid-story guides; provide for expansion every 30' (9144 mm)[#]
CPVC	Solvent cemented	1" (25 mm) and smaller, 3' (914 mm), 1¼" (32 mm) and larger, 4' (1219 mm)	Base and each floor; provide mid-story guides[#]
Lead	Wiped or burned	Continuous support	Maximum of 4' (1219 mm)
Copper	Mechanical	In accordance with standards acceptable to the Authority Having Jurisdiction	
Steel and brass	Mechanical	In accordance with standards acceptable to the Authority Having Jurisdiction	
PEX	Metal insert and metal compression	32" (813 mm)	Base and each floor; provide mid-story guides
PEX-AL-PEX	Metal insert and metal compression	½" (15 mm) ¾" (20 mm) 1" (25 mm) } All sizes 98" (2489 mm)	Base and each floor; provide mid-story guides
PEX-AL-PE	Metal insert and metal compression	½" (15 mm) ¾" (20 mm) 1" (25 mm) } All sizes 98" (2489 mm)	Base and each floor; provide mid-story guides
Polypropylene (PP)	Fusion weld (socket, butt, saddle, electrofusion), threaded (metal threads only), or mechanical	1" (25 mm) and smaller, 32" (813 mm); 1¼" (32 mm) and larger, 4' (1219 mm)	Base and each floor; provide mid-story guides

IAPMO

[*] Support adjacent to joint, a maximum of eighteen (18) inches (457 mm).
[†] Brace at a maximum of forty (40) feet (12,192 mm) intervals to prevent horizontal movement.
[‡] Support at each horizontal branch connection.
[§] Hangers shall not be placed on the coupling.
[‖] Vertical water lines shall be permitted to be supported in accordance with recognized engineering principals with regard to expansion and contraction, when first approved by the Authority Having Jurisdiction.
[#] See the appropriate IAPMO Installation Standard for expansion and other special requirements.

Figure 10-38. Maximum hanger and support spacing is based on the piping material, type of joint, and the orientation of the piping.

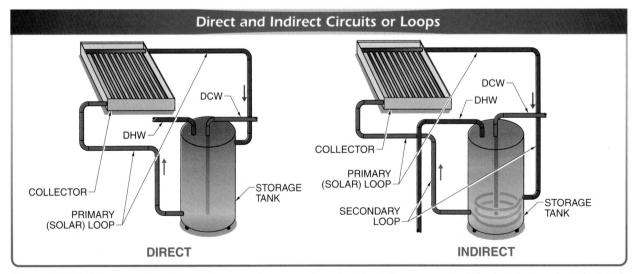

Figure 10-39. Solar water heating systems can be connected directly to the storage tank for heating pools or indirectly connected for DHW.

A solar water heating system with multiple solar collectors or collector arrays is typically used for larger buildings. In a multiple collector system, each collector must be piped together to obtain the highest amount of solar radiation as possible. The collector connecting piping must be configured properly for the solar water heating system to work efficiently. This configuration will depend on the type of solar collectors and the desired flow rate through the system. Multiple collector arrays can be connected together in parallel or in series. **See Figure 10-40.**

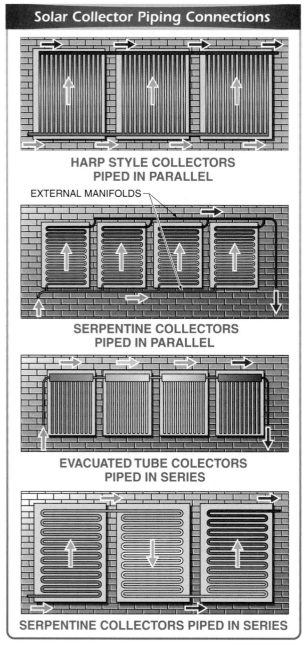

Figure 10-40. Solar collectors can be connected in parallel or series.

Parallel piping arrangements use a header piping technique to interconnect the collectors. HTF flows directly to each of the collector inlets through the header, providing for an even flow through the collectors. This can be accomplished by utilizing the collector's internal header if using the harp style collector, or the collectors can be connected using an external header and connecting each collector, such as a serpentine collector, separately from the header.

The parallel piping method is the most common method of interconnecting multiple solar collectors. The external header technique is almost exclusively the method used for piping multiple swimming pool collector arrays. This is due to the fact that this piping arrangement will present the least amount of head loss to the system, which will be critical in determining the size of pump to be used in an active system.

Series piping arrangements interconnect a solar collector by connecting the outlet of one collector to the inlet of the next collector. In this piping technique, HTF flows from one collector to the next until exiting the system through the last collector outlet. Evacuated tube collector arrays are easily piped this way due to the manifold system at the top of the collector.

The series piping method allows the HTF to reach the highest temperature possible. However, this method also presents the highest friction losses to the pump due to the HTF flowing through each collector. Friction loss through each collector will be combined to add to the friction loss through the system. This friction loss will determine the size of pump used in the system.

The solar collector manufacturer recommendations and instructions should be followed to determine the proper piping method to use. If external piping headers are used, the piping must be insulated, and the insulation must be protected from weathering. The header piping should be graded for complete drainage of the HTF from the collectors. The header piping should also be supported and secured to the roof.

Collector Connection

Each solar collector must be properly connected to the solar loop piping. Collector inlets and outlets may be supplied with copper tubing, copper thread adaptors, or other metal alloy thread adapters for piping connections. In each case, the proper attachment method should be used, and care must be taken to protect the collector's interior tubing and absorber from heat or torque damage. **See Figure 10-41.**

Solar Collector Piping Connectors

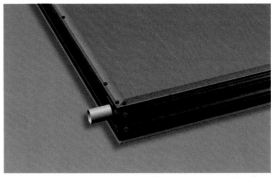

Alternate Energy Technology

**STRAIGHT COLLECTOR
TUBING END**

**THREADED ADAPTER
CONNECTORS**

**SOLDERED COPPER
COUPLINGS**

BRASS AND COPPER UNIONS

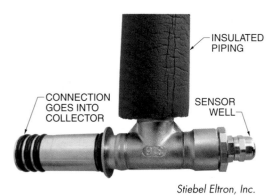

INSULATED
PIPING

CONNECTION
GOES INTO
COLLECTOR

SENSOR
WELL

Stiebel Eltron, Inc.

PROPRIETARY CONNECTORS

UNION NUT

THREADED
PIPE ADAPTER

ISOLATION
SLEEVE

BRASS
SWEAT
SOLDER
INSERT

Watts Water Technologies, Inc.

DIELECTRIC UNION

Figure 10-41. A large variety of solar collector piping connectors are available.

Collector copper tubing ends may be connected to system piping using copper couplings or copper/brass unions. Copper couplings or the union cup must be soldered to the tubing ends. Copper couplings will not allow for easy separation of the collector. However, copper/brass unions incorporate soldered copper joints and a threaded union for ease in separating the collector from system piping.

Collectors with copper or brass female thread adapter connections can be piped with copper male thread adapters soldered to system piping. Soldering should be done before inserting the male adapter into the female adapter. Proper thread sealant, either tape or paste, should be used to ensure a watertight installation. Care should be taken to not overtighten the adapter to the female end because this could damage the internal collector copper tubing.

Collectors with copper or brass threaded male ends can be connected with copper female adapters soldered to system piping. They can also be connected with brass unions. Proper thread sealant tape or paste should be used. Care should be taken to not overtighten the female adapter and the union nut because this could cause damage to the collector or system piping.

Collectors with aluminum, stainless steel, or other metallic ends should be connected to system copper tubing with dielectric unions. The dielectric union will keep dissimilar metals separated from each other and inhibit erosion of the copper tubing by galvanic action. Proper thread sealant tape or paste should be used. Care should be taken when tightening the female ends and the union nut to produce a secure and leak free joint without deforming the collector or system piping.

Components at the Solar Collector

Once the collector piping connection method is determined, the solar piping system components situated above the roof can be installed. **See Figure 10-42.** Depending on the system requirements, the solar piping system components above the roof may include the following:
- Sensor well—The sensor well must be inserted at the collector outlet to obtain the most accurate HTF temperature leaving the collector.
- Air vent—The air vent must be placed at the collector outlet in the uppermost portion of the piping system.
- Vacuum breaker or vacuum relief valve—The vacuum breaker or vacuum relief valve must be placed at the collector outlet adjacent to the air vent in the uppermost portion of the piping system.

- Pressure-relief valve—The pressure-relief valve should be installed in the outlet collector piping before it penetrates the roof, and the drain should be properly installed per code requirements.
- Freeze protection valve—The freeze protection valve should be installed in the outlet piping of the collector above the roof penetration. The drain should not flow over any roof penetration or cause a nuisance of any kind.
- Heat dump—The piping connection for the heat dump should go in the outlet piping of the collector in order to alleviate excess heat.
- Air separator—The air separator is usually installed in the collector inlet piping before the pump.
- Check Valve—The check valve is usually installed under the roof to prevent the fluid from flowing backward through the system.

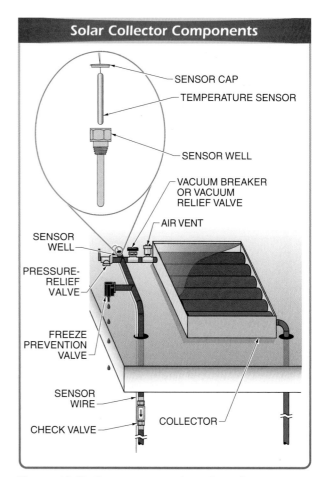

Figure 10-42. Components to the solar collector are attached after the collector is installed.

Roof Piping Penetration

Piping from the solar collector inlet and outlet will normally penetrate the roof close to the solar collector piping connections. When determining where these penetrations will be located, spacing for any system components must be accounted for. A few extra inches should be allowed to ensure there is enough room to install the needed components.

Piping penetrations through the roof are made in the same manner as collector support roof penetrations, except that the hole will be larger. The roof penetration should be large enough for the collector piping and insulation to snuggly fit through the hole. A hole saw normally is used to drill the hole. The hole must then be protected and waterproofed with flashing material and sealant. **See Figure 10-43.** The flashing material and sealant should be similar in style to those used in plumbing vent piping penetrations through the roof. The type and shape of flashing material used will depend upon the type of roofing material encountered.

In most cases, flashing material is placed under the roofing material and sealed using proper roofing sealant, such as the sealant used for collector mounting assembly penetrations. A critical element in waterproofing piping penetrations is obtaining a proper seal around the piping insulation. This can be accomplished with roof jacks with rubber booting or roof jacks using flashing caps or coolie caps.

The roof jack with rubber booting will slip over the piping insulation and present a watertight seal around the insulation. A flashing cap uses a larger piece of copper tubing, which is slipped over the collector copper tubing and is soldered in place just above the flashing. This creates a waterproof seal around the tube. However, it does not allow for full insulation of the piping within the flashing. A coolie cap flashing incorporating piping and sensor wiring protection can also be used allowing for one flashing for two penetrations. Flat roofs may require a pitch pocket installation. **See Figure 10-44.**

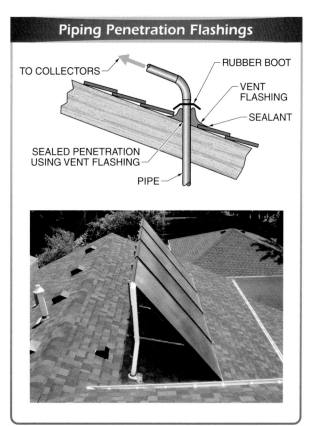

Solar Service Inc.

Figure 10-43. Flashing and sealant are installed around the piping penetrations.

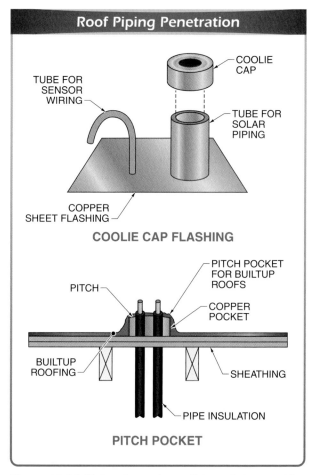

Figure 10-44. Flat roofs require a pitch pocket around the piping penetrations.

Remaining Solar Loop Piping

Once the solar piping roof penetrations are completed, the balance of the solar loop piping can be routed and installed. This piping should take the shortest route possible to the solar storage system. The route should also have as few turns as possible. This will not only use fewer fittings, but it will present less friction loss to the system pump. Each fitting that is used will add friction to the flow of HTF. The fewer fittings that are used in the piping system, the less friction loss there will be and the smaller the pump that will be needed.

Horizontal Grade. The horizontal piping from the solar collector to the storage system must be graded at ¼″ per foot. This will facilitate drainage of HTF not only in drain back or drain down systems but also for draining the HTF when necessary for maintenance or replacement.

Thermal Expansion of Piping Material. *Thermal expansion of piping material* is the expansion or gain in the length of piping when the system is heated. This occurs when there is a heated fluid flowing through a piping system. This gain is especially noticeable in long runs of piping. Relatively high-temperature HTF flowing in the piping will cause an even greater amount of expansion. If this gain in length is not accounted for, it can lead to pipe damage and eventual leakage.

There are several methods to relieve thermal expansion in piping systems. Specially designed fittings that will expand and contract internally with the rise and fall of fluid temperatures are sometimes installed. These expansion fittings work very well, however they can be expensive. One of the most common and inexpensive methods used to account for piping thermal expansion is to create an expansion loop.

An *expansion loop* is an arrangement of pipe and fittings that allows the thermal expansion gain to be absorbed within the loop rather than along the length of tubing or piping where it can cause pipe damage. **See Figure 10-45.** The expansion loop in copper tubing can be a short length of soft copper tube in the shape of a U inserted in a relatively long piping run. An arrangement of fittings and hard copper tube in a U shape utilizing four 90° fittings or two or three offsets in the piping can also be used to absorb thermal expansion. In each case, the gain in pipe length occurs along the run of tubing, while the U flexes with the expansion and contraction of the tubing run. In a solar loop, the expansion loop is normally installed in the horizontal piping run from the collector to the storage system. This run of piping is usually the longest run in the system and the area where thermal expansion will affect the system the most.

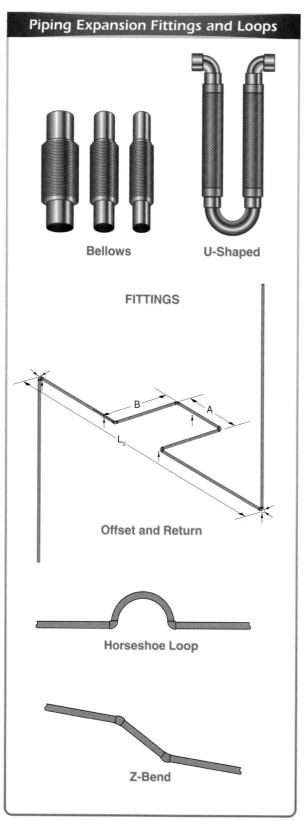

Figure 10-45. Expansion fittings or loops allow thermal expansion of the piping system.

Solar loop piping normally extends from the solar collector through attic areas and enters into the solar storage area from the ceiling and as close as possible to the storage system. It is then piped and connected to the storage system. System components are added to system piping where necessary. This can be accomplished using individual components or using a pump station.

Installing Individual Components

In an individual component system, the solar collector outlet piping or solar hot water supply is connected to the solar storage inlet piping. The solar collector inlet piping or solar cool return is connected to the solar storage outlet piping. In direct solar water heating systems, the storage inlet connection will be the water heater/solar storage DCW inlet, and the storage outlet connection will be the water heater/solar storage DHW outlet.

Individual system components are installed either as the piping is being run or cut in after piping is completed. The method used will depend on the preference of the installer and the type of connection used when installing the component. Some components, such as heat exchangers and pumps, may have flanges that are easier to install after piping is completed. Some components are soldered or threaded and will be easier to install while system piping is being run. **See Figure 10-46.**

In passive direct solar water heating systems, there may be no additional components besides two isolation valves at the water heater/solar storage inlet and outlet. **See Figure 10-47.** However, most passive direct systems use the solar system as a preheater for DHW. In this system, an additional three-way valve is installed at the water heater inlet to provide for isolation of the solar collector and still allow flow of DCW into the tank. Additionally, boiler drain valves may be added to facilitate draining down the system. The water heater should be supplied with a T&P relief valve per code requirements. The water heater outlet (DHW outlet) should have a tempering valve installed and connected to the DCW to eliminate scalding temperatures in DHW supplied to demand fixtures.

In active direct solar water heating systems, a circulator or pump is added to the collector inlet piping. **See Figure 10-48.** The pump then pushes the DHW through the solar collector and back to the water heater/solar storage inlet. A check valve should be placed in the solar collector/water heater inlet piping above the isolation valve and the three-way valve. This will eliminate reverse thermosiphoning.

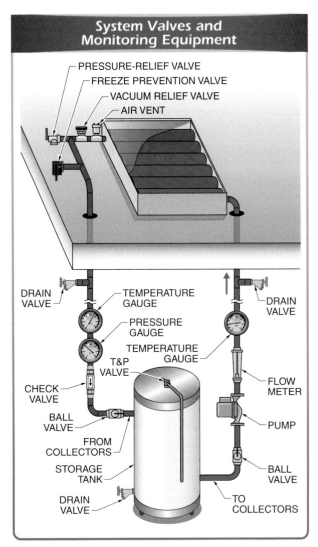

Figure 10-46. Individual components are installed with the piping.

A solar water heating installation company should be well-stocked with various system components.

Passive Direct System Components

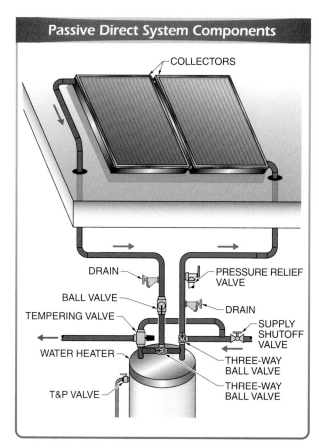

Figure 10-47. Passive direct solar water heating systems have fewer components.

Active Direct System Components

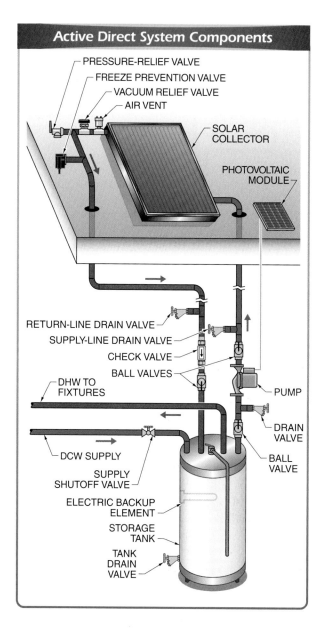

Figure 10-48. Active direct solar water heating systems use a pump to circulate the water.

In indirect solar water heating systems, the solar collector piping will be connected to a heat exchanger, which is in turn connected to the solar storage system. The heat exchanger can be either integrally immersed within the solar storage tank or an externally installed heat exchanger. In either case, the solar collector outlet hot water supply is connected to the heat exchanger inlet, and the solar collector inlet cool return is connected to the heat exchanger outlet. When an integrally contained heat exchanger is used, the solar loop piping is completed at the heat exchanger connection. **See Figure 10-49.**

The installation of an external heat exchanger will include mounting or supporting the heat exchanger and additional piping to the solar storage inlet and outlet. In some cases, such as with the plate heat exchanger, the piping may be able to support the heat exchanger. If the filled heat exchanger is too heavy to be supported by piping only, supports should be provided to carry the weight.

The external heat exchanger is normally piped in a counterflow method. **See Figure 10-50.** In the counterflow method, the highest temperature HTF flows in the opposite direction through the heat exchanger of the cooler storage system HTF. This allows the cooler DHW to flow against the highest temperature HTF creating a more efficient heat transfer from hot HTF to cool DHW. However, the heat exchanger manufacturer's installation instructions must always be followed.

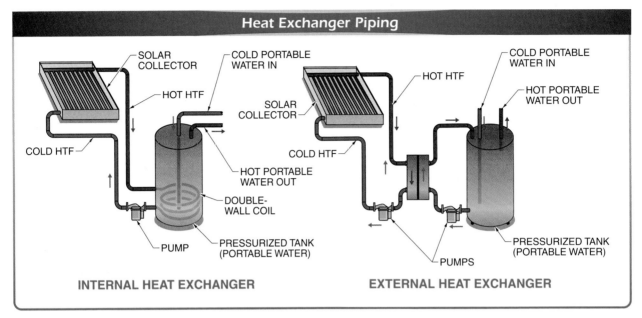

Figure 10-49. Indirect solar water heating systems use a heat exchanger to heat the DHW.

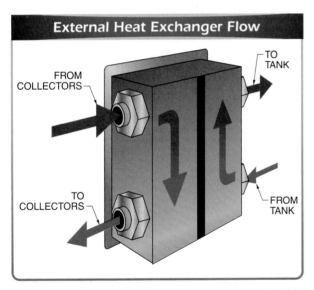

Figure 10-50. External heat exchangers are piped in a counterflow method to increase the amount of heat exchanged.

The heat exchanger must also be the correct size for the system to operate properly and efficiently. The heat exchanger should be sized for the highest Btu hourly rating of the solar collector or collector array. For example, if the solar collector or collector array is capable

of 100,000 Btu/hr, the heat exchanger should be capable of transferring 100,000 Btu/hr. If this is not the case, excess heat will build up in the solar collectors, which wastes solar radiation. It is also better to have too much heat exchanger capacity than not enough capacity. When a total system is ordered from a manufacturer, the configuration of the system, including the heat exchanger size, is calculated by the manufacturer. If the system is an OG-300 certified system, the OG-300 specifications should be followed.

A passive indirect solar water heating system will normally be a nonseparable thermosiphon system mounted on the roof. Once the heat exchanger piping and system components are installed at the collector, the system is ready for testing and control wiring.

An active indirect solar water heating system requires that a circulator or pump be installed in the collector inlet piping along with an expansion tank and a pressure-relief valve. **See Figure 10-51.** The pump should be placed in the collector inlet piping and configured so that maintenance or replacement will be relatively easy. Isolation valves should be placed before and after the pump so the pump can be removed without loss of HTF. Some pumps are lightweight and can be supported by system piping. However, large systems have large pumps and the pumps should be supported by hangers rather than system piping.

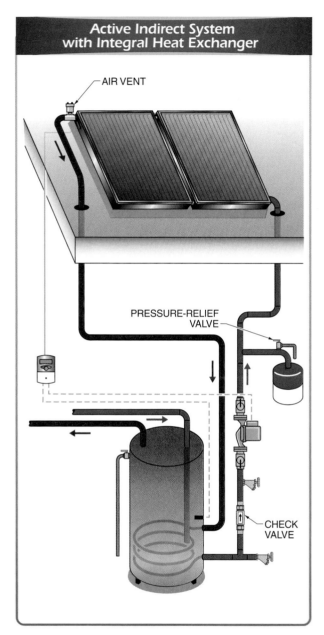

Active Indirect System with Integral Heat Exchanger

AIR VENT

PRESSURE-RELIEF VALVE

CHECK VALVE

Figure 10-51. Active indirect solar water heating systems use a pump and require an expansion tank and pressure-relief valve.

Boiler drain valves should be installed at the lowest portion of the collector inlet piping from the heat exchanger. This will facilitate drainage of HTF when needed. In some cases, a check valve may be installed upstream of the pump so that HTF cannot flow backwards into the heat exchanger when the pump shuts off. This will also eliminate reverse thermosiphoning. An additional boiler drain can be

installed downstream of the check valve to drain the downstream piping if needed.

The solar loop of the active indirect system is a closed system. Because the loop contains high temperature HTF, thermal expansion of the HTF will occur. The thermal expansion must be relieved by the installation of an expansion tank. The expansion tank should be installed downstream of the system pump. No valves should be installed before the expansion tank. The tank will contain HTF and should be supported per the manufacturer's recommendations. **See Figure 10-52.**

The expansion tank is sized by the total volume of HTF in the system and by the amount of thermal expansion expected within the system. If steam back is expected, the expansion tank should be sized to accept that volume of HTF. If a total package system is used, the manufacturer will have calculated the correct size of the expansion tank. If a customized system is installed, it is recommended that the expansion tank manufacturer complete the sizing calculations. The installation of an incorrectly sized expansion tank could cause severe damage to the solar system and the customer's property.

Another safety component that must be installed within the closed system solar loop is a pressure-relief valve (PRV). The PRV should be installed in a convenient area in the collector inlet piping. The PRV can be installed in the short line, which is teed off the collector inlet piping for the expansion tank. The PRV should have a pressure-relief setting that is below the other system component's maximum pressure rating. This setting is normally 125 psi.

Other components that are installed in the solar loop piping include temperature and pressure gauges. Temperature gauges should be installed on the collector outlet piping close to the storage tank or heat exchanger, on the collector outlet piping after the pump, and at the top of the storage tank at the DHW outlet. These gauges will allow monitoring of the system to ensure it is operating properly. A pressure gauge should be installed in the solar loop to monitor system pressure. It should be installed close to any flow control valve for easy viewing when adjustments to the system flow are needed.

System components should only be installed in areas that are readily accessible. The home owner or maintenance repairman should be able to reach all system components, especially valves, without having to remove panels or doors. This will create a convenient maintenance friendly installation.

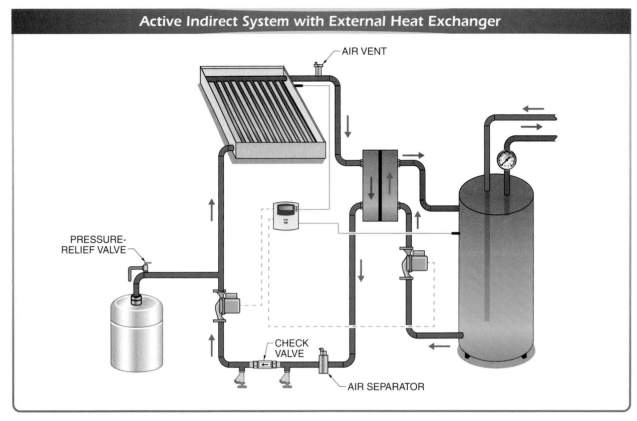

Active Indirect System with External Heat Exchanger

AIR VENT

PRESSURE-RELIEF VALVE

CHECK VALVE

AIR SEPARATOR

Figure 10-52. Active indirect solar water heating systems are closed loop systems with high temperature HTF.

Installing Pump Stations

A pump station places most of the solar water heating system components in a compact control package. Most pump stations contain all of the components, including the pump, gauges, valves, and expansion tank connection. **See Figure 10-53.** In many cases, all that remains in piping the solar loop is connecting piping to and from the pump station.

In the pump station, the solar collector outlet piping is connected to the supply inlet port on the station. The supply outlet port is then connected to the heat exchanger inlet. The heat exchanger outlet is connected to the collector inlet or return inlet port on the station. The station return outlet port is then connected to the solar collector inlet piping. The expansion tank is then mounted and connected to the pump station, and the solar loop piping is complete and ready for testing.

The pump station control component package provides a compact and complete system that does not need a lot of space to install the solar water heating piping. **See Figure 10-54.** This system also saves

time and effort in installation. However, these package systems are only designed for moderate-sized systems. Large solar water heating systems require an individual component system.

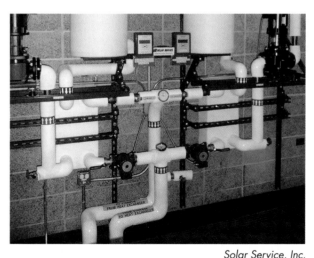

Solar Service, Inc.

Multiple pumping and control systems may be required on larger residential or commercial solar water heating systems.

Pump Stations

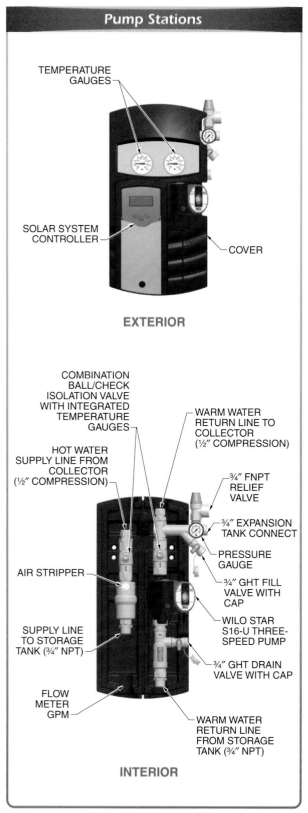

TEMPERATURE GAUGES

SOLAR SYSTEM CONTROLLER

COVER

EXTERIOR

COMBINATION BALL/CHECK ISOLATION VALVE WITH INTEGRATED TEMPERATURE GAUGES

HOT WATER SUPPLY LINE FROM COLLECTOR (½" COMPRESSION)

AIR STRIPPER

SUPPLY LINE TO STORAGE TANK (¾" NPT)

FLOW METER GPM

WARM WATER RETURN LINE TO COLLECTOR (½" COMPRESSION)

¾" FNPT RELIEF VALVE

¾" EXPANSION TANK CONNECT

PRESSURE GAUGE

¾" GHT FILL VALVE WITH CAP

WILO STAR S16-U THREE-SPEED PUMP

¾" GHT DRAIN VALVE WITH CAP

WARM WATER RETURN LINE FROM STORAGE TANK (¾" NPT)

INTERIOR

A. O. Smith

Figure 10-53. Pump stations are compact control packages.

Pump Station Control Component Packages

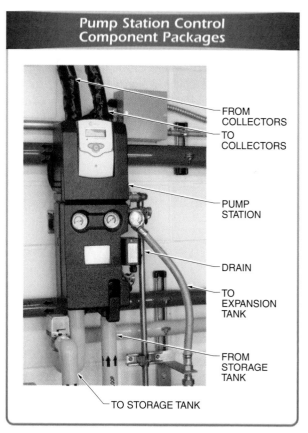

FROM COLLECTORS

TO COLLECTORS

PUMP STATION

DRAIN

TO EXPANSION TANK

FROM STORAGE TANK

TO STORAGE TANK

Figure 10-54. Pump stations are connected to the system piping to control the flow of HTF.

Installing Drainback Tanks

The drainback system, which is normally an active indirect system, will require that the additional components of the drain back tank and flow meter are installed in the solar loop piping. The drainback system can function properly in the individual component configuration or the pump station system. **See Figure 10-55.**

The drainback tank is normally installed in the collector outlet piping, while the flow meter is installed in the collector inlet piping. These two components must be installed at the same level so that the flow meter is mounted so that the level of HTF is visible when the system pump is off and the collectors have drained back to the tank.

Component and Tank Port Piping Connections

Each component and tank piping connection must be made with the proper fittings. System components, such as valves, PRVs, and flow meters, are made with brass fitting ends that are either solder cups or male or female threads. If copper tubing is used for the solar loop piping, copper fitting adapters can be used to connect the component, or copper tubing can be soldered to the brass fitting cup. If CSST is used, the components should be ordered with female thread connections. The brass male adapter ends of the CSST fitting can be used to thread into the brass female ends of the components.

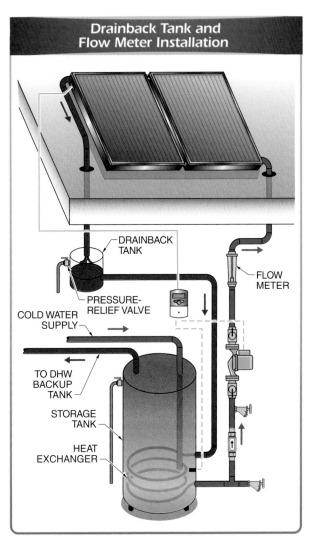

Figure 10-55. Drainback tanks are installed at the same level as the flow meter so that the flow meter can be used to determine the HTF level in the drainback tank when the system is not operating.

Water heater/storage tank port connections to copper tubing should be made with dielectric unions. This may entail adding a male nipple to a female port. Brass nipples can be used for this installation rather than galvanized or steel nipples. This will help to inhibit galvanic action. The dielectric union should still be used with the brass nipples but must be used when connecting to galvanized or steel nipples. Proper thread sealant tape or paste should be used to ensure a water tight installation. Care should be taken to not over-tighten the nipples or the union nut because this could damage the port or nipple threads and lead to leaking.

DCW and DHW Piping Connections

The secondary loop of a solar water heating system contains the fluid that is being heated by the solar heat exchanger. This is normally the DCW. The DCW is contained within the water heater or solar storage tank, is heated through the heat exchanger system, and exits the water heater storage tank as DHW. This DHW is also potable water, and piping must meet the requirements of the local plumbing code. Two of the most important code considerations for this secondary loop are the quality of piping materials and tempering of the DHW to demand fixtures and appliances.

The DCW and DHW piping connections and additions made by the solar water heating installer must be made with materials suitable for potable water. The piping materials should not be mixed. For example, if the existing DCW and DHW is piped in copper tubing, then copper tubing should be used for all connections and piping. Only new piping materials should be used for the DCW and DHW piping.

The DHW that will be leaving the water heater/solar storage tank may be at a very high temperature. Most water heaters are set to 120°F (49°C) outlet temperature. Solar systems have the ability to heat the DHW to much higher temperatures. For this reason, and in case other system temperature limiting devices fail, a tempering valve must be installed at the DHW outlet of the storage system. This tempering valve will be connected to the DCW so that DCW and DHW can be mixed to bring the outgoing DHW from the tempering valve to the safe and desired temperature. **See Figure 10-56.**

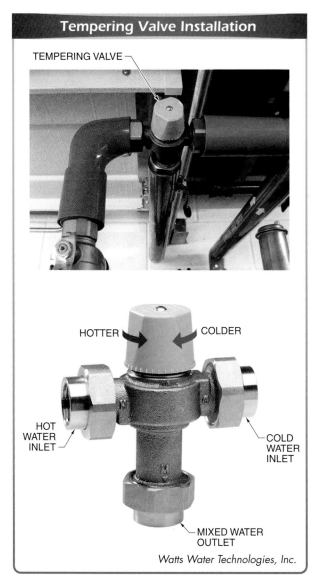

Figure 10-56. Tempering valves are used to control the discharge temperature used as DHW.

Installing PRV, T&P Valve, and Heat Exchanger Drains

All PRV and T&P relief valve drains should be piped per the local plumbing code requirements. For example, the 2012 *Uniform Plumbing Code* states the following:

608.5 Drains. Relief valves located inside a building shall be provided with a drain, not smaller than the relief valve outlet, of galvanized steel, hard-drawn copper piping and fittings, CPVC, PP, or listed relief valve drain tube with fittings that will not reduce the internal bore of the pipe or tubing (straight lengths as opposed to coils) and shall extend from the valve to the outside of the building, with the end of the pipe not more than 2 feet (610 mm) nor less than 6 inches (152 mm) aboveground or the flood level of the area receiving the discharge and pointing downward.

Such drains shall be permitted to terminate at other approved locations. Relief valve drains shall not terminate in a building's crawl space. No part of such drain pipe shall be trapped or subject to freezing. The terminal end of the drain pipe shall not be threaded.

Heat Exchanger Drains. Double-walled heat exchangers will also have a drain. This drain connects to the interior space between the two walls. This space is provided so that if a leak occurs in the interior wall of the exchanger, the HTF will leak into the space and drain out of the heat exchanger. This drain should be piped to an area that is open and visible rather than to the outside or a sink or floor drain. The HTF leaking from the heat exchanger drain is an indication that the heat exchanger has failed. The heat exchanger should be removed and replaced as soon as possible. Once all of the system piping connections are made and tightened and all drains are run, the system is ready for control wiring and pressure testing.

Chapter 10 Review and Resources

Installing and Wiring Operational Control Systems

Unless the system is a very simple passive direct system, the temperatures of the heat transfer fluid must be monitored at various points within the solar water heating system. If the system is an active solar water heating system, a controller is needed to energize and deenergize the pump when necessary. There may also be a need to monitor the efficiency of the system and communicate this information in real time. An operational control system can be used to accomplish one or all of these tasks. Once the solar water heating system piping is installed, the operational controls can be installed and wired.

OPERATIONAL CONTROLLERS

An *operational controller* is an electronic device used to monitor and control the heat transfer fluid (HTF) in a solar water heating system. Operational controllers are needed for active solar water heating systems that require pumps and valves to be activated or HTF temperatures to be monitored.

The operational controllers of active systems are required to energize and deenergize circulators or pumps. More sophisticated systems require that temperature sensors be installed within the system to relay HTF temperatures to the controls. **See Figure 11-1.** Active systems may also include individual solar water heating components that require monitoring or activation, such as motorized valves, solenoids, and relay switches. Therefore, a separate control module is needed to operate the system. In some active systems, a packaged component system or pump station system incorporates a controller into a single module that contains a pump, valves, gauges, and the controller.

Passive solar water heating systems do not require pump activation and may not require temperature monitoring. However, the passive systems that require temperature monitoring also require some type of control or monitoring device.

There are three basic types of operational controllers used in the solar water heating industry. These are timer

controllers, photovoltaic controllers, and differential controllers. The type of control device that should be used for a specific system depends on the type and sophistication of the solar water heating system and building owner needs.

Proper backing, such as plywood, should be in place to provide a secure mounting surface for an operational controller or other system components.

Active System Control Systems

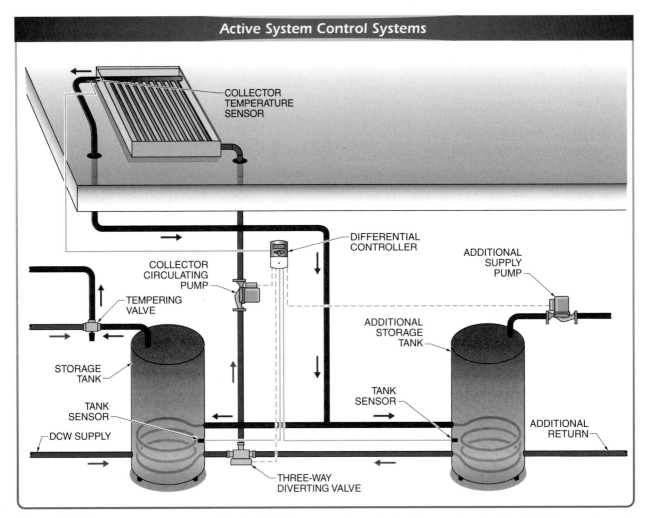

Figure 11-1. Temperature sensors must be installed in some solar water heating systems to relay HTF temperatures to the controls for the operation of individual components like pumps or valves.

Timer Controllers

A *timer controller* is a basic operational controller that switches an electrically powered device on and off based on the time of day. The timer controller is normally set to energize the pump of a solar water heating system during the highest solar insolation times of 9:00 AM to 4:00 PM each day. Time is the only parameter on which the controller is based. Therefore, a timer controller is used only for solar water heating systems located in mild temperature locations where freezing does not occur.

The type of solar water heating system that normally incorporates a timer controller is a retrofitted system. A retrofitted system may use a solar conversion valve. A *solar conversion valve (SCV)* is a specialized control valve that allows existing hot water tanks to be retrofitted into solar storage tanks. **See Figure 11-2.**

The SCV is installed into the bottom inlet, or drain port, of the tank and allows solar supply and return flow. It also retains the original function of the port ("cold in" or drain). The copper return pipe ensures delivery of hot water at a higher level in the tank than the "cold out," thus preventing disruption of the thermal stratification layers within the tank. This also ensures that the coolest water is always delivered to the solar collector for heating, thus maximizing solar harvesting efficiency. **See Figure 11-3.**

The pump used in this system is usually a small pump of $\frac{1}{100}$ HP to $\frac{1}{250}$ HP. This allows sufficient flow throughout the system without disturbing the temperature stratification within the tank. The only wiring required for this controller is from the timer controller to the pump. The timer controller need only be wired to the pump.

Solar Conversion Valves (SCVs)

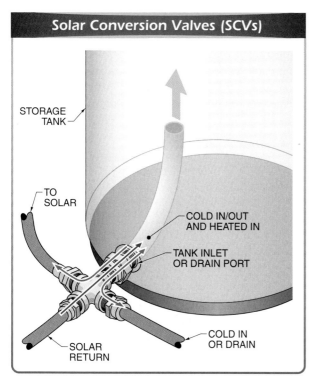

Figure 11-2. An SCV is a specialized control valve that allows existing hot water tanks to be retrofitted into solar storage tanks.

Timer-Controlled Retrofit Systems with SCVs

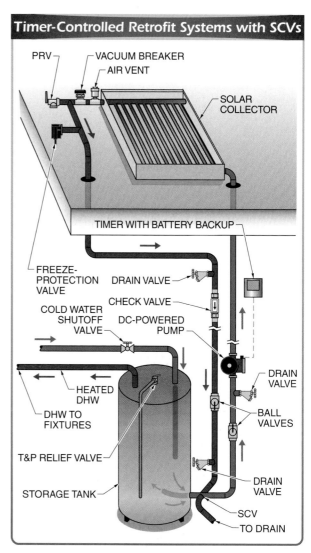

Figure 11-3. The SCV ensures the delivery of hot water at a higher level in the tank than the cold out. The coolest water is always delivered to the solar collector for heating.

There are some disadvantages of the timer controller. A major disadvantage of the timer controller is that it will activate the pump when the system does not need heating or when there is not adequate insolation. For example, the pump will continue to run when the tank reaches its maximum temperature. The timer will also continue to energize the pump during a low insolation day. Energy is wasted when the pump runs unnecessarily or HTF circulates through a collector that is cooler than the stored HTF. Another disadvantage is that the timer controller requires a back-up battery for continued operation during a power failure.

Photovoltaic Controllers

A *photovoltaic (PV) controller* is an operational controller that uses energy generated in the form of direct current electricity via a PV module or panel to energize the pump of a solar water heating system. The PV module generates electricity only when insolation is available for solar water heating. The PV module and pump are usually packaged together so that the panel size will match the energy needed to run a specific sized pump.

In order to gain even more system efficiency, the PV controller can be used in conjunction with a differential controller. This allows the pump to be energized by the PV module while the system is controlled by the temperature monitoring differential controller. **See Figure 11-4.**

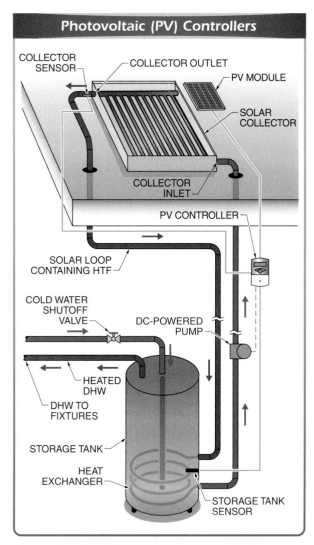

Figure 11-4. Many PV controllers contain a simple temperature differential relay that energizes the pump when the collector temperature is higher than the storage tank temperature.

The pump used with the PV controller is a direct current (DC) variable-speed pump. The PV controller and pump are designed so that the pump speed will be proportional to the amount of energy generated from the PV module. For example, when there is high insolation, the PV module will generate higher amounts of energy. Therefore, the pump will run at a faster rate and send HTF through the solar collector at a higher volume. This allows for higher efficiency in transferring heat to the HTF. In some cases, the PV controller may be used with a linear current booster, which will increase the energy output at low insolation times and allow the pump to initially run at a faster rate.

Although the PV-controlled system can be very efficient, the drawback to this system is that the pump can be energized when the tank has reached its maximum temperatures. Because of this, some installers install a manual or timer-controlled switch at the pump to shut off the pump if necessary.

Differential Controllers

A *differential controller,* also known as a differential temperature controller, is an operational controller used to energize an electrically powered component based on a setpoint difference between two temperature values. **See Figure 11-5.** Unlike the timer and PV controllers, the differential controller operates a solar water heating system based on temperatures within the system rather than external conditions, such as time, amount of insolation, or temperature.

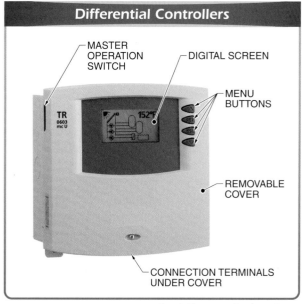

Bosch Thermotechnology Corp.

Figure 11-5. A differential controller operates a solar water heating system based on temperatures within the system rather than external conditions, such as time, amount of insolation, or temperature.

The differential controller has several advantages, which is why it is the most widely used control device in the solar water heating industry today. When properly installed and set, the differential controller is the most efficient operational control device and provides the best results possible from the solar water heating system. It is also the most versatile of controllers. For example, it can control a simple retrofitted single-collector domestic hot water (DHW) system or a multiple-collector/multiple-storage-tank system with real-time computer monitoring capabilities.

Several manufacturers in North America produce and supply differential controllers. Often solar product retailers license and rebrand these controllers and place their logos on the product. The solar product retailers may also install proprietary software functions within the controller. Each differential controller manufacturer offers several controller models with varying functions, such as storage tank charging (heating the tank) and shutdown, temperature monitoring, freeze protection, a holiday function, and an evacuated tube collector function. Regardless of the type or brand of controller used, manufacturer installation instructions and recommendations must be followed precisely.

Differential Controller Functions

The main function of the differential controller is to energize or deenergize the system pump or circulator when a temperature differential setpoint is reached by monitoring sensors installed in the system. The differential controller continuously monitors the temperatures at the two primary temperature sensors within the solar loop. It compares those temperature values to parameters, or differential setpoints, programmed into the controller. The differential setpoint for an individual function could be the difference between the temperatures at the two primary temperature sensors. It could also be the difference between the temperature at one of the sensors and a temperature value programmed as one of the controller parameters.

The primary temperature sensors in the solar water heating system are installed at the solar collector outlet (T1) and the solar storage tank/water heater (T2) in the lower portion of the tank. **See Figure 11-6.** T1 monitors HTF temperature at the point where the highest temperatures will be reached in the system as the result of solar radiation capture. This usually occurs at the top of the collector outlet. T2 monitors HTF temperature at the point where the coolest temperatures will be reached within the solar storage/water heater tank. *Note:* Depending on the manufacturer, the temperature sensor at the solar collector outlet may also be known as S1. The temperature sensor at the storage tank/water heater may also be known as S2.

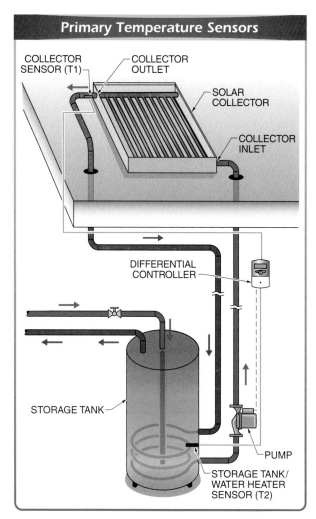

Figure 11-6. The primary temperature sensors of a solar water heating system are installed at the solar collector outlet (T1) and the storage tank/water heater (T2) in the lower portion of the tank.

The differential controller compares the temperature values at T1 and T2 to the parameters programmed into the various functions of the controller. Then, it either closes or opens a relay switch connected to the system pump. A power source is connected to one side of the relay switch, and the other side of the switch is wired to the pump. When the relay switch closes, the pump is energized. When the relay switch opens, the pump is deenergized.

Functions can be programmed by using the digital screen interface of the controller. The controller interface varies by manufacturer. Most controllers allow the installer to scroll through the function menu and turn on or off each function. In many instances, the installer can also adjust the function setpoint temperature value.

For example, on the Buderus TR 0301 U controller, the digital screen will display the operational condition of the solar water heating system including temperature values. **See Figure 11-7.**

Menu buttons can be used to scroll through the functions or access the status of function values if they are not shown on the default screen. The typical entry-level differential controller may perform the following operational and safety functions:

- operation switching
- automatic storage tank charging
- storage tank shutdown
- collector temperature monitoring, such as for evacuated tube collectors
- storage tank temperature monitoring (holiday function)
- freeze protection

Master Operation Switches. Most controllers are equipped with a master operation switch. The master operation switch can be used to turn the controller to one of three modes of operation: on, off, or auto. Switching the controller to the OFF position shuts down the controller and all functions. Switching the controller to the ON position activates the primary controller functions but does not activate other functions, such as the holiday function. Switching the controller to auto allows the controller to recognize and apply all functions that are programmed and activated and allows the system to be automatically operated by the controller.

Automatic Storage Tank Charging. During automatic storage tank charging, the differential controller constantly compares the temperature values between the collector outlet sensor (T1) and the storage tank sensor (T2). The controller then energizes or deenergizes the system pump when the set temperature values are reached. When the temperature at T1 is 16°F higher than the temperature at T2, the pump energizes. The pump will run, heating or charging the tank, if the safety functions of the controller do not prohibit it from energizing. The differential value of 16°F can be either a constant fixed value or a programmable value, depending on the complexity of the controller. The digital display on the Buderus controller will show the sun and the pump symbol rotating when the system is running.

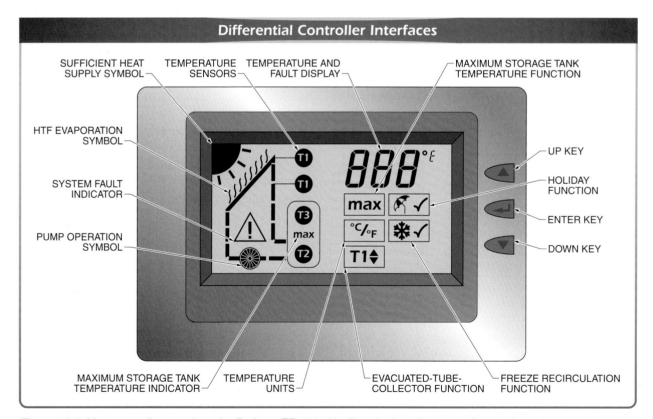

Figure 11-7. Most controllers, such as the Buderus TR 0301 U, allow the installer to scroll through the function menu and turn on or off each function using the digital screen interface of the controller.

Solar Service Inc.
Commercial solar water heating systems may require the use of multiple controllers.

The pump will continue to run, send HTF through the collector, and charge the storage tank until the controller senses a temperature differential between T1 and T2 of 8°F. At this point, the controller will deenergize the pump and stop the flow of HTF. The pump symbol will stop rotating.

Maximum Temperature Storage Tank Shutdown. The storage tank shutdown function deenergizes the pump when the storage tank reaches a maximum storage, or charging, temperature. Some controllers have an adjustable temperature value for this function, while other controllers have a factory-set fixed temperature value. For example, the Buderus controller has a fixed setpoint of 140°F. When this temperature is reached, the pump stops, the pump symbol stops rotating, the sun symbol is visible, and the MAX symbol appears in the storage tank symbol. When the controller senses a differential of 6°F below the maximum set temperature, in this case 134°F, the pump energizes again, if not prevented from doing so by the other safety functions. When the pump is energized, the pump symbol rotates and the MAX symbol is not present.

Collector Temperature Monitoring. The collector temperature monitoring function is a safety function. It prevents the pump from energizing if the temperature at the collector outlet sensor (T1) reaches a maximum set temperature. When there is high insolation and low DHW usage, the temperature in the collector can reach above 270°F. This will lead to steamback conditions. Therefore, the pump should not be energized until lower temperatures in the collector are reached. For example, the Buderus TR 0301 U controller has a maximum temperature value at T1 of 266°F. The controller will prevent the pump from energizing once this temperature is reached. When

the temperature at T1 falls below 261°F, the pump may be energized if not prevented by other safety functions.

Evacuated-Tube-Collector Function. The evacuated-tube-collector function provides an accurate method of sensing temperatures at the T1 sensor of an evacuated tube collector. Most evacuated tube collectors have a heat exchanger incorporated into the manifold at the top of the collector. The temperature sensed at the outlet of the collector does not necessarily reflect the accurate outlet HTF temperature if there is no flow in the system. Therefore, the function allows the pump to be energized for a short period of time at regular intervals to create flow. For example, a Buderus TR 0301 U controller using the evacuated-tube-collector function energizes the pump for 30 sec every 30 min. Therefore, the temperature of the HTF as it flows through the heat exchanger is more accurately reflected. When this function is selected, the T1 symbol will be displayed below the normal T1 symbol.

Holiday Function. The holiday function is a safety function that allows the storage tank to be cooled if the HTF temperature within the tank becomes warmer than the HTF temperature in the collector. This condition can occur if DHW is not used and there is high insolation. For example, when a Buderus TR 0301 U controller senses temperature at the T2 sensor that is 16°F warmer than at the T1 sensor and the temperature is 20°F below the maximum storage tank temperature of 140°F, the pump will energize. This will occur as long as other safety functions do not prevent the pump from energizing. This function allows the flow of cooler HTF within the collector to cool the storage tank. This process will continue until the temperature at T2 falls to 95°F or the temperature differential between T1 and T2 falls to 8°F. When the holiday function is activated, a small umbrella symbol will appear in the display.

Freeze Protection. The freeze protection function is a safety function that energizes the pump when the temperature at T1 reaches 41°F. This protects the HTF within the collector from freezing by allowing it to flow. Once the temperature at T1 reaches 45°F, the pump deenergizes. When this function is activated, a snowflake symbol will appear in the display.

The freeze protection function is not fail-safe. The controller will only energize the pump if there is power to the controller. During a power failure, the controller will not function unless it has a battery backup. Therefore, if long periods of cold weather are expected, the HTF should be an antifreeze solution mixed to the proper concentration for the expected temperatures.

Many differential controllers allow additional sensors to be monitored. These sensors are commonly referred to as T3 or T4 (or S3 or S4, depending on the controller manufacturer). The sensors are typically placed at the top of the storage tank, on the outlet of the storage tank, or in a secondary storage tank. The sensors monitor the HTF temperature when the HTF is at its warmest within the solar storage tank or water heater. They may also monitor the HTF or DHW temperature in the secondary tank. Some controllers do not have functions related to the additional sensors. The additional sensors are only for monitoring purposes.

Advanced Differential Controllers

Advanced differential controllers perform basic differential controller functions plus provide operational control for a more sophisticated solar water heating system. For example, functions for drainback systems, multiple-tank charging, and pump speed control are available with advanced controllers. These controllers can accommodate more temperature sensors and more relay switches. The relay switches can also be used to energize motorized valves and the pump. **See Figure 11-8.**

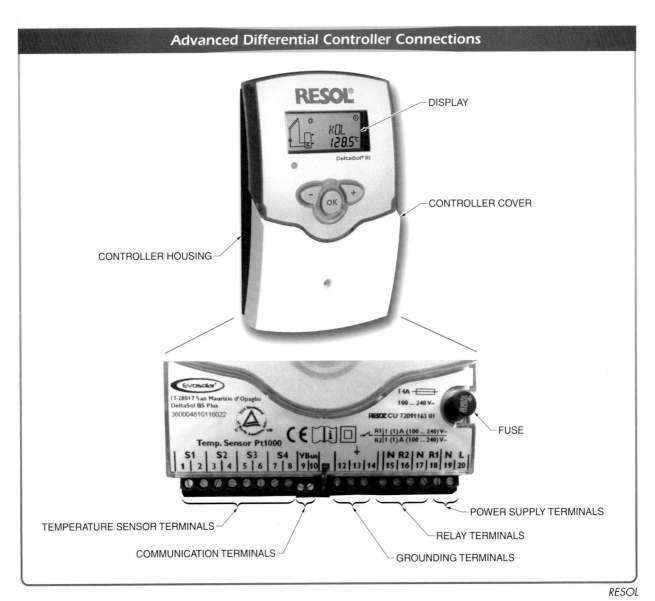

Advanced Differential Controller Connections

Figure 11-8. Advanced differential controllers perform basic functions plus provide operational control for a more sophisticated solar water heating system.

For example, the RESOL® DeltaSol® BS Plus controller is designed for four temperature sensors and two relay switches and can be used to control one of nine different system configurations. These configurations can include monitoring and controlling components, such as multiple storage tanks, collectors, heat exchangers, and pumps.

Many advanced differential controllers offer real-time Internet monitoring of the system. With the addition of an Internet LAN connection or a wireless connection, the solar water heating system can be operated and monitored remotely via the Internet. **See Figure 11-9.** This is not only a convenience factor for the building owner or for the installation/maintenance contractor, but a safety factor and allows for immediate adjustments to system operation when needed.

Advanced differential controllers may also allow for the addition of data loggers, which track the performance of the solar water heating system. Data can be collected, such as total energy harvested, temperature values at all sensor points, and HTF flow rates. Some data loggers allow the data to be stored over the lifetime of the system in order to provide historical data. Historical data can be used to verify the efficiency of system operation for subsidy programs and for system maintenance and

troubleshooting. Data from the data logger can be saved to a computer memory card or sent over the Internet and saved on a dedicated server. The data may then be viewed and used to operate a variety of devices in real time. **See Figure 11-10.**

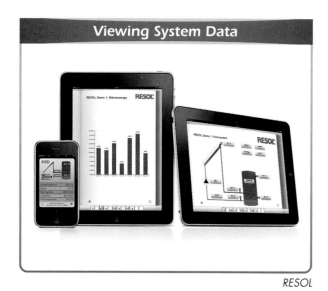

RESOL

Figure 11-10. Data received from controllers and data loggers can be viewed and used to operate a variety of electronic devices in real time.

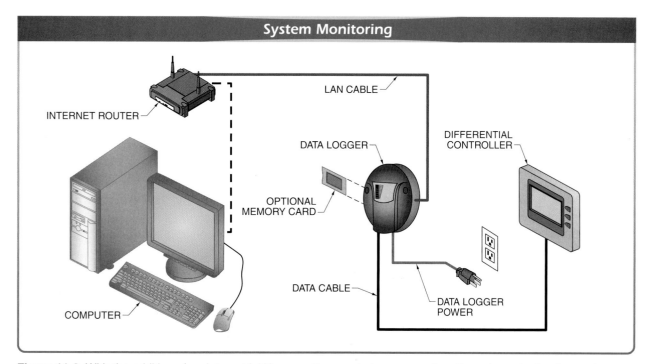

Figure 11-9. With the addition of an Internet LAN connection or wireless connection, the solar water heating system can be operated and monitored remotely via the Internet. The building owner can adjust the system parameters while away from the system.

Pump Stations

The various configurations of differential controllers are also available for pump station systems. The controllers that are integral to the pump station are in most instances identical to the separate differential controllers. The differential controller included with the pump station will operate as the controllers described above. Normally, the pump station is used for small to moderate solar water heating systems, while larger systems use the individual component system. **See Figure 11-11.**

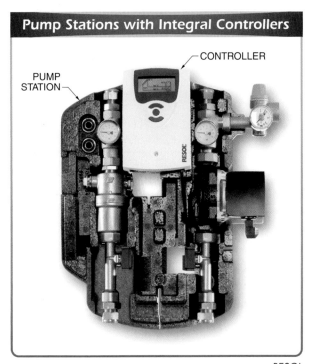

Pump Stations with Integral Controllers

CONTROLLER

PUMP STATION

RESOL

Figure 11-11. The controllers that are integral to the pump station are in most instances identical to the separate differential controllers.

CONTROLLER INSTALLATION

Controllers and pump stations for solar water heating systems must be properly and securely installed. The placement of the controller must be carefully considered before installation. For example, there must be ample room to access the controller once the rest of the system has been installed. There are general mounting considerations for all controllers and more specific considerations for timer, PV, and differential controllers. The electrical needs of the controller and the location of the nearest accessible electrical wall outlet must be taken into consideration as well. In some cases, a new electrical wall outlet must be installed.

Controller Mounting Considerations

During controller installation, the body of the controller or pump station should be mounted to a solid, flat surface. If the wall is made of drywall, a piece of plywood should be mounted to the wall studs as a mounting board. This allows for the use of wood screws to solidly mount the controller rather than using plastic anchors or molly screws in the drywall.

Most controller bodies have two or more mounting holes. Some manufacturers include a paper mounting template with the controller. The mounting template or the controller itself can be used for marking the layout of the mounting holes. **See Figure 11-12.** Once the controller mounting holes have been marked, small pilot holes should be drilled, and the controller can be mounted to the board with screws. Once mounted, the controller is ready to be wired. However, timer-controlled, PV-controlled, and differential-controlled systems have different installation considerations that must be taken into account before being wired.

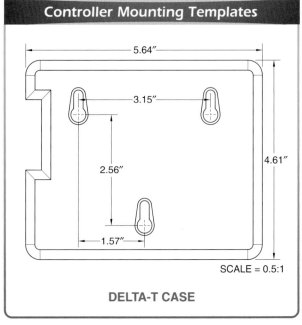

Controller Mounting Templates

5.64″

3.15″

2.56″

4.61″

1.57″

SCALE = 0.5:1

DELTA-T CASE

Figure 11-12. Mounting templates allow for the marking and drilling of pilot holes used to mount the controller to the board.

Mounting Timer Controllers. A timer controller is typically a small device with a timer wheel and set tab. **See Figure 11-13.** The timer controller should be mounted securely on a wall in close proximity to the pump and an electrical wall outlet. In some cases, the timer controller can be plugged directly into an electrical wall outlet and the pump plugged into or attached to the timer.

Intermatic, Inc.

Figure 11-13. A timer controller is typically a small device with a timer wheel and set tab. It should be mounted securely on a wall in close proximity to the pump and an electrical wall outlet.

Mounting PV Controllers and Modules. In order for a PV-controlled system to operate properly, the PV module must be mounted at the solar collector. The module should be mounted at the same tilt and orientation as the solar collector for the module to receive the same amount of insolation as the solar collector. The PV module should also be installed where it will receive the same shading pattern as the collector. **See Figure 11-14.** If the PV module is mounted where it will receive insolation when the solar collector does not, the pump will be energized and send HTF through a cool collector, which wastes heat. Once the PV module is mounted, it should be covered so that the pump does not energize until necessary.

The PV controller should be mounted securely on a wall in close proximity to the pump. In addition, there should be enough room to access the controller in order to connect the wiring and service the controller as necessary. The PV controller usually has connections for the PV module, collector and tank sensors, and the pump.

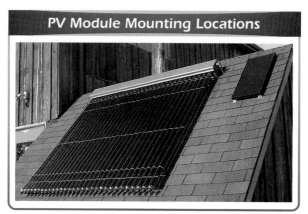

ThermoTech

Figure 11-14. A PV module should be mounted at the same tilt and orientation as the solar collector and should also be installed where it will receive the same shading pattern as the collector.

In certain configurations, the solar collector, PV module, and pump are combined in a packaged system. If this is the case, the PV module is attached to the solar collector using either screws or bolts. There is no separate controller module to be mounted in the solar storage area. The entire system will run off of a differential controller or the PV module can be directly wired to the pump.

Mounting Differential Controllers. The differential controller is typically a relatively small, plastic, and lightweight device. Its design allows for easy installation on a wall within the solar storage area. In most cases, the controller case, which consists of the controller body and the cover panel, is manufactured in two pieces.

The controller should be mounted in a dry indoor area in close proximity to the solar storage tank / water heater and the system pump. There should be enough room to access the controller to connect the wiring, adjust the settings, and service the controller. Manufacturer installation instructions should be followed in regard to specific mounting procedures.

Mounting Pump Stations. The control module is usually incorporated within the body of the pump station frame. Therefore, the controller must be mounted with the pump station.

Pump station placement is critical since both the system wiring and system piping will be connected to the station. Therefore, the control module must be mounted close to the storage tank / water heater, in-line with the piping coming from the solar collector, and in a dry area.

Because of the weight of the pump station, a solid mounting board should be used if the existing wall cannot properly support the pump station. **See Figure 11-15.** Sturdy screws designed for penetration of the mounting surface should always be used. The screws should be securely fastened to the wall. However, care should be taken to not overtighten and damage the pump station case. Once the station is mounted, the necessary piping and wiring can be installed.

Pump Station Mounting

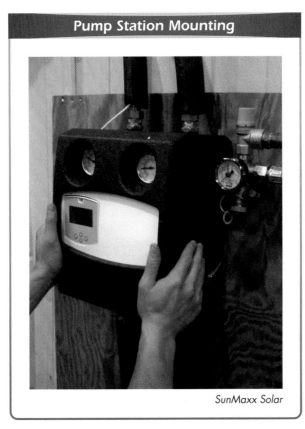

SunMaxx Solar

Figure 11-15. A solid mounting board should be used if the existing wall cannot properly support the pump station.

SENSORS

A *sensor* is an electrical component that is used to measure a property or condition, such as fluid temperature within a solar water heating system. The sensors in the solar water heating system used to measure fluid temperature are connected to the operational controller and should be wired and connected to the controller before the controller power wiring is connected. This will preclude any possible electrical shock that could occur. These sensors include the thermistor and the resistance temperature detector (RTD).

The electrical resistance of thermistors and RTDs varies with temperature. The resistance of the thermistor decreases as the temperature increases and resistance increases as the temperature decreases. RTD resistance properties are just the opposite. The resistance of the RTD decreases as the temperature decreases and resistance increases as the temperature increases. Resistance is measured in ohms (Ω).

The thermistor is comprised of two metal leads embedded in a ceramic or polymer disk surrounded by a metal sheath. Electrical resistance of the thermistor is created by the ceramic or polymer material between the two leads. There are two types of thermistors based on their resistance at 77°F (25°C). The 10K thermistor has a resistance of 10,000 Ω at 77°F (25°C), and the 3K thermistor has a resistance of 3,000 Ω at 77°F (25°C). Because of this difference in resistance, these two thermistors should not be interchanged. Each differential controller is designed to measure the resistance of one type of thermistor only. Therefore, only the thermistor/sensor specified by the controller manufacturer must be used. In most cases when ordering the controller, the number of sensors needed can be supplied.

There are two types of RTDs: wire-wound and thin-film. A wire-wound RTD consists of an element made of platinum, nickel, or copper wire that is wound around a ceramic core. A thin-film RTD consists of a thin film of platinum that has been deposited on a ceramic substrate. Both types of RTDs are covered with a protective metal sheath. Electrical resistance of the RTD is created by the element or thin film of platinum.

The main difference between thermistors and RTDs is their temperature response. RTDs are capable of sensing higher temperature ranges and are, therefore, more accurate than thermistors. Therefore, RTDs can be used in systems that reach higher temperatures than thermistors. Since thermistors and RTDs respond differently to temperature changes, they cannot be interchanged.

Thermistors and RTDs look similar, are two-wire sensors, and attach to system components in the same manner but operate differently. **See Figure 11-16.** Because of the similarities between the thermistor and RTD, it is advised that when working with one type of sensor that the other type of sensor be kept away from the job site.

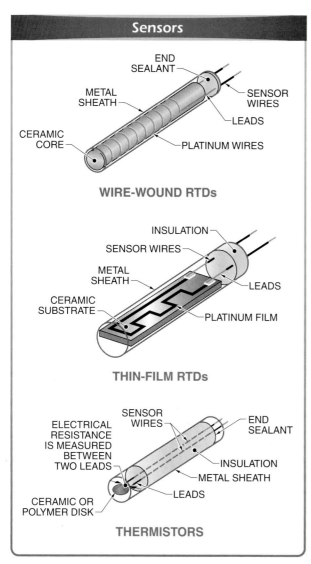

Figure 11-16. Thermistors and RTDs look similar, are two-wire sensors, and attach to system components in the same manner but operate differently.

Sensor Installation

Thermistor and RTD sensors used in the solar water heating system are manufactured in various designs, based on how they are to contact the piping or storage system. Sensor types based on design include the immersion and flat-tab sensor.

An *immersion sensor* is a long, cylindrical sensor that is installed in the sensor well placed in system piping, a collector, or a storage tank. **See Figure 11-17.** A *sensor well,* also known as an immersion well or

thermowell, is a closed-end threaded fitting that is designed to hold and protect an immersion sensor. The sensor well allows the immersion sensor to be surrounded by fluid without the risk of leakage. The immersion sensor should be coated with thermal grease to increase the conductance between the sensor and the walls of the sensor well. A strain relief nut can be used to hold the sensor securely in the sensor well. A *flat-tab sensor,* also known as a clip-on sensor, is a sensor with a flat or slightly curved end that is typically secured to the surface of a pipe or tank using a screw or stainless steel clamp.

RESOL

Figure 11-17. One way in which sensors differ is how they contact the piping or storage system.

Regardless of the type of sensor used, the connection points should be coated with thermal grease. Also, the sensor should be tightly secured for the best conduction of heat and to provide the most accurate temperature measurement.

At least two primary sensors must be installed in order for a differential controller to properly monitor a solar water heating system. The two primary sensors that should be installed are the collector sensor and the storage tank/water heater sensor. Secondary sensors may also be installed in other areas, such as additional storage tanks or DHW piping to allow the controller to monitor the system more closely.

Collector Sensors. The collector sensor, sometimes known as the T1 or S1 sensor, should be installed at the collector outlet. Many solar collector manufacturers provide a connection cross fitting containing a sensor well that can be installed at the collector outlet for this purpose. **See Figure 11-18.** The cross fitting fastens to the collector outlet. The top of the cross fitting can be either plugged or provide a port for the installation of an air vent. An immersion sensor should be installed in the sensor well of the cross so that the sensor will sit in the flow of HTF exiting the collector absorber.

Before installation into the sensor well, the sensor should be coated with thermal grease. Some collectors are designed with a threaded hole for the installation of the sensor well and sensor. Others have an integrated sensor well. **See Figure 11-19.**

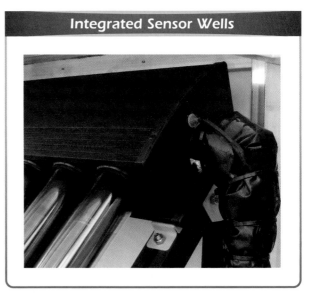

Integrated Sensor Wells

Figure 11-19. Some collectors are designed with an integrated sensor well for the installation of the sensor.

If the installation of an immersion sensor is not possible, a flat-tab sensor may be installed by attaching it to the collector outlet piping. The optimal position for this sensor is on the collector outlet pipe as it exits the absorber. The sensor can be secured by a stainless steel band. **See Figure 11-20.** The pipe and the sensor must then be insulated so that the ambient air will not affect the sensor.

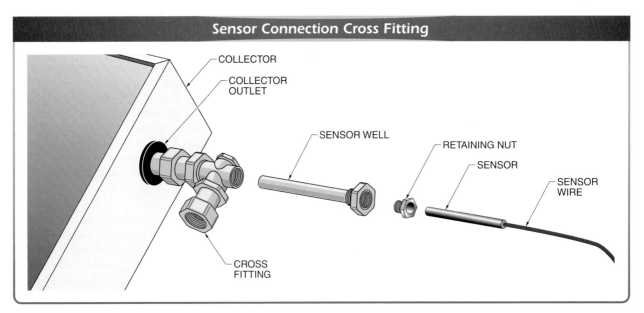

Sensor Connection Cross Fitting

COLLECTOR
COLLECTOR OUTLET
SENSOR WELL
RETAINING NUT
SENSOR
SENSOR WIRE
CROSS FITTING

Figure 11-18. Many solar collector manufacturers provide a connection cross fitting containing a sensor well that can be placed at the collector outlet for the installation of the sensor.

Flat-Tab Sensor Installation

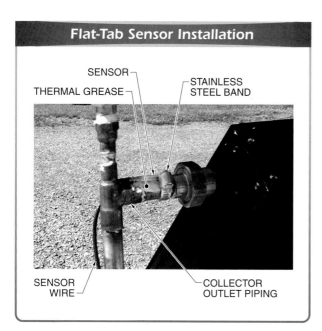

Figure 11-20. The optimal position for a flat-tab sensor is on the collector outlet pipe as it exits the absorber.

Storage Tank/Water Heater Sensors. The storage tank/water heater sensor, sometimes known as the T2 or S2 sensor, must be installed in the lower portion of the storage tank/water heater. Some storage tanks are manufactured with sensor wells at predetermined points. **See Figure 11-21.** A brass elbow with a sensor well, similar to a cross fitting used with solar collectors, is sometimes used to allow the placement of the sensor within the heat exchanger of the tank. However, some storage tanks and water heaters require surface-mounted sensors.

It is important to ensure good contact of the sensor with the surface of the tank. Both surfaces should be coated with thermal grease. It is acceptable to place the sensor between the tank insulation and the tank. However, the sensor must be permanently secured. Also, any insulation removed must be replaced so that the ambient air will not affect the sensor.

Secondary Sensors. Secondary sensors, which are sensors other than the collector sensor and the storage tank/water heater sensor, may be installed within the solar piping system. Secondary sensors may be used to measure temperatures in areas of the system other than the storage tank or water heater. For example, a secondary sensor is typically installed on the DHW outlet. Also, a secondary sensor may be installed in the lower portion of the second storage tank in a two-tank system. **See Figure 11-22.**

Storage Tank Sensor Installation

Heat Transfer Products, Inc.

Figure 11-21. The storage tank/water heater sensor, sometimes known as the T2 or S2 sensor, must be installed in the lower portion of the storage tank/water heater.

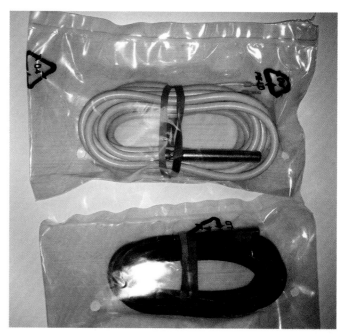

Most controllers come prepackaged with the necessary primary sensors.

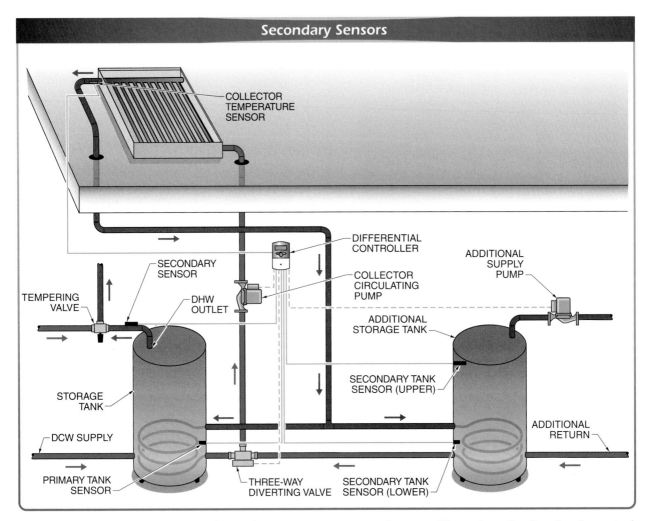

Figure 11-22. Secondary sensors may be used to measure temperatures in areas of the system other than the storage tank or water heater. For example, secondary sensors may be installed on the DHW outlet or the lower portion of the second storage tank in a two-tank system.

Sensor Wiring

Sensor wiring is used to transmit the temperature at the sensor to the controller. Typically, dual No. 18/20 AWG stranded insulated wire is used to extend sensor wiring. Solid wiring is not used for this purpose. Depending on the distance from the sensor to the controller, a heavier gauge wire may be required. *Note:* Lower gauge wires are thicker or heavier. There may also be a maximum wiring distance recommended by the manufacturer. For example, the Buderus controller limits the maximum distance of No. 18 wire to 165′. The maximum distance for No. 14 sensor wire is 330′. Regardless of the type of controller used, the manufacturer installation instructions for the particular controller must be followed.

The sensor may come prewired with a predetermined length of wire. Since the sensor may be a considerable distance from the controller, it may require longer lengths of wiring to reach the proper terminals at the controller. If this is the case, the sensor wiring may need to be spliced, or extended, to reach the controller. **See Figure 11-23.** It may be possible to order the sensor with the proper length of wiring so that sensor wires do not need to be spliced.

It is good practice to splice sensor wire so that the factory ends are at the controller terminal. The factory ends are normally tinned (soldered to prevent fraying) for a good connection at the controller terminal. The ends to be spliced are first stripped of their insulation and then soldered together. The splice should be covered with a heat-shrink wire covering.

Splicing Sensor Wires

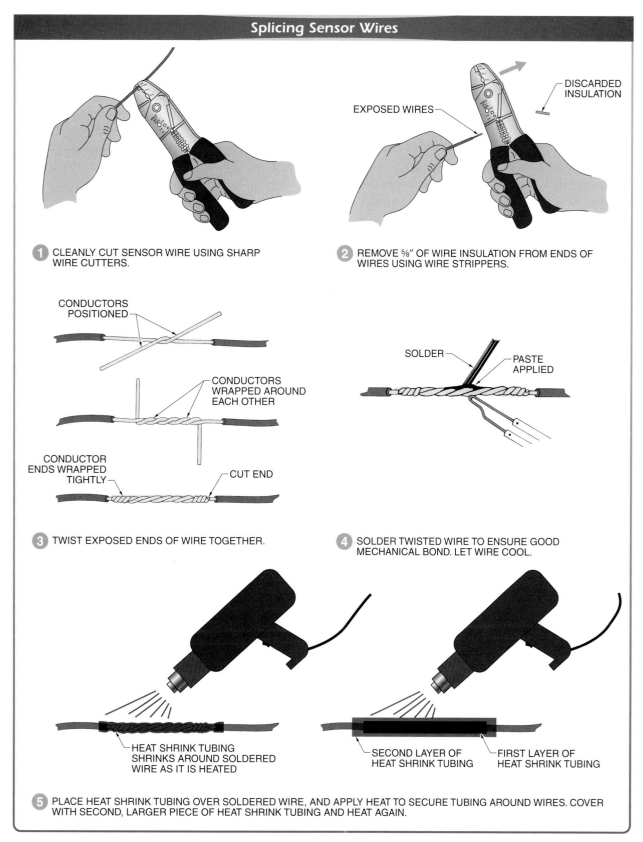

① CLEANLY CUT SENSOR WIRE USING SHARP WIRE CUTTERS.

② REMOVE ⅝″ OF WIRE INSULATION FROM ENDS OF WIRES USING WIRE STRIPPERS.

DISCARDED INSULATION

EXPOSED WIRES

CONDUCTORS POSITIONED

CONDUCTORS WRAPPED AROUND EACH OTHER

CONDUCTOR ENDS WRAPPED TIGHTLY

CUT END

SOLDER

PASTE APPLIED

③ TWIST EXPOSED ENDS OF WIRE TOGETHER.

④ SOLDER TWISTED WIRE TO ENSURE GOOD MECHANICAL BOND. LET WIRE COOL.

HEAT SHRINK TUBING SHRINKS AROUND SOLDERED WIRE AS IT IS HEATED

SECOND LAYER OF HEAT SHRINK TUBING

FIRST LAYER OF HEAT SHRINK TUBING

⑤ PLACE HEAT SHRINK TUBING OVER SOLDERED WIRE, AND APPLY HEAT TO SECURE TUBING AROUND WIRES. COVER WITH SECOND, LARGER PIECE OF HEAT SHRINK TUBING AND HEAT AGAIN.

Figure 11-23. Sensor wire may need to be spliced in order for the wire to reach from the sensor to the controller.

It is not good practice to use wire nuts or crimp connectors to splice the wire. The wire nut connection may oxidize over time and lose the ability to conduct electricity. This causes incorrect temperature readings. The crimp connection may loosen and cause the wire to slip out of the connection.

Sensor Wire Installation

In most installations, the T1 sensor wiring from the collector must penetrate the roof. Roof penetrations for sensor wiring are made similarly to collector piping penetrations. The roof penetration must be protected by flashing. Wiring must be protected as it penetrates the roof and flashing by conduit of some type. The roof penetration and the opening around the sensor wiring must be sealed with the proper sealant for protection from the elements and UV degradation.

A coolie cap is commonly used to route the wire. Sensor wiring should be routed away from sharp objects that can chafe or pierce the wiring. Sensor wiring should also be routed away from other wiring containing voltages of over 110 V, which could cause electrical interference. Sensor wiring can be installed in PVC sheathing or conduit, where allowed, to protect the wire from physical damage and provide a neat and professional appearance. **See Figure 11-24.** However, commercial applications may require that the wiring be installed in metal conduit.

Before sensor wiring is connected to the controller terminal, it should be ensured that the controller is deenergized and properly locked/tagged out. The two ends of the sensor wires for each sensor can then be securely attached to the proper controller terminals. Since there is no polarity in sensor wiring, either end of the wire can be attached to either side of the terminals. The wiring should be run through the strain relief clamps provided on the controller to prevent the wires from accidently being pulled out of the terminals. **See Figure 11-25.**

CONTROLLER AND SENSOR TESTING

The controller and pump should be tested once the controller, sensors, and sensor wiring have been installed and connected and the piping system has been pressure tested and is fully charged. This ensures that the controller and pump will not be damaged when the system operates dry. The manufacturer instructions for system testing should be followed.

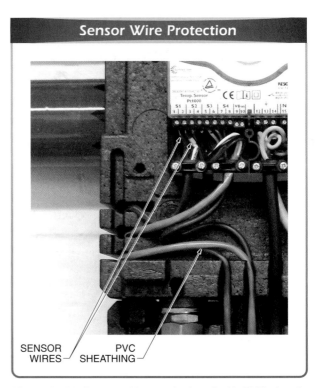

Figure 11-24. Sensor wiring can be installed in PVC sheathing or conduit, where allowed, to protect the wire from physical damage and provide a neat and professional appearance.

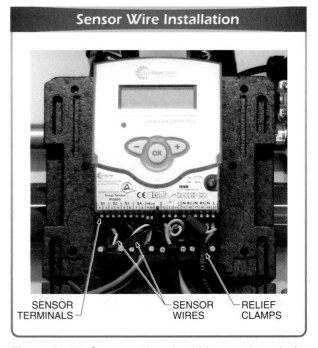

Figure 11-25. Sensor wiring should be run through the strain relief clamps provided on the controller to prevent the wires from accidently being pulled out of the terminals.

The connectivity and accuracy of a solar water heating sensor can be determined by testing the resistance of the sensor with an ohmmeter. First, the temperature reading of the piping or tank must be taken at the sensor location. The sensor wiring must be disconnected from the controller terminals. The resistance reading must then be taken across the two wires of the sensor. **See Figure 11-26.** The sensor temperature and resistance can then be compared to the appropriate table for sensor temperature and resistance.

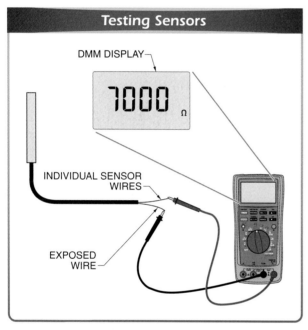

Figure 11-26. In order to test a sensor, the temperature reading of the piping or tank must be taken first at the sensor location. Once the sensor wiring is disconnected from the controller terminals, the resistance reading must be taken across the two wires of the sensor.

Facts

Since the negative temperature coefficient of silver sulphide was first observed by Michael Faraday in 1833, there has been a continual improvement in thermistor technology. The most important characteristic of a thermistor is its extremely high temperature coefficient of resistance. Also, modern thermistor technology has resulted in the production of devices with extremely precise resistance versus temperature characteristics, making them the most advantageous sensors for a wide variety of applications.

For example, if testing a 10K thermistor at 92°F, the resistance level should read 7000 Ω, according to the proper thermistor table. **See Appendix.** If this reading closely matches the ohmmeter reading, the thermistor wiring is intact and the thermistor is accurate. If not, the wiring and thermistor should be checked and replaced if not functional.

Care must be taken to use the correct table for the thermistor or RTD being tested. Typically, the correct tables are provided in the controller and sensor installation manuals. When testing sensors, the instructions provided by the manufacturer should always be used.

Wiring Controller Power

In some cases, the controller may be equipped with an electrical power cord long enough to be plugged into a wall outlet. If a wall outlet is not close enough, a new wall outlet may have to be installed. However, the controller may need to be hardwired using wiring run from the electrical panel to the controller. It may also be necessary to run electrical wiring from the pump to the controller. It is recommended that a qualified electrician install wiring. In many areas in North America, a licensed electrical contractor may be required to do this electrical work. An unqualified solar water heating installer should not attempt to run the electrical power wiring to the control system.

Chapter 11 Review and Resources

Testing, Startup, Maintenance, and Troubleshooting

The startup of the solar water heating system may begin after the system has been pressure tested, inspected, and insulated. Any leaks in the system must be repaired before the system can be inspected by the authority having jurisdiction. Most authorities require a 24 hr notice for inspections and may also require that the installer be present for the inspection. The installer must ensure that there is a means for the inspector to access all areas of installation including roof and attic areas. Once the system has passed inspection, the piping can be insulated and the system can be prepared for startup.

SYSTEM TESTING

Once the entire solar water heating system is installed, the system is ready for testing. The piping system, wiring system, and controllers must be tested. The piping system should be pressure tested according to the codes adopted in the local area. The sensors should be tested for proper resistance. The controllers should be tested using the control manufacturer's installation and operation procedures.

The piping for a solar water heating system should be pressure tested per local code requirements and inspected by the local authority having jurisdiction (AHJ). The pressure test parameters vary for different types of systems, such as open loop, closed loop, and swimming pool systems. Testing parameters are outlined in plumbing, mechanical, or solar energy codes. As an example of possible pressure test requirements, the following are the testing requirements contained in the *Uniform Solar Energy Code* published by the International Association of Plumbing and Mechanical Officials (IAPMO):

316.0 Testing.

316.1 Media. The piping of the solar system shall be tested with water, air, heat transfer fluid, or as recommended by the manufacturer except that plastic pipe shall not be tested with air. The authority having jurisdiction shall be permitted to require the removal of any plugs, etc., to ascertain if the pressure has reached all parts of the system.

316.2 Liquid Solar System. Upon completion, the system, including piping, collectors, heat exchangers, and other related equipment, shall be tested and proved tight.

316.2.1 Open Loop Systems. Open loop systems directly connected to the potable water system shall be tested under a water pressure not less than the maximum working pressure under which it is to be used. The water used for tests shall be obtained from a potable source of supply. A fifty (50) pound per square inch (344.5 kPa) air pressure test shall be permitted to be substituted for the water test.

316.2.2 Other Open Loop Systems. Systems operating at atmospheric pressure shall be tested under actual operating conditions.

316.2.3 Closed Loop Systems. Closed loop or other type pressure systems shall be tested at one and one-half (1½) times maximum designed operating pressure. All systems shall withstand the test without leaking for a period of not less than fifteen (15) minutes.

These testing requirements are fairly straight forward. The atmospheric, open, and closed loop systems have differing test values because each system will be under different pressure levels. Therefore, they should be tested according to the testing requirement for the pressure level.

All portions of the new piping system must be tested. The testing apparatus must have the proper gauge indicator for monitoring the test. The pressure test apparatus can be connected to the boiler drain. The piping being tested should be isolated from other connected piping. In addition, the expansion tank should be isolated or removed for the test. During the test, the piping and components should be visually inspected for leaks while the pressure gauge is monitored over the appropriate amount of time.

Open Loop Piping Tests

Open loop piping is rarely pressurized over atmospheric pressure. Therefore, open loop piping must be tested under operating conditions. This typically involves a running water test under normal line pressure.

If it is not possible to perform a running water test, the piping loop can be isolated and air tested. The local building or plumbing inspector should be consulted for permission before changing any testing parameter.

In most cases, open loop direct systems use potable water as the heat transfer fluid (HTF). If using a liquid test for the piping, only potable water should be used for the test. Open loop direct systems are normally pressurized at the same pressure as the DCW. Therefore, the pressure test would consist of purging air out of the system by filling it with DCW. The system is then left under pressure for the duration of the pressure test, which is a minimum of 15 min.

Closed Loop Piping Tests

Closed loop piping must be tested at 1½ times the operating pressure. Most closed loop systems are not pressurized over 30 psi, which makes the required test pressure approximately 45 psi.

A closed loop piping test may be performed by filling the system with water from the potable water system to purge air out of the system. In many cases, the normal line pressure will exceed the required 45 psi. When filling the system, the proper method of backflow prevention must be used.

Swimming Pool Piping Tests

Solar swimming pool heating systems are also open loop systems. They require running water tests under normal line pressure. If pressure testing is necessary, it must be performed with water because air testing of PVC piping is prohibited by codes and PVC manufacturers. This

is because PVC piping under air pressure can develop fractures and shatter. This can cause injury or property damage. If PVC piping is required to be pressure tested, it should be hydrostatically tested with water.

Air Tests

In the winter, stagnant water that could freeze should not be left in piping. An air test allows for piping to be tested without filling the system with liquid. **See Figure 12-1.** The air test allows the piping system to be tested without the possibility of liquid freezing under cold conditions or water damage from leaking joints. It is far easier to repair system leaks when testing with air than water.

Air Testing

Figure 12-1. An air test allows for piping to be tested without filling the system with liquid.

An air test requires a test setup that is connected to a boiler drain and uses compressed air or nitrogen as the test medium. The system, such as an open loop direct system, must be flushed, cleaned, and possibly disinfected per plumbing code requirements when the air test is completed.

Alternative Tests

Advanced solar water heating systems may require specialized fluids other than propylene glycol (PG). Many fluids may not be mixed with water. If this is the case and water is used for testing, all water must be removed from the system and the piping dried. A pressurized dry-air purge may be used to dry the pipes. Typically, the AHJ will allow the 50 psi air test to be substituted for the liquid test. Requirements and allowances must be confirmed with the AHJ before an air test is used to test the system.

STARTUP PROCEDURES

Once a solar water heating system has been pressure tested it can be prepared for startup. However, a number of tasks must be completed before starting the system for the first time. A system startup checklist is a valuable tool for streamlining the startup process. The startup procedure for a solar water heating system includes the following:

- flushing system piping
- cleaning system piping
- charging the system
- energizing the system
- checking HTF flow
- programming the controller
- verifying the operation of system functions
- inspection by the AHJ

Flushing System Piping

Newly installed piping for a solar water heating system should be flushed to remove any debris, such as dirt, dust, deburring shavings, solder, and flux residue. If debris is left in the system piping, the check valves and control valves may malfunction. Debris that has built up in the piping or collector can also inhibit or stop HTF flow.

In most cases, it is only necessary to flush the solar loop piping. Solar loop piping should be flushed with clean water. Line pressure should be sufficient to flush out any debris within the system. If line pressure is not available, clean water should be transported to the site and pumped through the system. This is especially important for direct solar water heating systems.

Most storage tank / water heaters have been flushed at the factory. However, storage tanks built at a job site should be washed down and flushed by the installer. The specific procedures for flushing direct, indirect, and drainback systems vary. Therefore, manufacturer recommendations should always be followed.

Direct Systems. Direct systems contain potable hot water and the lines must not be fouled with polluted or contaminated water. Direct systems can be flushed by connecting a hose from the DCW to the collector outlet or return side boiler drain. Another hose should be connected to the collector inlet or supply side boiler drain, which will be used as the flush drain hose. This hose should be placed where it can easily drain away the fluid and debris into a floor drain, floor sink, or bucket. **See Figure 12-2.** The flush water should drain away without ponding or creating a hazard. This configuration flushes the system opposite of normal flow.

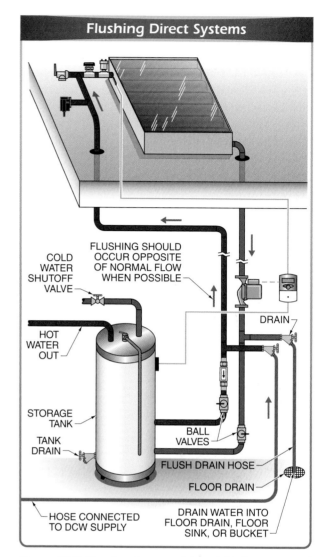

Figure 12-2. Direct systems contain potable hot water and the lines must not be fouled with polluted or contaminated water. The contaminated fluid and debris flushed from a direct system should be drained into a floor drain, floor sink, or bucket.

If a check valve is installed on one of the lines and cannot be removed, the solar loop should be flushed in the direction of flow. The end of the flush drain hose should be visible to monitor the cleanliness of the water. Once debris is no longer visible in the flush water, the DCW should be turned off. The short runs from the tank to the boiler drain can be flushed by opening the isolation valves and running water out of the boiler drain. The solar loop should not be filled before the lines are cleaned.

Indirect Systems. Flushing an indirect system is similar to flushing a direct system. In most individual component systems, boiler drains are installed on the lowest system piping on either side of the system check valve. **See Figure 12-3.** A hose connected to the DCW can be attached to the collector outlet boiler drain and a flush drain hose can be attached to the other boiler drain. The expansion tank should be removed beforehand to prevent flushing debris into the tank. Isolation valves should be opened to allow flow throughout the solar loop. The DCW can be turned on and the piping flushed until no sign of debris is observed in the flush drain water.

Flushing an indirect system can be made easier by removing the check valves and replacing them with a spool piece made of union ends and a brass nipple or copper pipe. Other components such as the circulating pump can be replaced with a spool piece made with flanges and pipe or tubing. This ensures that system flushing is completed without fouling the check valve or pump. **See Figure 12-4.**

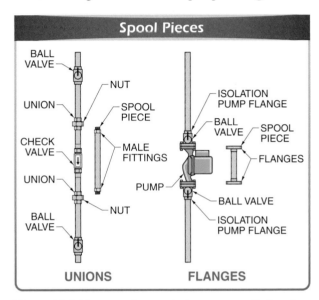

Figure 12-4. The use of a spool piece ensures that system flushing is completed without fouling the check valve or pump.

Drainback Systems. To flush a drainback system, the drainback tank may need to be removed if there is no drain on the tank. The system should not be flushed with the drainback tank in-line if there is no way to remove debris from the tank. If the tank must be removed, the system can be flushed from both directions.

Systems with Pump Stations. Systems that incorporate a pump station should be flushed per the manufacturer's instructions. A pump station is compact and individual components cannot be removed easily for replacement with spools or for cleaning if debris becomes lodged in the system. Therefore, manufacturer recommendations should be followed carefully when flushing the system. **See Figure 12-5.** For example, the A. O. Smith Cirrex pump station should be flushed using the following steps as specified in the manufacturer-provided installation manual:

1. Connect a water hose from a water source with at least 40 psi supply pressure to the fill valve (P).
2. Connect a discharge hose to the drain valve (T) and open the valve. Place the other end of the discharge hose into an appropriate drain.
3. Turn the supply ball valve (Q) to 90° (horizontal) so that the valve is closed.

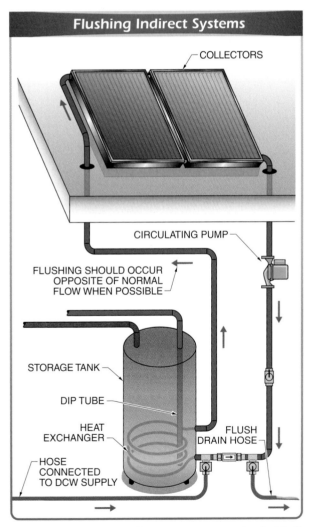

Figure 12-3. In most indirect systems comprised of individual components, boiler drains are installed on the lowest system piping on either side of the system check valve to allow for proper flushing of the system.

4. Turn the slot of the flow restrictor (S) to the horizontal position using a flat head screwdriver. The integrated ball valve is now closed and will prevent debris from entering the pump.

5. Turn on the supply water and open the fill valve (P) then leave running for 15 min to flush debris out of the solar loop.

6. After the 15 min flush, turn off the supply water at the source and allow the system to drain. When water stops flowing out of the discharge hose, close the fill valve (P).

7. Close the drain valve (T) on the pump station, and move this hose to the drain valve that was installed at the lowest point of the solar loop near the bottom of the tank.

8. Open this drain valve with a flat head screwdriver, open the fill valve (P) slowly to vent the system, and allow the remaining water to drain out. In order to vent the system completely, it is necessary to open the flow restrictor (S) by turning with the screwdriver until vertical.

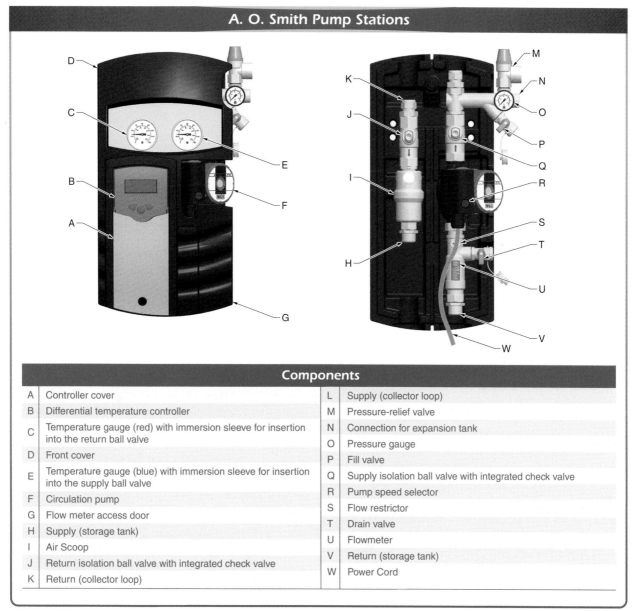

A. O. Smith Pump Stations

Components

A	Controller cover	L	Supply (collector loop)
B	Differential temperature controller	M	Pressure-relief valve
C	Temperature gauge (red) with immersion sleeve for insertion into the return ball valve	N	Connection for expansion tank
		O	Pressure gauge
D	Front cover	P	Fill valve
E	Temperature gauge (blue) with immersion sleeve for insertion into the supply ball valve	Q	Supply isolation ball valve with integrated check valve
		R	Pump speed selector
F	Circulation pump	S	Flow restrictor
G	Flow meter access door	T	Drain valve
H	Supply (storage tank)	U	Flowmeter
I	Air Scoop	V	Return (storage tank)
J	Return isolation ball valve with integrated check valve	W	Power Cord
K	Return (collector loop)		

A. O. Smith

Figure 12-5. A system that incorporates a pump station, such as the A. O. Smith Cirrex pump station, has individual components that cannot be removed easily. Therefore, manufacturer recommendations should be followed when flushing the system.

Flushing Considerations. Care must be taken after the solar water heating system is flushed. It may be impossible to drain all water from the system. Sags and dead ends may collect enough water that could freeze if low temperatures are encountered before the system is cleaned and charged with HTF. It may be possible to blow out the system using an air compressor. However, this still may not remove all water from the system. The solar water heating installer should plan on testing, flushing, cleaning, and charging the system in the same day.

Any domestic cold and hot water piping that has been installed should also be flushed and cleaned. It may be necessary to open the domestic cold water (DCW) and domestic hot water (DHW) isolation valves, flush this piping back into the solar loop lines, and then flush the solar loop. Once the domestic water lines are in operation, such as lavatories and kitchen sinks, the aerators of various fixtures should be removed and cleaned. A small amount of water should be run out of the faucets to remove any large debris or air that might have remained in the line.

Cleaning System Piping

Simply flushing the system piping may not remove all polluting or contaminating agents from the piping. Flux residue, oils, or excess pipe lubricant may still be entrained along piping walls even after flushing. It is especially important to properly clean the potable water piping of a direct system and any new DCW or DHW piping. In order to remove these substances, the piping should be cleaned with a 1% to 2% solution of trisodium phosphate and water.

Trisodium phosphate (TSP), also referred to as sodium phosphate, is a white powder used as a cleaning agent or degreaser. TSP in a 1% to 2% solution with water has a pH of 12 and is, therefore, a very effective degreaser for removing grease or oil from surfaces, such as the interior walls of piping. **See Figure 12-6.** Because TSP is very alkaline, the proper personal protective equipment (PPE), such as eye and skin protection, must be worn when using TSP. The material safety data sheet (MSDS) should be consulted for more detailed information on TSP.

The same configuration that is used for flushing the system can be used to clean the system with TSP. It is necessary to dilute the TSP with water before introducing it to the piping system. This can be done in a large bucket or a barrel. Once the lines are full, the system should be circulated with the TSP solution for 1 hr to 2 hr. The system can then be drained just as the flush water was drained from the system. The diluted TSP is not hazardous and can be drained away through drainage piping.

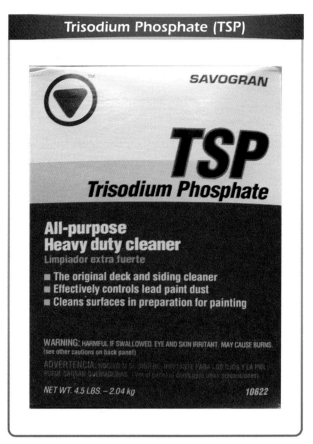

Figure 12-6. TSP is used as a cleaning agent for removing stains and as a degreaser for removing oil or grease from surfaces, such as the interior walls of piping.

Charging Systems

The solar water heating system can be charged after it has been flushed and cleaned. *Charging* is the process of purging air from a solar water heating system while filling the system with the appropriate HTF at the proper pressure. All air within the system must be completely removed as the system is filled. Drainback systems do not require the purging of all air from the system, but the system must be checked to ensure that the proper amount of HTF and air is in the system.

Direct Systems. Direct systems use potable cold water as the HTF. If the storage tank is new or not yet filled, the tank should be filled first while the air inside the tank is removed via the temperature and pressure (T&P) valve test lever. This relieves air out of the T&P drain line and does not introduce that air to the solar loop piping. This will also flush and clear any debris from the drain and ensure that the drain is functioning.

The tank should be filled by slowly opening the DCW supply valve to the storage tank while opening the test valve on the T&P valve until the tank is full. **See Figure 12-7.** When the tank is full, the solar loop can be filled by slowly opening the collector inlet or supply isolation valve. The opposite-side solar loop boiler drain can be opened until there is water at the drain. The other isolation valve can then be opened to fill the rest of the loop.

There still may be air in the system and the DHW piping. The DHW side of the fixtures should be run until there is no air in the system. If the air vent at the collector outlet is a manual air vent, the air must be relieved from the system and the vent should be closed. If the system is an active system, the pump should be run for a while and the air vent opened again to ensure that all air is out of the system. *Note:* The solar collectors should still be covered while flushing, cleaning, and charging the system. The collector should not be uncovered until the system is ready to generate heat.

Indirect Systems. To charge indirect systems, an HTF is typically used rather than potable water. This is due to the possibility of freezing temperatures throughout most of North America and because of the amount of hardness in most potable water sources. Indirect drainback systems may use distilled water as an HTF. However, this should

only be done where mild temperatures are typical. When water is used as the HTF, it should be distilled rather than potable. Distilled water does not have caustic chemicals or minerals that may build up in or damage the system over time. Distilled water is also an ideal HTF because it has superior heat transfer properties over potable water, PG solutions, or synthetic oils. However, most indirect solar water heating systems use a glycol-based solution, such as distilled water and PG, as the HTF because of its temperature properties and protection characteristics.

The distilled water and PG should be mixed prior to filling the system. This can be done in the same barrel or bucket that was used for cleaning the system. After the liquids are mixed together, the HTF mixture should be pumped into the piping system. This can be done using either a transfer pump and hoses or a filling station. **See Figure 12-8.**

Charging the system can be accomplished using the same method as cleaning the system. The expansion tank, check valve, and pump should be installed and all spool pieces should be removed. The manufacturer instructions for the expansion tank should be followed when initially filling the system. Some expansion tank manufacturers may require the pressure within the tank to be reduced on initial charging.

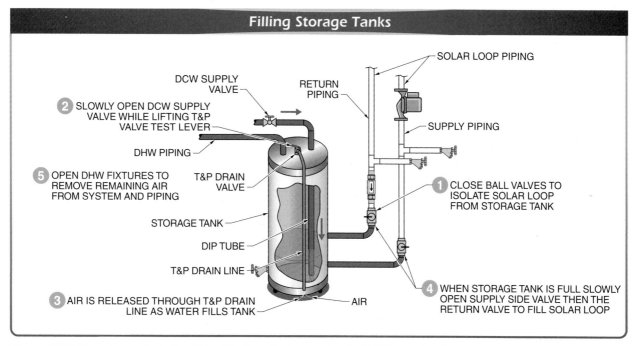

Figure 12-7. The storage tank should be filled by slowly opening the DCW supply valve to the storage tank while opening the test valve on the T&P valve until the tank is full.

HTF Filling Stations

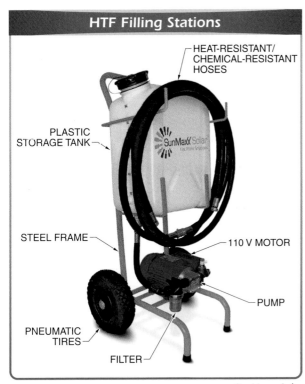

HEAT-RESISTANT/
CHEMICAL-RESISTANT
HOSES

PLASTIC
STORAGE TANK

STEEL FRAME

110 V MOTOR

PUMP

PNEUMATIC
TIRES

FILTER

SunMaxx Solar

Figure 12-8. A filling station allows for easy mixing, transport, and transfer of an HTF mixture.

To fill the system, the hose from the pump outlet should be connected from the pump outlet to the opened collector inlet or supply boiler drain. Another hose should be connected to the opened collector outlet or return boiler drain. The other hose end should be placed in the HTF bucket or barrel along with the suction hose end of the pump. When energized, the pump will transfer fluid to the system piping while purging air out of the system. As the system is filled, an HTF and air mixture will be pumped out of the drain hose. The HTF should be allowed to circulate through the system until the HTF from the drain hose is clear of air bubbles. If the system has an air vent, the air vent should be opened until all air is out of the system.

Once the HTF is flowing without air bubbles, the collector return drain should be slowly closed. The pump should be allowed to continue to add HTF until the system operating pressure is reached. The system pressure gauge should be monitored closely so that the system design pressure is not exceeded. Then the collector supply boiler drain should be closed and the pump shut off. During this process, the system should be carefully monitored so that it does not run dry or the fill/return hose is not allowed to come out of the container and add air into the system

as a result. If a filling station is used, the manufacturer instructions for charging the system should be followed. If a pump/controller station is used in the solar water heating system, the manufacturer filling instructions for the pump station should be followed.

Determining HTF Requirements. Most indirect solar water heating systems will use a glycol-based solution as the HTF, although there are other types of HTF such as synthetic and silicon oils. However, other products are rarely used in the solar water heating system. Glycol-based products are normally used due the temperature properties and the protection characteristics of glycol-based solutions. There are two common glycol-based solutions used in hydronic or solar water heating systems. The solutions are ethylene glycol and propylene glycol (PG) solutions.

There are two major characteristic differences between propylene and ethylene glycol HTFs. The characteristic differences are the viscosity and toxicity. Ethylene glycol-based HTF is less viscous than PG-based HTF. It generally provides a more efficient heat transfer medium and better low-temperature protection than a PG-based solution and is preferred for most hydronic applications. However, PG-based HTF has a lower toxicity rating than the toxicity rating of ethylene glycol-based HTF and is preferred for use in solar water heating systems over ethylene glycol solutions.

Indirect solar water heating systems provide heat transfer through heat exchangers. The heat is normally transferred through the exchanger to potable water. Most model plumbing codes require a double-wall heat exchanger when potable water is used in the other side of the heat exchanger, unless the HTF is a Class 1 (potable water) or Class 2 (essentially nontoxic) transfer fluid. Essentially nontoxic HTF can be used in single-wall heat exchangers, which are less expensive than double-wall heat exchangers. However, because there may be a large volume of HTF within the system that can increase toxicity in the potable water if there is a leak in the heat exchanger wall, most AHJs will require the use of double-wall heat exchangers. The local AHJ should be consulted before using single-wall heat exchangers with potable water and PG-based solutions.

PG-based HTF solutions are the norm for most indirect solar water heating systems. PG-based products such as DOWFROST and NOBURST are readily available at local plumbing and mechanical supply houses. **See Figure 12-9.** PG alone will be too viscous to pump through a system. Therefore, PG is used in solution with water as the HTF of choice for solar water heating systems.

Noble Company

Figure 12-9. PG is available in 5 gal. containers.

PG manufacturers recommend that PG be mixed with distilled water rather than tap water or mineral water. Tap water or mineral water may have too high of a total hardness or contain chemicals such as chloride, sulfate, or iron that can deplete the effectiveness of the PG solution. Most PG products also contain inhibitors to protect the piping system from corrosion, and many of the chemicals that can be found in tap water can degrade the inhibitors leaving the piping prone to corrosion.

Most PG products are made in two formulations that are ideal for use in solar water heating systems. For example, the two variations that DOWFROST PG can be found in are DOWFROST and DOWFROST HD. The main difference between these two variations is the low- and high-temperature protection values. DOWFROST is an industrially corrosion inhibited fluid with a maximum operating range of –50°F to 250°F (–46°C to 121°C). DOWFROST HD is an industrially corrosion inhibited fluid with a maximum operating range of –50°F to 325°F (–46°C to 163°C).

The HD product is an ideal HTF for solar water heating systems because of the high-temperature protection that the HD product is capable of providing. In fact, the listed ratings are for undiluted PG solutions at atmospheric pressure. The PG product will have varying protection properties based on various dilution percentages and at differing pressures.

Two other protection properties of PG solutions that are important to a solar water heating system are freeze protection and burst protection. *Freeze protection* is the temperature rating at which a PG product will begin to create ice crystals and become slushy and almost impossible to pump through a system. *Burst protection* is the temperature rating at which a PG solution will totally

freeze and burst system piping. Each manufacturer creates dilution and pressure tables that allow the installer to determine the proper temperature rating needed in the local area and design the system around these properties. The determination should be based on 5° higher and lower than the highest and lowest ambient temperatures that will be encountered. The concentration of PG and the pressure within the system can be designed based on these properties. For example, this principle can be seen in the freeze temperatures and the boiling points at various concentrations of DOWFROST HD. **See Figure 12-10.**

Freezing and Boiling Points of DOWFROST HD Solutions		
Glycol by Volume*	Freezing Temperature[†]	Boiling Temperature[†]
10	26 (–3)	212 (100)
20	19 (–7)	213 (101)
30	8 (–13)	216 (102)
40	–7 (–22)	219 (104)
50	–28 (–33)	222 (106)
60	–60 (–51)	225 (107)
70	<–60 (–51)	230 (110)

* in %
[†] in °F (C)

Figure 12-10. The freezing and boiling temperatures of HTF depend on the concentration of propylene glycol in the HTF.

If temperatures above those in the table are reached for any length of time, the PG solution will deteriorate. The inhibitors contained in the PG solution that prevent oxidation within piping will also deteriorate, and the piping may then corrode. However, most PG solutions will be rated for increased temperatures above the table values but only for brief periods of time.

Manufacturer recommendations should be followed regarding the proper HTF, HTF dilution percentage, and system pressure for each installation. The use of a product other than the manufacturer-specified HTF and HTF solution may void any system warranty. For example, the A. O. Smith Cirrex system recommendations are for a 50% solution of PG with distilled or demineralized water at an operating pressure of 30 psi. This will provide freeze protection down to –30°F (–34°C) and burst protection down to –60°F (–51°C). The system will also have a boiling point temperature of 282°F (139°C) and high-temperature protection under stagnant conditions.

The PG and distilled water should be mixed to the correct percentage and amount before being pumped into the system. This involves finding the total amount of HTF needed and then determining the amount of PG and water to be mixed. The amount of solution needed can be calculated by determining the amount of HTF needed to fill the solar collectors, all piping, and any tanks or vessels that contain HTF, including heat exchanger coils.

The required amount of HTF for a solar collector can be determined by using the SRCC collector rating sheet or documentation from the manufacturer. For example, the fluid capacity listed on the rating sheet for a Chromagen flat-plate collector is 1.4 gal. **See Figure 12-11.** Therefore, the fluid capacity of two Chromagen flat-plate collectors is 2.8 gal.

The amount of HTF in piping must be calculated using piping charts. **See Figure 12-12.** For 1′ of ¾″ copper tube, the amount of HTF is 0.0251 gal. A system containing 100′ of ¾″ Type L copper tubing requires 2.51 gal. of HTF.

If the system contains a heat exchanger coil, the fluid capacity of the coil must also be accounted for. For example, the fluid capacity of the heat exchanger coil in the SUNX 80 solar water heater is 2.8 gal. **See Figure 12-13.**

The total HTF capacity of a system that contains two Chromagen flat-plate collectors, 100′ of ¾″ Type L copper tubing, and a SUNX 80 solar water heater is as follows:

- two Chromagen flat-plate collectors = 2.8 gal.
- 100′ of ¾″ Type L copper tube = 2.51 gal.
- heat exchanger coil capacity = 2.8 gal.
- total HTF capacity of system = 8.11 gal.

Therefore, a 50% solution will consist of 4.05 gal. of PG and 4.05 gal. of distilled water.

Energizing Systems

After the system is charged and before being energized, the pressure in the solar water heating system should be monitored for pressure drops. A pressure drop indicates a leak in the system. System piping should also be reinspected to ensure that all valves are at the correct orientation and all sensor wiring is connected and secured. If the system shows no sign of leaks and all valves have the correct orientation, the system can be energized. However, the back-up heating source for the solar water heating system should not be energized until the operational functions of the system are verified.

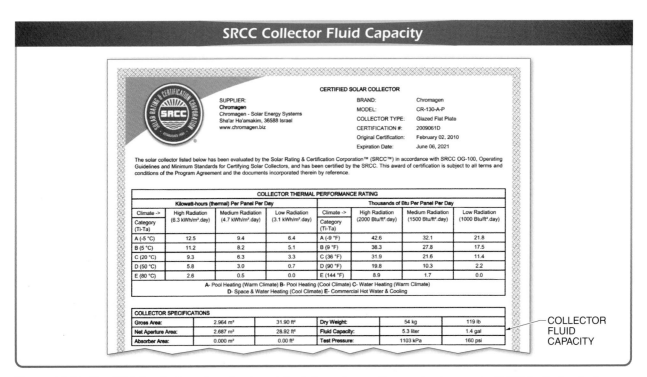

Figure 12-11. The SRCC collector rating sheet contains information on the required amount of HTF for that collector.

Dimensions and Physical Characteristics of Copper Tube — Type L								
Nominal or Standard Size*	Nominal Dimensions*			Physical Characteristics				
	Outside Diameter	Inside Diameter	Wall Thickness	Cross-Sectional Area of Bore†	Weight of Tube‡	Weight of Tube and Water‡	Tube Contents	
							Volume§	Capacity‖
¼	0.375	0.315	0.030	0.078	0.126	0.160	0.0054	0.00405
⅜	0.500	0.430	0.035	0.145	0.198	0.261	0.00101	0.00753
½	0.625	0.545	0.040	0.233	0.285	0.386	0.00162	0.0121
⅝	0.750	0.666	0.042	0.348	0.362	0.512	0.00242	0.0181
¾	0.875	0.785	0.045	0.484	0.455	0.664	0.00336	0.0251
1	1.125	1.025	0.050	0.825	0.655	1.01	0.00573	0.0429
1¼	1.375	1.265	0.055	1.26	0.884	0.43	0.00875	0.0655
1½	1.625	1.505	0.060	1.78	1.14	1.91	0.0124	0.0925

* in in.
† in sq. in.
‡ in lb/ft
§ in cu ft/ft
‖ in gal./ft

Copper Development Association

Figure 12-12. Piping charts are used to determine the amount of fluid for a given diameter and length of pipe.

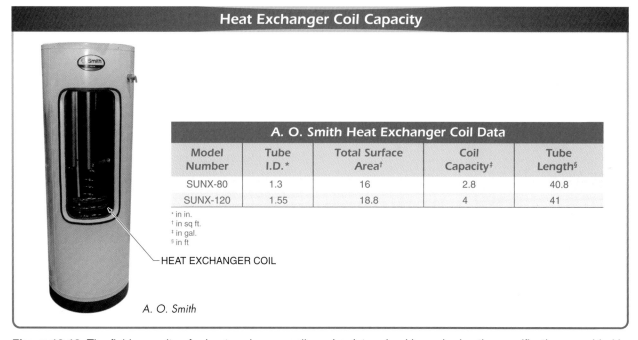

Figure 12-13. The fluid capacity of a heat exchanger coil can be determined by reviewing the specifications provided by the manufacturer.

The pump should be checked to ensure that it is wired or plugged into the controller. The pump can be checked with a digital multimeter to verify a good power connection. Pump operation can also be checked without the controller by simply plugging the pump into a wall outlet if possible.

If energizing the pump without the controller is not possible, the controller can be energized by either plugging the controller into a power outlet or energizing the proper electrical panel circuit breaker. The controller should be set to manual so that the system will not operate automatically until the installer is ready. The controller power indicator LEDs should be ON. If there is no manual setting, the system should register a low temperature in the storage tank, which should energize the pump. When the pump is energized, a humming sound can be heard or the pump will vibrate.

Checking HTF Flow

Once the pump is energized, HTF flow can be checked. Checking HTF flow is easy if a flow meter is installed in the system. The flow meter float indicates whether there is HTF flow in the system and the flow rate (in gpm or lpm) when the pump is ON. **See Figure 12-14.** Flow can also be checked by removing the cover from the solar collectors and allowing the system to operate.

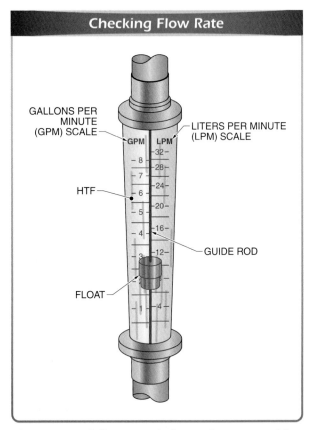

Figure 12-14. A flow meter indicates the amount of flow when a pump is activated.

Initially, the HTF temperatures in the collector without insolation should register similar temperatures at the collector sensor (T1 or S1) and the storage tank sensor (T2 or S2). The cover over the collector should be temporarily removed to allow the collector to function. As HTF flows through the collector on a sunny day, the temperature will rise at T1 while the temperature will be lower at T2. This indicates that there is flow through the collector and heat is being transferred. On a cold or cloudy day, the temperatures may be opposite as the sensors or temperature gauges reflect cooler temperature HTF from the collectors entering the storage tank.

A rough estimate of flow rate can be made without the use of a flow meter. This can be done by timing the change in temperature at the storage tank after allowing the temperature to rise in the collectors for a brief period of time and then restarting the pump. The stagnant HTF in the collector outlet piping at the tank should cool during this waiting period. The system can be energized again. When the temperature changes at the tank, there should be a volume of highly heated HTF from the collector that passes on to the tank. When all of the HTF volume in the collector passes T2, the temperature should drop as cooler circulated HTF flows from the collector.

The gpm rate is determined by dividing the capacity of the collectors by the time it takes for the temperature at S2 to change from hot to warm temperature. For example, if the flow rate in the system is 1 gpm, then it should take close to 3 min for the 2.8 gal. volume of hot HTF in the collectors to flow through to the storage tank. If the actual flow matches that time, then the flow rate in the system is correct. The solar thermal collector is then recovered until the installer is ready to verify controller functions.

Programming Controllers

Once the solar water heating system is energized and it has been verified that the HTF is flowing, the controller can be programmed. Each controller type has a slightly different layout for controller functions and different screen symbols. The manufacturer operation instructions should always be followed for controller programming and troubleshooting.

Each controller function and parameters (values) should be checked. In most cases, the controller is preprogrammed with default settings. The controller parameters should be checked against the system design parameters. If the controller function parameters do not match the design parameters, the function parameters should be reset. For example, the maximum storage tank temperature setting of the controller should be checked and matched to the system design parameter. If the system is designed for 140°F (60°C) maximum tank temperature and the controller is set at 120°F (49°C), the controller should be reset for 140°F (60°C).

Verifying System Functions

It should be verified that the controller safety and operational functions, such as pump startup/shutdown and pump freeze protection, are working before the collector covers are removed. Proper controller operation of the pump startup function can be easily verified when the sensor at T2 is at a low temperature. At low temperatures, the controller should energize the pump. The pump shutdown function at maximum tank temperature can be checked by removing the T2 sensor wires and shorting the two contacts with a jumper wire. This will cause the controller to deenergize the pump, and the pump should stop. The freeze protection function can be checked by temporarily removing sensor T1 and placing it in ice water. The temperature should lower past the setpoint. This will cause the controller to energize the pump, and the pump should run. *Note:* When verifying the proper operation of system functions, the back-up heat source should remain off.

All of the solar water heating functions should be monitored. The system gauges and the controller should be monitored to verify proper HTF flow. Once it is verified that the HTF is flowing properly, the back-up heat source can be energized. It is best to verify system operation on a day of high insolation because less time is required for the process.

Once the controller functions have been verified to be operating properly, the solar water heating system is ready to operate. The controller should be set to automatic and the collector cover can be removed. The installer should return periodically after the system is running to ensure proper functioning of the system before it is turned over to the owner.

System Inspection

Once proper system operation has been verified, the solar water heating system is ready for final inspection by the authority having jurisdiction (AHJ). During final inspection, an inspector will verify that the system meets all pertinent codes and is functioning properly.

The installer should ensure that the system is ready for inspection. The system should be fully functioning as designed. All system piping should be labeled. The piping content and direction of flow should be on each label. **See Figure 12-15.** All valves should be tagged for function and normal operation orientation. System documentation and any prints or drawings should be on-hand for the inspector to reference if needed. The installer must be available to provide access to the system, answer any questions about the system, and demonstrate system operation. Once the inspector completes the final inspection and the system is approved, the system can resume normal operation.

Piping Labels

Figure 12-15. All system piping should be labeled. The piping content and direction of flow should be on each label.

Commissioning of the solar water heating system may be required in some jurisdictions after the final inspection. *Commissioning* is a formal assessment of the functionality and efficiency of a system. Commissioning may also be required for incentive or rebate verification. The commissioning process is sometimes conducted by an independent third-party commissioning service. The third-party agent attests to the viability and efficiency of the system and provides certification documentation to the owner.

Facts

Plumbing and electrical work is usually checked by separate plumbing and electrical inspectors. If inspectors discover a violation of the building code, they have the legal authority to have the work demolished and reconstructed properly, which can be costly to the owner or installation contractor.

DOCUMENTATION AND OWNER TRAINING

The installer should prepare an operation manual that should be given to the property owner or facilities operator. The operation manual should be prominently placed in the storage area of the solar water heating system and should include the following:
- site information, including facility description, system history, and requirements
- installation contractor and installer contact information
- as-built design drawings
- installation manuals
- equipment and component inventory and maintenance log
- equipment and component warranty documentation, including model and serial number
- system events log
- recommended maintenance requirements
- basic troubleshooting guide
- other relevant resources and documentation

The property owner or facilities operator should be thoroughly trained in the operation and maintenance of the solar water heating system. This step is critical to the ongoing functionality and efficiency of the system. The owner or operator must understand how the system operates and shuts down or the system may be damaged, harm occupants, or damage the building. Owner training should include the following:
- an overview of the operation and function of the system, including startup and shutdown procedures
- the identification of all equipment and components of the system
- an explanation of system labeling and valve operation
- an overview of controller programming and normal system operation settings
- a review of the equipment and component operation manuals and warranties
- an overview of maintenance and simple troubleshooting procedures
- a review of safety issues and procedures

MAINTENANCE

A professionally installed and programmed solar water heating system should function properly for many years without problems. To ensure years of trouble-free operation, the system must be properly maintained. System maintenance includes periodic inspections of system equipment and components as well as verification of proper HTF quality, levels, and solution concentration.

Maintenance Inspections

System maintenance should include a monthly inspection of the collectors, piping insulation, and HTF by the property owner or facility manager. For example, the solar collector or collector array should be visually inspected monthly for debris such as leaves or branches, damaged or dirty glazing, and broken evacuated tubes if used. The outdoor piping insulation should be checked for damaged or deteriorated insulation. Proper HTF flow should be verified. Drainback systems should be checked for proper HTF levels. If any problems are found, the system should be immediately repaired.

An annual maintenance inspection, which should be much more thorough than the monthly inspection, also should be conducted. In addition to the collector, insulation, and HTF flow, the following should be checked during the annual maintenance inspection:
- all valves (for leaks)
- motorized equipment to ensure smooth operation
- PG solutions (for fluid quality, proper pH levels, and proper PG concentration)
- expansion tank air pressure
- controller function settings

Swimming pool water heating systems require far less maintenance than DHW solar water heating systems. The collector of a swimming pool heating system should be checked for debris, which should be removed from around the collector if found. If a controller is used, its functions should be checked. A flow test and a valve operation test should be performed before the pool season begins.

HTF Solutions

When checking the quality of the HTF PG solution, a sample of the solution should be checked for sludge, grit, and discoloration. The sample should be taken from the middle area of the system if possible so that sediment from the lower areas of the piping system does not alter results. If the HTF has undergone degradation, the fluid should be flushed and replaced.

PG concentration should be checked with a refractometer or glycol test kit. **See Figure 12-16.** The refractometer or test kit can be used to determine the actual level of freeze and corrosion protection of a glycol and water solution with a glycol concentration of up to 55%. The refractometer or test kit instructions for testing the fluid should be closely followed.

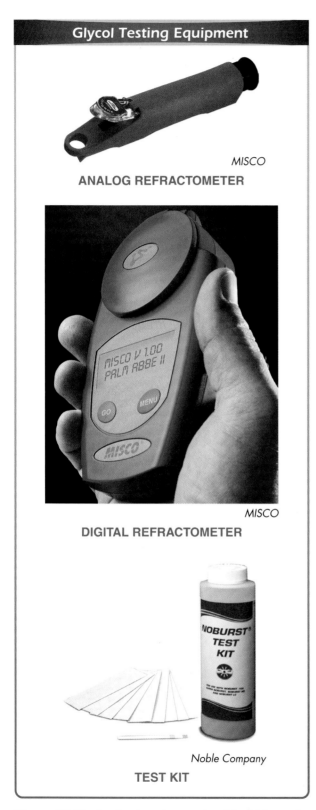

Glycol Testing Equipment

MISCO
ANALOG REFRACTOMETER

MISCO
DIGITAL REFRACTOMETER

Noble Company
TEST KIT

Figure 12-16. A refractometer or glycol test kit can be used to determine the actual level of freeze and corrosion protection of a glycol and water solution with a glycol concentration of up to 55%.

The pH level is a good indicator of the condition of the PG and is best measured with litmus paper or, for more accurate measurements, a field pH meter. The pH level of a 30% to 50% solution of PG and water should be between 8.3 and 9.0. Readings below 8.3 indicate that the HTF has been degraded. PG solution with a pH level of 7 or less should be removed from the system as soon as possible. The system should be flushed and then filled with new solution.

TROUBLESHOOTING AND REPAIR

A properly installed and maintained solar water heating system should last for many years. Many of the system components, such as the pump, expansion tank, and motorized valves, should last about 20 years. However, even the most carefully installed systems will develop problems and/or wear out. A major indicator of problems with the solar water heating system is DHW that is either too hot or not hot enough.

Determining the causes of system problems and their remedies can be difficult when the installer has installed the system and is familiar with the system design. It is even more challenging to determine the cause of system problems when the installer is called upon to troubleshoot and repair a solar water heating system someone else has installed. A service call may involve something as simple as loose pump or controller wiring or air in the system. However, the problem could be quite complicated and may require the installer or repair technician to review specifics of the entire system including DHW usage, collector orientation, system design, and component and piping installation.

Fortunately, manufacturers include troubleshooting information in their installation and operation manuals to assist in troubleshooting their systems. **See Figure 12-17.** Installation and operation manuals should be consulted first when looking for solutions to system problems. Most manufacturers post these manuals on their website. For this reason, a smartphone or laptop computer with wireless Internet capability is an essential tool for the installer or repair technician to have available during a service call. When repairing solar water heating systems, all safety precautions should be taken.

System problems typically include electrical, equipment/component, or HTF flow problems. These problems can occur individually or in tandem with other problems. It is up to the installer or repair technician to troubleshoot and determine the cause of the problem and then to repair the system.

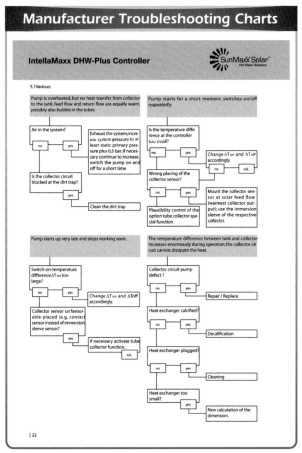

SunMaxx Solar

Figure 12-17. Manufacturers usually include troubleshooting charts in their installation and operation manuals to assist in troubleshooting their systems.

Electrical Problems. Electrical problems can be caused by faulty or deteriorating line power wiring. This could cause pump, back-up system, and controller failure. Electrical failure could occur in the controller itself caused by other problems such as dirty contacts, loose wiring, or failure of the digital process programming. In addition, sensors could fail or become dislodged, or sensor wiring could deteriorate or develop interference problems.

Equipment and Component Problems. Equipment and component problems include pump failure, expansion tank flooding, sticking of the valve, stem breaks, and motorized or check valve failure. The back-up heating source may also develop problems that can affect the system. In addition, the solar collector could develop problems caused by dirty glazing, condensation within the collector, or broken evacuated tubes. An external problem that can occur is shading of the collector by vegetation or buildings that were not present at the time of installation.

HTF Flow Problems. HTF flow problems include air in the system that causes air lock, plugged absorber tubing, HTF deterioration, and reverse thermosiphoning of the HTF. These flow problems can be a result of another type of failure and, therefore, may make it difficult to discover the true cause of the problem. Sometimes there is no installation or operation manual available, or the problem is not with any piece of equipment or component. Therefore, a generic troubleshooting guide with possible remedies to system problems, such as the one provided on the following pages, can be consulted.

Solar Water Heating System Troubleshooting . . .		
Problem	**Possible Cause**	**Possible Repair**
No hot water	No power to electric heating element	Check high temperature protection and reset thermostat
	Gas-fired water heater fails to ignite	Check pilot mechanism
	Malfunctioning safety switch	Check and replace
	Defective automatic gas pilot valve	Check and replace
	Too much primary air	Adjust pilot shutter
	Debris in pilot orifice	Clean and open orifice
	Defective pilot valve	Replace
	Loose thermocouple connection	Tighten
	Defective thermocouple	Replace
	Improper pilot gas adjustment	Adjust
	Defective thermostat	Replace
	Improperly adjusted mixing/antiscald valve	Check water temperature at fixture and adjust valve setting
	Defective mixing and/or antiscald valve	Replace or remove from system plumbing
	Leak in distribution piping	Locate and correct
Not enough hot water	Undersized auxiliary water heater	Replace
	Auxiliary water heater tank heat loss	Insulate tank
	Auxiliary water heater thermostat set too low	Increase setpoint temperature
	Auxiliary water heater element failure	Replace element
	Auxiliary water heater thermostat failure	Replace thermostat
	Auxiliary water heater lower element disconnected	Reconnect element and set thermostat to low temperature
	Heat loss due to defective or improperly installed check valve	Inspect valve and repair or replace
	Solar return dip tube is missing, in wrong location, or defective	Replace dip tube
	Cold water supply dip tube is missing or in wrong location	Install in proper location
	Degraded collector absorber coating	Recoat or replace absorber (contact manufacturer)
	Undersized collector area	Increase collector area
	Excessive collector condensation	Inspect and repair glazing seal, pipe gaskets, weep holes, and vents
	Dirty collector glazing	Clean as required
	Collector leaks	Repair

. . . Solar Water Heating System Troubleshooting . . .		
Problem	**Possible Cause**	**Possible Repair**
Not enough hot water	Collector at improper orientation	Check orientation; face collector ±45° east or west of south
	Outgassing inside collector glazing	Clean surface and contact manufacturer
	Degraded plastic collector glazing	Replace
	Hazy collector glazing	Replace
	Shaded collector(s)	Remove shading or relocate collector(s)
	Improperly tilted collector	Check tilt; set to within 15° of latitude
	Improperly plumbed collector	Compare with system schematic
	Improper controller operation (cycling, late turn on)	Check sensor placement and insulation from ambient conditions
	Defective sensors or controller	Measure sensor resistance or check controls by placing sensors against hot- and cold-water glasses and watching pump function; replace defective units
	Improper or loose wiring	Compare with system schematic; check for proper connections; seal all splices against moisture
	Shorted sensor wiring	Check wiring for breaks, metal contact, water exposure, and corrosion
	Undersized heat exchanger	Replace with properly sized heat exchanger
	Scaling/clogged heat exchanger	Back flush and clean
	Closed isolation valves	Open
	Improperly adjusted mixing/ antiscald valve	Reset temperature indicator
	High water usage	Check system size and discuss solar system and owner's lifestyle
	Clogged piping (corrosion or sediment)	Replace excessively corroded components
	Insufficiently insulated piping	Add insulation where required
	High heat loss from piping	Check pipe insulation for splits, deterioration, absence
	Nighttime thermosiphoning	Check for pump operation at night
	Improper plumbing	Compare with system schematic; check flow direction
	Isolation valve failure after closing	Replace valve
	Flow blockage	Flush system; check effluent for dirt/scaling
	Low system pressure	Check pressure gauge; refer to owner's manual for correct pressure

. . . Solar Water Heating System Troubleshooting . . .		
Problem	**Possible Cause**	**Possible Repair**
Not enough hot water	Pump has no power	Check breaker, pump, and controller; repair or replace
	Pump flow rate is too high or low	Adjust flow rate
	Defective pump	Listen for irregular noises in pump operation; feel collector feed and return pipes for temperature difference; check and replace
	Pump activation switch is OFF	Check switch on PV to pump wiring
	Pump runs continuously	Check control system for opens and shorts
	Improperly installed pump	Compare with system schematic
	Improper or damaged sensor wiring	Check for damage or loose connections and correct
	Shorted sensor wiring	Check and repair
	Undersized storage tank	Install larger tank
	Storage tank heat loss	Insulate tank
No hot water in morning	Check valve is stuck open or does not seat	Replace check valve
	Sensor wires are reversed	Check wiring and reconnect
	Water heater circuit breaker is off	Switch breaker on
	Recirculation freeze protection tripped and back-up power off	Switch back-up power on
	Excessive consumption	Discuss hot water usage; check system size and auxiliary heater status
Water too hot	Auxiliary water heater thermostat setpoint is too high	Reduce temperature setpoint
	High limit sensor is miscalibrated	Check, recalibrate, or replace
	Lack of hot water use	Run hot water to reduce tank temperature
	Mixing or antiscald valve temperature set too high	Adjust
	Mixing/antiscald valve failure	Replace valve
No water	Cold water supply valve closed	Open valve
High electric use	Pipe leak	Repair leak
	Tank thermostat set too high	Check setting and adjust to desired temperature
	Tank thermostat is inaccurate	Recalibrate or replace
	Collector return is above tank thermostat	Increase length of dip tube
	Collector return is above tank	Check tank plumbing

. . . Solar Water Heating System Troubleshooting . . .		
Problem	**Possible Cause**	**Possible Repair**
Pump does not start	Controller in OFF position	Switch to automatic or normal operating mode
	Unplugged controller	Return power to controller
	ON/OFF temperature differential setpoints too high	Reset according to specifications
	Loose contacts	Clean and tighten connections
	Defective controller	Replace controller
	Sensors connected to wrong terminal	Correct per manufacturer's recommendations
	Loss of circuit continuity	Check and repair or replace
	Power switch is OFF	Switch to ON
	Blown fuse or tripped breaker	Determine cause and replace fuse or reset breaker
	Brownout	Await resumption of utility power
	Shaded photovoltaic module	Relocate or eliminate shade
	Defective photovoltaic module	Check and replace
	Improper, damaged, or loose photovoltaic wiring	Check and repair
	Pump motor failure	Check brush holders and other mechanical components that may be loose, worn, dirty, or corroded; replace as appropriate; check for thermal overload
	Pump motor runs when started by hand; capacitor failure	Replace capacitor
	No power to pump	Check breaker, wiring, and controller
	Stuck shaft, impeller, or coupling	Replace
	Frozen pump bearings	Replace
	Defective sensor(s)	Replace
	Improper sensor installation	Clean and reinstall properly
	Sensors out of calibration	Recalibrate or replace
	Defective sensor wiring	Repair or replace
	Open in collector sensor wiring	Check wiring; repair or replace
	Short in tank sensor wiring	Check wiring; repair or replace
	Defective timer	Replace
	Incorrect ON and/or OFF time	Reset and check battery
Pump cycles continuously	ON/OFF temperature differential setpoints are too close together	Reset according to specifications
	Freeze protection setting is too high	Reset according to specifications
	Defective controller	Perform controller operation check; repair or replace
	Reversed collector plumbing	Reconnect properly

... Solar Water Heating System Troubleshooting ...		
Problem	**Possible Cause**	**Possible Repair**
Pump cycles continuously	Sensors in wrong location	Relocate sensors as per system design or manufacturer's requirements
	Sensors not properly secured	Secure properly
	Defective sensors	Check and replace, if necessary
Pump cycles after dark	Radio frequency interference in sensor wiring	Use shielded sensor wire
	Check valve does not seat	Replace
Pump runs continuously	OFF temperature differential setpoint is too low	Reset according to specifications
	Controller lightning damage	Replace controller
	Controller in ON mode	Switch to automatic or normal mode
	Sensors out of calibration	Recalibrate
	Defective sensor	Replace
	Improper sensor installation	Reinstall
	Radio frequency interference in sensor wiring	Use shielded sensor wire
	Short in collector sensor wiring	Check wiring; repair or replace
	Open in tank sensor wiring	Check wiring; repair or replace
Pump operates but no fluid flows from collector	System air-locked; air vents closed	Disassemble and clean seat and seal; replace if necessary
	Air vent in improper location	Install at the highest point; install at all high points if possible air trap locations exist; install in true vertical position
	Air vent cap is too tight	Loosen ¼ turn
	Improperly installed check valve	Check flow arrow on valve to ensure direction is per system design
	Collector flow tubes clogged	Flush collector tubing
	Drain-down valve stuck in drain position	Clean; check power to valve
	No fluid in direct system	Open cold water supply valve
	Loss of fluid in indirect system	Locate leak and refill
	Loss of fluid in drain-back system	Cool system, locate leak, and refill properly
	Closed isolation valves	Open valves
	Clogged or damaged piping	Unblock piping or repair damaged piping
	Pump impeller broken or separated from shaft	Replace impeller, shaft, or pump
	Improperly installed pump	Install to ensure correct flow

. . . Solar Water Heating System Troubleshooting . . .		
Problem	**Possible Cause**	**Possible Repair**
Pump operates but no fluid flows from collector	Improperly vented pump	Install in correct orientation
	Undersized pump	Check pump specifications; change pump if required
	Valves closed	Open valves
Pump cycles on and off after dark	Corroded or defective check valve	Repair or replace
Pump runs after dark, but eventually shuts off	Defective sensors	Change sensor
	Improper sensor location	Relocate
	Uninsulated sensor	Insulate
No power to pump with switch on	Open in sensor circuits	Check wiring; repair or replace
	Weak or failed controller output relay	Replace relay or controller
Noisy pump	Air trapped in system	Open automatic air vent
	Dry or worn pump bearings	Lubricate or replace
	Loose pump impeller	Tighten or replace impeller
	Pump enclosed in small room	None
	Pump attached to wall (wall acts as amplifier)	Relocate pump if noise is unacceptable
	Corrosion or particles in pump volute	Clean volute and impeller; replace if required
	Air trapped in pump	Open vent port and/or vent valve and bleed air
Noisy system	Air locked	Bleed air
	Entrapped air (direct systems only)	Purge system by running water up supply pipe and out drain on return line (isolation valves closed)
	Pipe vibration	Isolate piping from walls
Controller does not switch ON or OFF in the automatic mode but operates in the manual mode	Defective controller	Conduct function check and repair or replace
	Defective sensors	Check resistance with DMM; correct or replace sensors
	Improper sensor contact or insulation	Ensure proper contact is made; insulate sensors
	Improper sensor location	Relocate
	Short or open in wiring	Check wiring; repair or replace
System shuts off at wrong high limit or continues to run	Defective controller	Conduct function check and repair or replace
	Defective sensors	Check resistance with DMM; correct or replace sensors
	Improper sensor location	Relocate
System leaks	Pipe burst due to freeze or defective joint	Repair or replace; check freeze-protection mechanisms
	Freeze-protection valve did not reseal after opening	Recalibrate, if required, or replace

. . . Solar Water Heating System Troubleshooting		
Problem	**Possible Cause**	**Possible Repair**
System leaks	Freeze-protection valve out of calibration	Recalibrate, if required, or replace
	Normal operation of freeze-protection valve (most open at 42°F)	No action
	Hose clamps not tightly secured	Tighten clamps; replace clamp or hose
	Thermal expansion and contraction of pipe joints	Replace and provide for flexibility
	Improperly made pipe joint	Reassemble
	Improper seal in system using glycol solution	Make a good seal; use recommended sealer *Note:* Glycol leaks through joints where water does not
	T&P relief valve did not reseal after opening	Replace
	Defective T&P relief valve	Replace
	Improper setting on relief valve	Reset, if possible, or replace
	Loose valve gland nuts	Tighten nuts; replace seal or packing if necessary
	Deteriorated valve seats	Replace seat washers; redress seat; replace valve if necessary
Water comes off the roof	Defective T&P relief valve seal	Replace
	T&P relief valve activates due to no circulation through collector(s)	Check flow in collector loop (See "Pump does not start" and "Pump operational but no flow to collector")
	Freeze-protection valve out of calibration	Recalibrate per manufacturer's instructions or replace
	Defective freeze-protection valve seal	Repair or replace
	Normal operation of freeze-protection valve (most open at 42°F)	No action required
	Ruptured collector piping due to freeze	Repair or replace
	Defective piping	Repair or replace
	Power loss; pressure loss from well	No action required
System does not drain	Collector (parallel) fins and tubes are horizontal and sagging, preventing drainage from occurring	Straighten fins and tubes; realign collector so it will drain completely
	Collector (serpentine) design prevents drainage	Blow water out of tubes with air compressor
	Insufficient piping slope for drainage	Check and ensure piping slopes ¼" per foot of piping
	Defective vacuum breaker does not open	Clean or replace

Chapter 12 Review and Resources

Appendix

Snow Loads

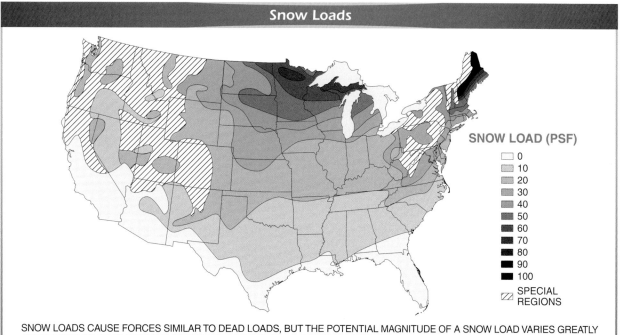

SNOW LOADS CAUSE FORCES SIMILAR TO DEAD LOADS, BUT THE POTENTIAL MAGNITUDE OF A SNOW LOAD VARIES GREATLY AMONG GEOGRAPHIC REGIONS OF NORTH AMERICA. IF A COLLECTOR IS TILTED AT A HIGH ANGLE OR 30° TO 45° THE SNOW WILL USUALLY FALL OFF EASILY AND IT MAY NOT BE NECESSARY TO TAKE THE SNOW LOAD INTO CONSIDERATION. IF THE COLLECTOR IS AT A LOW ANGLE OR FLAT, THE LOAD AMOUNT SHOULD BE ADDED TO ROOF LOADING CALCULATIONS.

Company Logo	# 1 Solar Heating Company (555) 555-5555

SOLAR WATER HEATING SITE ASSESSMENT REPORT

Date of Assessment: 1/31/12   **Installer/Assessor:** Mike Jones

Time of Inspection: 1:30 PM   **Weather Conditions:** 52°F overcast, no precipitation

BUILDING OWNER AND LOCATION

Name: Joe Smith

Address: 000 Main Street, Topeka, KS 66601

Home Phone: (708) 555-1200

Mobile Phone: (708) 555-1400

Email: jsmith@internet.com

LOCATION INFORMATION

Latitude: 39.1°N   **Longitude:** 95.6°W   **Declination:** 3.14°E

Record Low Temperature: −32°F   **Record High Temperature:** 109°F

Property Type: ☒ Existing Construction   ☐ New Construction   ☐ Seasonal Use Only

                      ☒ One-Family Dwelling   ☐ Multifamily Dwelling   ☐ Apartments

                      ☐ Commercial

Solar System Use: ☒ Domestic Hot Water   ☐ Swimming Pool Heating   ☐ Space Heating

Additional Comments: _____

_____

_____

_____

EXTERIOR INSPECTION

UTILITIES LOCATION

Power/Telephone Lines: Underground from Main St. 10′ off north side of building no lines aboveground

Water: Underground from Main St. at center of east side of building

Gas: Underground from Main St. 3′ off south side of building entering at center of south side

Electric: Underground from Main St. 2′ off south side of building entering at southeast corner on south wall

Exterior Safety Assessment: Gas meter at south wall, high trees on west and north side of building, good access from front to south side of building, ground at south side firm, collectors can be manually lifted or mechanically lifted; care needs to be taken with grass around building, harness support from peak or crown

Existing Damage to Exterior of Building: None noted at time of inspection

COLLECTOR LOCATION

Roof Mounting: Good location available at south roof

Ground Mounting: None available too much shading, only front yard available

SHADING

Visual Roof: Shading on west edge of south roof, possible east in early morning

Visual Ground: No areas on ground acceptable

ROOF CONSTRUCTION

Roof Type: Asphalt Shingle

Roof Orientation (Azimuth): 11°E from true south

Roof Angle: 7:12 pitch, 30.26° angle

Optimum Angle at Location: 39°   **Winter:** 54°

Roof Height: 10′ to eave, 18′-6″ to peak

Size (length × width): Triangle shape, from 12″ at peak to 25′ at eave

Obstruction on Roof: None

Condition of Roof: Good, no weathering of shingles, roof firm

Mounting Type Recommended: ☒ Stand-off Mount   ☐ Tilt Mount   ☐ Flush Mount

Additional Comments:

SITE DRAWING

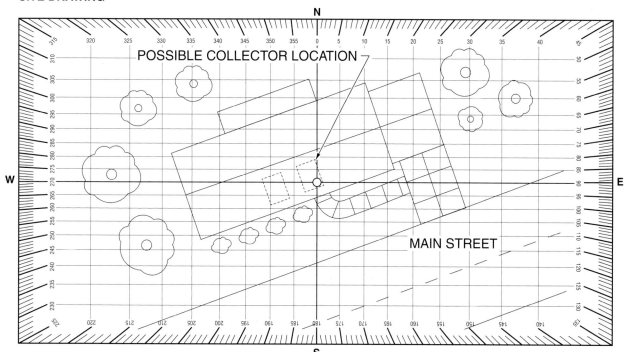

INTERIOR INSPECTION

Roof Collector Location Inspection: Roof sheeting looks good, rafters look good, 4′ of work space in area where collectors will mount, piping path is clear

Interior Safety Assessment: Exposed fiberglass insulation, some exposed nails overhead, will need plywood sheeting for work area–3 pcs. 2′ × 6′, firewatch needed if soldering in attic space

Rafter or Truss Size and Spacing: 2 × 6 – 24″ OC

MECHANICAL ROOM OR SPACE

Size: 10′ × 8′   **Comments:** Enough space for storage tank and other system equipment

Door Size: 29¼″ opening   **Comments:** Clear path for tank entry, one turn 3′ wide hall, no pad needed

Electric/Gas Availability and Location: Yes from existing, electrical panel 6′ away, outlets available but may have to add for controller or pump or change to four-gang outlet

Piping Access to Roof: Yes   **Distance to Collectors:** 25′

Additional Comments: Installation of two flat-plate collectors mounted vertically on south roof section is feasible. No obstructions in attic areas for piping path. Piping can be installed easily from collectors travelling east in attic to water heater room approximately 12′ from east edge of collectors. Tank installation may require removal of mechanical room door frames if tank is wider than 29″.

EXISTING WATER HEATER

Size: 50 gal.   **Fuel Type:** Gas   **Age:** 7 yr   **Condition:** Fair

Retrofit Possible: WH too old and might be too small

Piping Type: Copper

Clothes Washer: Yes, old   **Dishwasher:** Yes, Energy Star rated

Large Volume Fixtures: Whirlpool BT

Gallons per Day Estimate: 80 gal. to 90 gal.

Desired Hot Water Temperature: 125°F

Incoming Cold Water Temperature: 55°F

Estimated Solar Loop Length: 50′

Additional Comments:

DOMESTIC HOT WATER USE

Water Source: ☒ City   ☐ Well   **Incoming Water Pressure:** 60 psi

Occupants: 4   **Guests/Future Occupants:** 0

Number of Bedrooms: 4   **Future:** 0   **Number of Bathrooms:** 3

Additional Comments:

SWIMMING POOL HEATING

Pool Location: ☐ Outdoor   ☐ Indoor

Pool Operating Period: _____

Open Date: _____   **Close Date:** _____

Desired Pool Temperature: _____

Existing Heating System: ☐ Yes   ☐ No   **Type of Fuel:** _____

Storage System Needed: ☐ Yes   ☐ No   **If Yes, Size Required:** _____

Storage Location: _____

Estimated Solar Loop Length: _____

Additional Comments: _____

_____

_____

_____

SPACE HEATING

Existing Heating System: ☐ Forced Air   ☐ Radiant Floor   ☐ Radiant Baseboard   ☐ Other

A/C: ☐ Yes   ☐ No

Size of Heated Area: _____

Describe the type of space heating system that will use the solar heating system: _____

_____

_____

_____

Desired Hot Water Temperature: _____

Existing Cold Water Temperature: _____

Storage System Needed: ☐ Yes   ☐ No   **If Yes, Size Required:** _____

Storage Location: _____

Estimated Solar Loop Length: _____

Additional Comments: _____

_____

_____

_____

Solar Instructor
Training Network

Northeast Region
(Solar Heating & Cooling)

LESSON – PUMP SIZING

In a closed loop system, the pump is responsible for circulating the heat transfer fluid through the solar circuit. The pump (also referred to as a circulator in this type application) must be sized appropriately so that it is able to circulate the fluid at the appropriate flow rate through the piping, fittings, collector(s), and heat exchanger without consuming an inordinate amount of electricity.

The following lesson details the steps for:
- Determining the appropriate flow rate for a system
- Estimating the frictional head loss – also referred to as *pressure drop* – in a solar circuit, and
- Selecting an appropriate pump

> *Important note concerning the use of glycol as a heat transfer fluid:* References for flow rates and frictional head loss are typically based on using water as a heat transfer fluid. As a result, adjustments must be made to account for the use of glycol as a heat transfer fluid. One adjustment is made to the *flow rate* to account for the decreased heat transfer capability of glycol; the other adjustment is made to the *frictional head loss* to account for the increased viscosity of glycol. Details about these adjustment factors are discussed within this lesson.

See the *Detailed Procedure for Pump Selection in a Closed Loop Solar Thermal System* on page 2 for a summary of the step-by-step process described in this lesson.

1

DETAILED PROCEDURE FOR PUMP SELECTION
IN A CLOSED LOOP SOLAR THERMAL SYSTEM

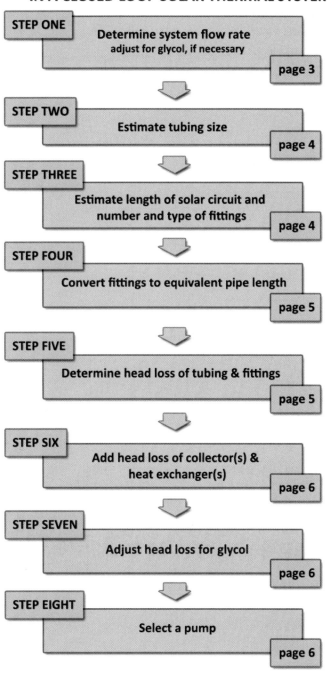

STEP ONE
Determine system flow rate
adjust for glycol, if necessary
page 3

STEP TWO
Estimate tubing size
page 4

STEP THREE
Estimate length of solar circuit and
number and type of fittings
page 4

STEP FOUR
Convert fittings to equivalent pipe length
page 5

STEP FIVE
Determine head loss of tubing & fittings
page 5

STEP SIX
Add head loss of collector(s) &
heat exchanger(s)
page 6

STEP SEVEN
Adjust head loss for glycol
page 6

STEP EIGHT
Select a pump
page 6

2

STEP ONE | Determine the appropriate flow rate for the system

Manufacturers typically have a recommended flow rate for their specific collector. ***These flow rates should be obtained directly from the manufacturer***. If this information is unavailable, the Solar Rating & Certification Corporation (SRCC) uses standard flow rates based upon the type and size of collector. Collector ratings are based upon those flow rates. Figure 1 shows the portion of the SRCC OG-100 Rating that discusses the test flow rate for a rated collector (in this case, the Sunda Seido 1-16 evacuated tube collector).

TECHNICAL INFORMATION

Efficiency Equation [NOTE: Based on gross area and (P)=Ti-Ta]

				Y INTERCEPT	SLOPE
S I UNITS:	η= 0.526	-1.32530 (P)/I	-0.00422 (P)^2/I	0.529	-1.697 W/m^2.°C
I P UNITS:	η= 0.526	-0.23345 (P)/I	-0.00041 (P)^2/I	0.529	-0.299 Btu/hr.ft^2.°F

Incident Angle Modifier [(S)=1/cosθ - 1, 0°<θ<=60°]

			Model Tested:	SEIDO1-8
K$\tau\alpha$ = 1	0.302 (S)	-0.306 (S)^2	Test Fluid:	Water
K$\tau\alpha$ = 1	0.00 (S)	Linear Fit	Test Flow Rate:	9.0 ml /s.m^2 0.0133 gpm/ft^2

Figure 1 - Excerpt from SRCC OG-100 Collector Rating

The SRCC test flow rates are based on aperture area. The Seido 1-16 collector has an aperture area of 33.15sf, thus the test flow rate is:

$$Q_{test} = (0.0133 gpm/ft^2)(33.15 ft^2);$$

$$Q_{test} = 0.44 gpm$$

It is important to note that this flow rate is *per collector* and *when using water as a heat transfer fluid*.

Flow rates for a multiple collector arrays

If the collector array consists of two (2) Seido 1-16 collectors, the flow rate based on SRCC testing would be:

$$2 \text{ collectors} \times 0.44 gpm/collector = 0.88 gpm$$

Again, it is important to consult the manufacturer to obtain recommended flow rates for their specific collectors and for multiple collector arrays.

Adjusting flow rates based on the type of heat transfer fluid

Water is an efficient medium for transferring heat, but it is a freeze hazard in our region of the country when used for solar heating and cooling. Typical pressurized SHC systems in the northeast region utilize a propylene glycol solution to provide freeze protection. Propylene glycol is not as effective a heat transfer fluid as water, thus it requires higher flow rates to transfer equivalent heat quantities.

To accommodate for this difference, a *glycol adjustment factor for flow rate* is utilized. A glycol adjustment factor for flow rate (f_g) of 1.1 is commonly used for 30-50% glycol mixtures. If a manufacturer's recommended flow rate for a flat plate collector is 1.0gpm *for water* (Q_w), then the recommended flow rate *for glycol* (Q_g) would be:

$$Q_g = f_g \times Q_w$$
$$Q_g = 1.1 \times 1.0 gpm$$
$$Q_g = 1.1 gpm$$

3

STEP TWO   **Estimate the tubing size in the solar circuit**

Once the flow rate is determined, the size of solar circuit piping must be selected to determine the amount of friction – or pressure drop – that occurs due to circulating fluid through piping, fittings, valves, collectors, and heat exchangers.

Pipe size selection is most dependent upon the flow rate. The relationship between flow rate and frictional head loss is exponential - if the flow rate is doubled, the frictional head loss may be quadrupled. The following rules of thumb can be used for initial attempts to size piping for the solar circuit:

Flow rate	Pipe size
≤ 1gpm	½"
≤ 4gpm	¾"
≤ 8gpm	1"

Since pipe size is also affected by the length of piping in the solar circuit, the number of fittings in the circuit, and head loss in collectors and heat exchangers, this rule of thumb may not apply to solar circuits that have long runs between the collectors and the utility room. By doing a more detailed analysis, it is possible to determine the approximate head loss for the specific solar circuit, determine the pump required to circulate fluid through the circuit, and decide whether the pipe size should be modified.

STEP THREE   **Estimate the length of solar circuit**

The length of pipe is determined by measuring (or approximating) the round trip distance the fluid must travel to make one complete cycle in the circuit. Some circuits may be comprised of different diameter piping. In this case, the two different lengths of each size pipe must be considered separately and then added together. For example, if a 200' solar circuit is comprised of 100' of ½" tubing and 100' of ¾" tubing, then the frictional head loss for the ½" piping and ¾" piping must be calculated separately. Once determined, the frictional head losses from the ½" and ¾" tubing are added together to determine the total frictional head loss for the solar circuit.

Parallel runs must be added separately, as well – though it is important to remember that when flow is split, the flow rate is reduced in each of the branches. Therefore, the parallel runs must be calculated at different flow rates than the primary circuit.

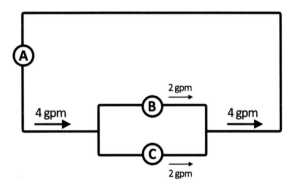

Figure 2 - Parallel runs in a closed loop system

An example of a parallel run in a closed loop circuit is shown in Figure 2. In this example, it would be necessary to determine the head loss in Circuit A at a flow rate of 4 gallons per minute, as well as the head loss in Branches B and C at 2 gallons per minute. These frictional head losses are then added together to determine total head loss for the circuit.

4

STEP FOUR   **Convert fittings to equivalent pipe length**

Once the total length of copper piping in the circuit has been determined, it is necessary to account for head loss in the various fittings in the circuit. This is accomplished by (a) determining the fluid velocity in the pipe and (b) utilizing Table 1 to determine the equivalent length of pipe for each fitting.

The fluid velocity can be determined by multiplying the flow rate by the following factors to determine fluid velocity:

For pipe size…	multiply the flow rate (in gpm) by…	to determine velocity in ft/s
½"	_____gpm x 1.634	= _____
¾"	_____gpm x 0.726	= _____
1"	_____gpm x 0.409	= _____

The equivalent length of pipe for 90-degree elbows is shown in Table 1. For example, in a solar circuit that utilizes ¾" copper and a fluid velocity of 3 feet per second, each elbow is equivalent in frictional head loss to 2.0 feet of copper.

It is also necessary to account for other fittings in the circuit. This is typically accomplished by comparing the fitting to a 90-degree elbow. Table 2 shows the equivalent head loss for various fittings. A 45-degree copper elbow has 70% of the head loss as a 90-degree elbow. Therefore, in the example where ¾" pipe is used to carry a fluid at 3 feet per second, the equivalent length of pipe for a 45-degree elbow would be 0.7 x 2.0 ft, or 1.4 ft of copper. A reducer coupling would be 0.4 x 2.0 ft, or 0.8 ft of copper.

Velocity,	Pipe size (in)				
fps	1/2	3/4	1	1-1/4	1-1/2
1	1.2	1.7	2.2	3.0	3.5
2	1.4	1.9	2.5	3.3	3.9
3	1.5	2.0	2.7	3.6	4.2
4	1.5	2.1	2.8	3.7	4.4
5	1.6	2.2	2.9	3.9	4.5
6	1.7	2.3	3.0	4.0	4.7
7	1.7	2.3	3.0	4.1	4.8
8	1.7	2.4	3.1	4.2	4.9
9	1.8	2.4	3.2	4.3	5.0
10	1.8	2.5	3.2	4.3	5.1

Table 2 - Equivalent length in feet of pipe for 90 degree elbows (from *ASHRAE Fundamentals*)

Fitting	Iron Pipe	Copper Tubing
Elbow, 90°	1.0	1.0
Elbow, 45°	0.7	0.7
Elbow, 90° long-radius	0.5	0.5
Elbow, welded 90°	0.5	0.5
Reduced coupling	0.4	0.4
Open return bend	1.0	1.0
Angle radiator valve	2.0	3.0
Radiator or convector	3.0	4.0
Boiler or heater	3.0	4.0
Open gate valve	0.5	0.7
Open globe valve	12.0	17.0

Table 1 - Iron and copper equivalents compared to 90-degree elbows (from *ASHRAE Fundamentals*)

STEP FIVE   **Determine head loss of tubing & fittings**

Figure 3 shows the relationship between flow rate and head loss for a variety of pipe sizes. It is critical to note that Figure 3 shows the head loss <u>for water</u>. Once the head loss is determined for water, it must then be adjusted for glycol in Step Five.

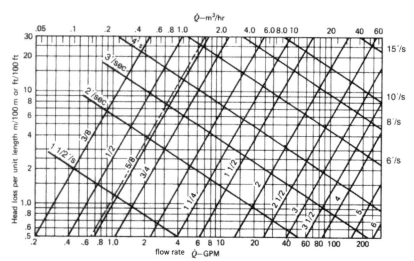

Figure 3 - Frictional Head Loss for Water in Type M Copper (from ASHRAE)

For ¾" copper at a flow rate of 4 gallons per minute, the head loss per 100' of copper is 5 feet. If the total equivalent length of the solar circuit were 235 feet, the total frictional head loss of the piping and fittings would be:

$$\Delta P = (5\text{ feet})(2.35) = 11.75\text{ft of water column } (for\ water)$$

Note that the units for head loss are in feet of water column. This can be converted to pounds per square inch by multiplying it by 0.43. In this example, the head loss is 11.75' or 5.1psi.

STEP SIX **Add head loss of collector(s) and heat exchanger(s)**

The head loss in the collectors and heat exchanger are unique to the specific products used. These properties can be obtained from the manufacturers. Alternately, the head loss in the collector may be available on the OG-100 certification.

The head loss from these components should be added to the head loss determined in Step Four.

STEP SEVEN **Utilize correction factor to account for increased viscosity of glycol**

Due to its higher viscosity, glycol causes higher frictional losses than water. A glycol correction factor of 1.23 is used to account for this phenomenon. If the total head loss for a solar circuit is 10.0' for water, it would have a head loss of 12.3' for 40/60 or 50/50 propylene glycol.

STEP EIGHT **Select a pump**

Once the flow rate and head loss have been determined, a pump can be selected for the circuit. Figure 4 shows the pump curves for a Wilo 3-Speed circulator.

6

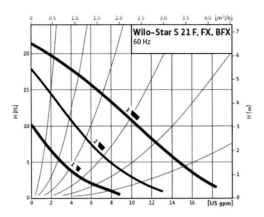

Figure 4 - Pump curves for Wilo 3-speed circulator

If a system had a total frictional head loss of 15' at a flow rate of 3.3gpm, the point on the pump curve would be above Speed 2 and below Speed 3. This means that Speed 3 should be capable of circulating 50/50 glycol in the solar circuit at 3.3gpm.

EXAMPLE

A domestic solar thermal system consists of two flat plate collectors and an indirect tank. It is a pressurized glycol system (50/50) with a recommended flow rate of 2gpm with water. The head loss in the collectors (for water) is equal to 0.2ft of water column, and the head loss at heat exchanger is 0.7ft of water column. The solar loop is 120' round trip and has the equivalent of forty 90-degree elbows.

Step One: Determine the appropriate flow rate for the system

The flow rate for this system should be:

$$Q_g = f_g \times Q_w$$
$$Q_g = 1.1 \times 2.0gpm$$
$$\boldsymbol{Q_g = 2.2gpm}$$

Step Two: Estimate the tubing size in the solar circuit

For a flow rate of 2.2 gpm, a pipe size of ¾" is selected.

Step Three: Estimate the length of the solar circuit

Because this is a simple looped system, the total length of pipe is 120'.

Step Four: Convert fittings to equivalent pipe length

For a flow rate of 2.2 gpm in ¾" pipe, the velocity is:
$$v = 0.726 \times Q$$

7

$$v = 0.726 \times 2.2gpm$$
$$\textit{v = 1.6fps}$$

At a velocity of 1.6 fps, the equivalent length of pipe for a single 90-degree elbow is 1.8'. For 40 elbows, the total equivalent length is:

$$L_e = (40)(1.8')$$
$$\textit{L}_e \textit{ = 72' of copper}$$

Step Five: Determine head loss of tubing and fittings

The total equivalent length of the circuit is:

$$L_t = 120' + 72'$$
$$\textit{L}_t \textit{ = 192' of copper}$$

At a flow rate of 2.2gpm, the head loss per 100' of ¾" copper pipe is approximately 1.7ft of water column (from Figure 3). Therefore the head loss in the solar circuit piping (for water) is:

$$\Delta P_{pipe} = 1.7\text{ft of water column} \times 1.92$$
$$\textit{\Delta P}_{pipe} \textit{ = 3.3ft of water column}$$

Step Six: Add head loss of the collector(s) and heat exchanger(s)

The head loss in the collectors and heat exchanger is now added. It is important to determine whether the head loss at the heat exchanger and collector are for the flow rate for water or the adjusted flow rate for glycol. In this case, the head loss given by the manufacturer is for a flow rate of 2.2gpm. Therefore,

$$\Delta P_{tw} = \Delta P_{pipe} + \Delta P_{collector} + \Delta P_{heat\ exchanger}$$
$$\textit{\Delta P}_{tw} \textit{ = 3.3ft + 0.2ft + 0.7ft = 4.2ft of water column}$$

Step Seven: Adjust head loss for glycol

The head loss must now be corrected for glycol. Therefore,

$$\Delta P_{tg} = \Delta P_{tw} \times 1.23$$
$$\Delta P_{tg} = 4.2\text{ft} \times 1.23$$
$$\textit{\Delta P}_{tg} \textit{ = 5.2ft of water column}$$

Step Eight: Select a pump

Based on the Wilo Pump curves, Speed 1 is capable of overcoming approximately 6ft of water column at a flow rate of 2.2gpm. Therefore, it should be adequate for this application.

8

1000 Ω RTD Temperature Sensor Resistance

Temperature*	Resistance†	Temperature*	Resistance†
30	861.8	35	876.1
40	890.5	45	905.1
50	919.2	55	933.9
60	948.7	65	963.7
70	978.2	75	993.4
80	1008.7	85	1024.0
90	1039.0	95	1054.6
100	1070.3	105	1086.1
110	1101.5	115	1117.5
120	1133.6	125	1149.9
130	1165.6	135	1182.1
140	1198.7	145	1215.4
150	1231.5	155	1248.4
160	1265.4	165	1282.5
170	1299.1	175	1316.5
180	1333.9	185	1351.4
190	1368.5	195	1386.2
200	1404.1	205	1422.0
210	1439.5	215	1457.7

* in °F
† in ohms

3000 Ω Thermistor Temperature Sensor Resistance

Temperature*	Resistance†	Temperature*	Resistance†	Temperature*	Resistance†
Thermistor open	Infinite	72	3390	116	1230
Thermistor shorted	0	74	3230	118	1180
30	10,400	76	3080	120	1130
32	9790	77	3000	124	1040
34	9260	78	2930	128	953
36	8700	80	2790	132	877
38	8280	82	2660	136	809
40	7830	84	2530	140	746
42	7410	86	2420	144	689
44	7020	88	2310	148	637
46	6650	90	2200	152	589
48	6300	92	2100	156	546
50	5970	94	2010	160	506
52	5660	96	1920	165	461
54	5370	98	1830	170	420
56	5100	100	1750	175	383
58	4840	102	1670	180	351
60	4590	104	1600	185	321
62	4360	106	1530	190	294
64	4150	108	1460	195	270
66	3940	110	1400	200	248
68	3750	112	1340	210	210
70	3570	114	1280	215	192

* in °F
† in ohms

10,000 Ω Thermistor Temperature Sensor Resistance

Temperature*	Resistance†	Temperature*	Resistance†	Temperature*	Resistance†
Thermistor open	Infinite	72	11,308	116	4094
Thermistor shorted	0	74	10,764	118	3922
30	34,565	76	10,248	120	3758
32	32,660	77	10,000	124	3453
34	30,864	78	9760	128	3177
36	29,179	80	9299	132	2925
38	27,597	82	8862	136	2697
40	26,109	84	8449	140	2488
42	24,712	86	8057	144	2298
44	23,399	88	7685	148	2124
46	22,163	90	7333	152	1966
48	21,000	92	7000	156	1820
50	19,906	94	6683	160	1688
52	18,876	96	6383	165	1537
54	17,905	98	6098	170	1402
56	16,990	100	5827	175	1280
58	16,128	102	5570	180	1170
60	15,315	104	5326	185	1071
62	14,548	106	5094	190	982
64	13,823	108	4873	195	901
66	13,140	110	4663	200	828
68	12,494	112	4464	210	702
70	11,885	114	4274	215	646

* in °F
† in ohms

Pipe Content—One Foot Length

Inside Diameter*	Volume†	Capacity‡	Weight of Water§
¼	0.5891	0.0026	0.0213
⅜	1.325	0.0057	0.0478
½	2.356	0.0102	0.0850
¾	5.301	0.0229	0.1913
1	9.425	0.0408	0.3400
1¼	14.726	0.0637	0.5313
1½	21.205	0.0918	0.7650
2	37.699	0.1632	1.3600
2½	58.905	0.2550	2.1250
3	84.823	0.3672	3.0600
4	150.797	0.6528	5.4400
5	235.620	1.020	8.5000
6	339.293	1.469	12.2400
8	603.187	2.611	21.7600
10	942.480	4.080	34.0000
12	1357.171	5.875	48.9600
15	2120.580	9.180	76.5000

* in in.   ‡ in gal.
† in cu in.   § in lb

English-to-Metric Conversion			
Quantity	**To Convert**	**To**	**Multiply By**
Length	inches	millimeters	25.4
	inches	centimeters	2.54
	feet	centimeters	30.48
	feet	meters	0.3048
	yards	centimeters	91.44
	yards	meters	0.9144
Area	square inches	square millileters	645.2
	square inches	square centimeters	6.452
	square feet	square centimeters	929.0
	square feet	square meters	0.0929
	square yards	square meters	0.8361
Volume	cubic inches	cubic millimeters	1639
	cubic inches	cubic centimeters	16.39
	cubic feet	cubic centimeters	2.832
	cubic feet	cubic meters	0.02832
	cubic yards	cubic meters	0.7646
Liquid Measure	pints	cubic centimeters	473.2
	pints	liters	0.4732
	quarts	cubic centimeters	946.3
	quarts	liters	0.9463
	gallon	cubic centimeters	3785
	gallon	liters	3.785
Weight	ounces	grams	28.35
	ounces	kilograms	0.02835
	pounds	grams	453.6
	pounds	kilograms	0.4536
	short tons (2000 lb)	kilograms	907.2
	short tons (2000 lb)	metric ton (1000kg)	0.9072
Pressure	inches of water column	kilopascals	0.2491
	feet of water column	kilopascals	2.989
	pounds per square inch	kilopascals	6.895
Temperature	degrees Fahrenheit (°F)	degrees Celsius (°C)	$\frac{5}{9}$ (°F − 32)

Glossary

A

absorbance: The amount of radiation absorbed by a solar collector.

absorber: A solar collector component that captures the radiant energy from the sun and converts that energy into thermal energy, or heat, which it transfers to a heat transfer medium.

absorptance: The rate at which a solar collector absorbs radiation.

absorption: The physical process of absorbing light.

active direct system: A solar water heating system that uses domestic potable water as an HTF and employs a pump or circulator to circulate the fluid through the solar collector.

active indirect system: A solar water heating system that uses a pump or circulator to force an HTF that is not the DHW through the solar loop.

active system: A solar water heating system that relies on pumps or circulators to create HTF flow through its solar collector and storage system.

air lock: A condition that occurs when trapped air surrounds a pump impeller and does not allow the pump to push any liquid through a system.

air mass: The amount of atmosphere that solar radiation must penetrate to reach Earth's surface.

air separator: A solar water heating component that uses mechanical means such as baffles or screens to separate entrained air from HTF.

angle of incidence: The angle at which solar radiation meets the surface of a solar collector.

aperture area: The unshaded opening of the collector that allows access to solar radiation.

asbestos: A mineral that has long, silky fibers in a crystal formation that was a component of many building materials, such as fireproofing materials, pipe insulation, siding, and tile, installed until the late 1980s.

asbestosis: A respiratory disease caused by the inhalation of asbestos fibers resulting in the formation of scar tissue (fibrosis) inside the lungs.

atmospheric storage tank: An unpressurized storage tank that is designed to hold large quantities of solar heated water. Also known as an open system tank.

atmospheric vacuum breaker (AVB): A device that allows air to enter a piping system to facilitate the draining of HTF out of the piping when necessary. Also known as a vacuum breaker.

automatic air vent: A valve that is controlled by a float that opens to allow air to escape from a closed piping system.

automatic draindown system: A solar water heating system that allows HTF to be drained out of the collector and vulnerable solar loop piping using a sensor-controlled automatic draindown valve.

automatic drain valve: A specialized drain valve placed at the bottom of a solar panel array that allows fluid in the array to drain when it is not in use.

B

balancing valve: A valve designed to regulate the rate of fluid flow through a solar water heating system to meet system requirements.

ball valve: A valve in which fluid flow is controlled by a ball that fits tightly against a resilient (pliable) seat in the valve body.

batch collector: A solar collector that consists of a single storage tank framed within an enclosure. Also known as a single-tank ICS collector.

bed section: The lower section of an extension ladder.

brazed joint: A method for joining copper tube and fittings using capillary action by heating the metals and filling the space between the tube and fittings with a filler metal at temperatures above 840°F (449°C).

brazed plate heat exchanger: An external heat exchanger that consists of stacked steel plates that are brazed together.

brazing: A copper tube joining method in which a nonferrous filler metal with a melting temperature of 840°F (449°C) or more is used.

butt spur: The notched, pointed, or spiked end of a ladder that helps prevent the ladder butt from slipping.

C

centrifugal pump: A pump in which a pressure is developed principally by the action of centrifugal force.

centrifugal force: The force that tends to impel a thing or parts of thing outward from a center of rotation.

certification: A voluntary system of standards, usually set by key stakeholders, that practitioners can choose to meet in order to demonstrate accomplishment or ability in their profession.

charging: The process of purging air from a solar water heating system while filling the system with the appropriate HTF at the proper pressure.

check valve: A valve that permits fluid flow in only one direction and closes automatically to prevent backflow.

chlorinated polyvinyl chloride (CPVC): A cream-colored thermoplastic material specially formulated to withstand higher temperatures than other plastic materials and used for potable water distribution, corrosive industrial fluid handling, and fire suppression systems.

circuit breaker: A device that opens and closes a circuit by nonautomatic means and automatically opens a circuit when a predetermined current overload is reached without damage to itself.

circulator: A line-mounted centrifugal pump used to move HTF or DHW through system piping and solar collectors.

clip-on sensor: *See* flat-tab sensor.

closed-loop system: *See* indirect system.

closed system: *See* indirect system.

code: A regulation or minimum requirement.

combination system: A solar water heating system that uses the heat harvested from collectors to heat the DHW supply and HTF used for other purposes. Also known as a combi system.

combi system: *See* combination system.

commissioning: A formal assessment of the functionality and efficiency of a system.

concentrated solar power (CSP) technology: Solar thermal technology that uses mirrors and lenses to reflect solar radiation from a large area onto a smaller target area.

conduction: The transfer of heat through a solid material resulting from a difference in temperature between different parts of the material.

convection: The mode of heat transfer in fluids (liquids and gases).

coolie cap flashing: Flashing typically made from copper that provides a watertight seal for solar piping and wiring.

cross-linked polyethylene (PEX): A thermosetting plastic made from medium- or high-density cross-linkable polyethylene that is used for water service piping and cold and hot water distribution piping.

D

daylighting: *See* solar lighting.

DC circulator: A small magnetic drive pump that can be powered by a photovoltaic module in a closed-loop solar water heating system.

deceleration distance: The additional vertical distance a falling worker travels before stopping, excluding lifeline elongation and free-fall distance, from the point at which a deceleration device begins to operate.

delta T (ΔT): The difference, or change in temperature either occurring or needed.

demand-type water heater: *See* tankless water heater.

dielectric union: A pipe fitting used in between two dissimilar metal piping materials.

differential controller: An operational controller used to energize an electrically powered component based on a setpoint difference between two temperature values. Also known as a differential temperature controller.

differential temperature controller: *See* differential controller.

diffuse radiation: The scattered solar radiation that eventually reaches the surface of a solar collector.

direct flow evacuated-tube collector (flooded collector): A solar collector that contains a flow pipe attached to an absorber in which heat transfer fluid flows through the manifold, down the pipe, and back up into the manifold within a glass enclosure.

direct mounting method: A solar collector mounting method that involves attaching the collector directly to the roof surface without any supports or blocking beneath it. Also known as the flush mounting method.

direct radiation: Solar radiation that reaches Earth in a direct line or beam from the Sun without any change in direction.

direct system: A solar water heating system that uses potable DHW as HTF. Also known as an open system or open-loop system.

direct vent system: A vent system that uses separate piping for combustion air and the products of combustion.

dish/Stirling system: A concentrating collector that uses a concave (dish-like) reflecting surface to concentrate direct solar radiation onto a receiver.

double-pole scaffold: A wood scaffold with both sides resting on the floor or ground and not structurally anchored to a building or other structure.

double-wall heat exchanger: A heat exchanger that has two walls separating the DHW and the HTF.

drainback system: A solar water heating system that drains heat transfer fluid from the collectors into a storage tank located inside the building to protect the system from freezing temperatures.

drainback tank: A small storage tank that is used to hold HTF when a drainback system is not in operation.

drain valve: A valve that contains a globe valve and hose threads and is used to drain or flush tank or system piping.

dry collector: A solar collector that uses air as the heat transfer medium. Also known as a solar air collector.

E

effective collector area: *See* net area.

electric shock: A condition that occurs when an individual comes in contact with two conductors of a circuit or when the body of an individual becomes part of an electrical circuit.

electric water heater: A storage water heater that uses heat produced by the flow of electricity through a heating element to heat water.

electromagnetic spectrum: A continuum of all electromagnetic waves arranged by frequency and wavelength.

electronic flow meter: A flow meter that uses electronic sensors to measure fluid flow in a system.

emissivity: The ability of a material to emit energy by radiation.

enclosure: A collector component that houses an absorber and piping, protects the absorber from ambient temperatures, and captures some of the radiant heat from the absorber.

evacuated-tube collector: A solar collector that uses a glass tube to enclose an absorber plate or fin and attached tubing.

expansion loop: An arrangement of pipe and fittings that allows the thermal expansion gain to be absorbed within the loop rather than along the length of tubing or piping where it can cause pipe damage.

expansion tank: A tank with an air-filled cushion that provides additional volume for water or HTF under thermal expansion.

extension ladder: An adjustable-height ladder with a fixed bed section and sliding, lockable fly sections.

external heat exchanger: A heat exchanger that resides outside of a storage tank.

F

fixed ladder: A ladder that is permanently attached to a structure.

flat-plate collector: A solar collector that contains a flat absorber mounted with attached flow tubes within an insulated and glazed framed enclosure.

flat-tab sensor: A sensor with a flat or slightly curved end that is typically secured to the surface of a pipe or tank using a screw or stainless steel clamp. Also known as a clip-on sensor.

flashing: A piece of metal or thermoplastic material placed around roof penetrations that are vulnerable to leakage.

flow meter: A solar water heating component that is used to measure the flow rate of fluid in piping.

flush mounting method: *See* direct mounting method.

fly section: The upper section of an extension ladder.

foot pad: A metal swivel attachment with a rubber or rubberlike tread that helps prevent a ladder butt from slipping on hard surfaces.

free-fall distance: The vertical distance of the fall-arrest attachment point on a body harness between the onset of a fall and the point at which a fall-arrest system applies force to arrest the fall.

full-way valve: A valve designed to be used in its fully open or fully closed position. Also known as a shutoff valve.

fuse: An electrical overcurrent protective device with a fusible portion that is heated and broken by the passage of excessive current.

G

gas-fired water heater: A storage water heater that uses heat produced by the combustion of natural or liquefied petroleum (LP) gas to heat water.

gate valve: A full-way valve used to regulate fluid flow in which a threaded stem raises and lowers a wedge-shaped disk that fits against a smooth machined surface, or valve seat, within the valve body.

glazed collector: A solar collector that encloses the absorber and piping in a frame and uses a glass or plastic cover.

glazing: The part of a collector that covers and protects the absorber, allows solar energy to reach the absorber, contains radiant heat from the absorber within the enclosure, and consists of glass or plastic materials.

globe valve: A valve used to control fluid flow by means of a pliable disk that is compressed against a valve seat surrounding the opening through which water flows.

gravity vent: *See* natural vent system.

gross collector area: The overall collector dimension.

ground: A safety feature designed to prevent shock due to a fault in an electrical system.

ground-fault circuit interrupter (GFCI) receptacle: A fast-acting receptacle that detects low levels of leakage current to ground and opens the circuit in response to the leakage (ground fault).

guardrail: A rail secured to uprights and erected along the exposed sides and ends of a platform.

H

hazardous material: Any material capable of posing a risk to health, safety, and property.

header serpentine design: A tube flow design in which one continuous tube snakes from one side to the other side for the length of the collector but uses a manifold at the top and bottom of the collector instead of one inlet and outlet.

heat capacity: *See* thermal mass.

heat dump: A solar water heating component that provides a method for heat to be transferred away from the solar collector.

heat exchanger: A component of a solar water heating system that is used to transfer solar-generated heat from one liquid to another.

heating element: An electrical device within a water heater or solar storage tank with an intentionally high resistance that produces heat when connected to a power supply.

heat pipe evacuated-tube collector: A solar collector that transfers heat from a set of pipes that are attached to a manifold that contains the heat transfer fluid.

heat taping: *See* heat tracing.

heat tracing: An electrical heating element embedded in flexible cable that runs along the length of vulnerable piping and activates when the ambient temperature drops below a predetermined setpoint in order to keep the pipe from freezing. Also known as heat taping.

high-temperature collector: A solar collector that collects and transfers heat in operating temperatures above 180°F (82°C).

hybrid flat-plate collector: *See* hybrid photovoltaic thermal (PV/T) collector.

hybrid photovoltaic thermal (PV/T) collector: A photovoltaic flat-plate collector that converts solar radiation into both electricity and thermal heat in one panel. Also known as a hybrid flat-plate collector.

I

immersed heat exchanger: *See* internal heat exchanger.

immersion heating element: A heating device within a water heater or solar storage tank that makes direct contact with the substance to be heated.

immersion sensor: A long, cylindrical sensor that is installed in the sensor well placed in system piping, a collector, or a storage tank.

immersion well: *See* sensor well.

indirect system: A solar water heating system that uses a heat exchange system to transfer the harvested heat from the collector to the DHW. Also known as a closed system or closed-loop system.

inline flow meter: A flow meter installed in a piping system that measures the displacement of a tapered piston or float.

insolation: The amount of solar irradiation reaching the surface of Earth.

instantaneous water heater: *See* tankless water heater.

insulation: A material used as a barrier to inhibit thermal transmission.

in-tank heat exchanger: *See* internal heat exchanger.

integral collector storage (ICS) collector: A solar collector that uses one or more storage tanks as its absorber.

integral collector storage (ICS) system: A passive solar water heating system that consists of either a single storage tank or a series of multiple large tubes or tanks mounted within a glazed enclosure.

integrally mounted collector: A collector that is recessed into a roof structure and in which the collector surface becomes the roof surface.

internal heat exchanger: A heat exchanger that resides in a solar storage tank and allows the transfer of solar-generated heat from the HTF to the potable water without permitting the two liquids to mix. Also known as an in-tank or immersed heat exchanger.

isolation flange: A combination isolation ball valve and companion flange for circulators.

J

J bolt: A fastener that hooks around a secure framing member and has a threaded end that is used with a nut and washer to secure a collector mounting assembly.

job hazard analysis: An analysis that focuses on the relationship between a worker, job tasks, the tools used, and the work environment in an effort to identify hazards before they occur.

L

ladder: A structure consisting of two side rails joined at intervals by steps or rungs for climbing up and down.

lag screw: A screw used to penetrate a roof truss and secure a collector mount to the roof.

lanyard: A flexible line of rope, wire rope, or strap that generally has a connector at each end for connecting a body harness to a deceleration device, lifeline, or anchorage point.

lead: A heavy and dense material with a low melting point, low strength, and high rate of expansion.

licensure: A mandatory system of standards, usually controlled by state government, to which a practitioner must conform in order to practice a given profession.

lift check valve: A check valve that prevents backflow through the use of a disk that moves vertically within the valve body.

lockout: The process of removing the source of electrical power and installing a lock, which prevents the power from being turned on.

low-temperature collector: A solar collector that collects and transfers heat in the operating temperature range of 70°F to 180°F (21°C to 82°C).

M

magnetic declination: The angle differential between magnetic north and geographic, or true, north.

manual draindown system: A solar water heating system that allows HTF to be drained out of the collector and vulnerable solar loop piping by an operator using isolation and drain valves when the ambient air temperature approaches freezing temperatures.

material safety data sheet (MSDS): A printed document used to relay hazardous material information from a manufacturer, distributor, or importer to a worker.

maximum intended load: The total of all loads, including the working load, the weight of the scaffold, and any other loads that may be anticipated.

midrail: A rail secured to uprights approximately midway between the guardrail and the platform.

multiple-point suspension scaffold: A suspension scaffold supported by four or more ropes.

multiple-tank ICS collector: A solar collector that consists of a series of small tanks or tubes that are also framed within a single enclosure.

multiple-tank ICS system: A passive solar water heating system in which the collector is a series of multiple large pipes or small tanks mounted within a glazed enclosure.

N

natural vent system: A vent system that uses the natural buoyancy of the products of combustion to carry exhaust to the outside. Also known as a gravity vent.

net area: The surface of the absorber that can accept radiation from the sun. Also known as the effective collector area.

nonseparable thermosiphon system: A passive solar water heating system that consists of a storage tank located above the collector and connected to the collector as one piece.

O

open-loop system: *See* direct system.

open system: *See* direct system.

open system tank: *See* atmospheric storage tank.

operational controller: An electronic device used to monitor and control the HTF in a solar water heating system.

overload: A small-magnitude overcurrent that, over a period of time, leads to an overcurrent, which may turn on an overcurrent protection device (fuse or circuit breaker).

P

parabolic trough system: A concentrating collector that uses a linear parabolic solar-energy reflector to concentrate energy on a receiver.

passive direct system: A solar water heating system that uses domestic potable water as an HTF but does not use pumps or other mechanical means to circulate the fluid through the collector.

passive indirect system: A solar water heating system that uses an HTF and heat exchanger to transfer heat to the DHW without the use of a pump or other mechanical means to circulate the HTF.

passive system: A solar water heating system that circulates HTF through the solar collector circuit or loop without the use of a pump or other mechanical means.

pawl lock: A pivoting hook mechanism attached to the fly sections of an extension ladder.

personal fall-arrest system: A safety system used to arrest (stop) a worker's fall.

PEX-AL-PEX tubing: PEX tubing that has a layer of aluminum sandwiched between an inner and outer layer of cross-linked polyethylene.

photovoltaic (PV) controller: An operational controller that uses energy generated in the form of direct current electricity via a PV module or panel to energize the pump of a solar water heating system.

photovoltaic (PV) system: An electrical system consisting of a PV array and other electrical components needed to convert solar energy into usable electricity.

photovoltaics: Solar energy technology that directly converts solar radiation into electricity using crystalline silicon wafers that are sensitive to sunlight.

pitch pan: *See* pitch pocket.

pitch pocket: A small reservoir made around a roof penetration to keep standing water away from the penetration. Also known as a pitch pan.

pole scaffold: A wood scaffold with one or two sides firmly resting on the floor or ground.

polyvinyl chloride (PVC): A plastic piping material used for swimming pool solar water heating, sanitary drainage and vent piping, aboveground and underground stormwater drainage, water mains, and water service lines.

pressure-reducing valve: An automatic device used to convert high and/or fluctuating inlet water pressure to a lower or constant outlet pressure. Also known as a pressure-regulating valve.

pressure-regulating valve: *See* pressure-reducing valve.

pressure relief valve (PRV): A safety device that is used to relieve excessive pressure in a solar water heating system.

pressurized storage tank: A storage tank built to withstand DCW pressure without rupturing.

pump curve: The relationship between flow rate and head.

pump flow rate: The amount of liquid a pump can move without any restrictions measured in gallons per hour (gal./hr) or liters per hour (L/hr).

R

rack mounting method: A solar collector mounting method that involves supporting the collector above a roof surface on a frame or rack.

radiation: The exchange of thermal radiation energy between two or more objects.

reflectance: The ratio of the total amount of radiation reflected by a surface to the total amount of radiation striking the surface.

reflection: The change in direction of a wave, such as a light or sound wave, away from a barrier the wave encounters.

reflectivity: A measure of the ability of a surface, such as an absorber and glazing combined, to reflect radiation.

reverse thermosiphoning: The movement of warm water up to a collector through convection when ambient temperatures are low enough to cool the HTF in the collector, thereby creating a reversal of intended flow.

rope grab: A deceleration device that travels on a lifeline to automatically engage a vertical or horizontal lifeline by friction to arrest the fall of a worker.

rotameter: A variable-area flow meter that consists of a clear tapered tube and a float with a fixed diameter.

S

sacrificial anode: A piece of metal that is more susceptible to galvanic corrosion than the metal structure to which it is attached.

safety plan: A comprehensive document that is developed by a company and distributed to its employees that includes the safety regulations and procedures that the company has instituted.

scaffold: A temporary or movable platform and structure for workers to stand on when working at a height above the floor.

sealed box system: A combustion system that uses the direct vent method and includes a sealed combustion chamber.

sectional metal-framed scaffold: A metal scaffold consisting of preformed tubes and components. Also known as a tube-and-coupler scaffold.

selective coating: A very thin layer of material applied to an absorber to increase absorptance efficiency.

self-pumping system: A passive solar water heating system that uses a phase change (liquid to vapor) or other passive means to cause the fluid in a collector to circulate and transfer heat from the collector to the storage system.

self-retracting lifeline: A type of vertical lifeline that contains a line that can be slowly extracted from or retracted onto its drum under slight tension during normal worker movement.

sensor: An electrical component that is used to measure a property or condition, such as fluid temperature within a solar water heating system.

sensor well: A closed-end threaded fitting that is designed to hold and protect an immersion sensor. Also known as an immersion well or thermowell.

serpentine design: A tube flow design in which one continuous tube snakes from one side to the other side for the length of the collector.

shell-and-tube heat exchanger: An external heat exchanger that consists of a large pressure vessel with multiple tubes that run through it.

shock-absorbing lanyard: A lanyard that has a specially woven, shock-absorbing inner core that reduces the forces of fall arrest.

shutoff valve: *See* full-way valve.

sight glass: A plumbing component that is composed of a transparent tube through which the level of liquid in a drainback tank or piping can be visually checked.

silent check valve: *See* spring check valve.

single ladder: A ladder of fixed length having only one section.

single-pole scaffold: A wood scaffold with one side resting on the floor or ground and the other side structurally anchored to the building.

single-tank ICS collector: *See* batch collector.

single-tank ICS system: A passive solar water heating system in which the collector is a single storage tank mounted within a glazed enclosure.

single-wall heat exchanger: A heat exchanger that has only one wall separating the DHW and the HTF.

sodium phosphate: *See* trisodium phosphate (TSP).

solar air collector: *See* dry collector.

solar conversion valve (SCV): A specialized control valve that allows existing hot water tanks to be retrofitted into solar storage tanks.

solar energy: Energy derived from the sun.

solar fraction: The amount of solar thermal energy produced divided by the water heating energy demand of a building; used to represent the annual amount of water heating energy produced by the solar water heating system.

solar irradiance: The power of solar radiation per unit area; commonly expressed in watts per square meter (W/m^2) or kilowatts per square meter (kW/m^2).

solar irradiation: A measure of solar irradiance over a period of time.

solar lighting: A method of capturing and redirecting natural light for use in the interior of a building using special equipment and techniques. Also known as daylighting.

solar loop: The piping that carries HTF into and out of a solar collector and to and from the storage tank or heat exchanger.

solar noon: The moment when the Sun crosses a point along a north-south longitudinal line.

solar radiation: Energy from the Sun in the form of electromagnetic radiation, which consists of waves with electric and magnetic properties.

solar storage tank: A pressurized storage tank specifically designed for use in a solar water heating system.

solar time: A timescale based on the apparent motion of the Sun crossing overhead.

solar tower plant: A concentrating collector that uses hundreds or thousands of tracking reflectors or mirrors to concentrate solar radiation on a central receiver.

solar window: The depicted elevation and path of the Sun at winter solstice (December 21) and at summer solstice (June 21) used to create a visual image of all possible elevations of the Sun throughout the year.

soldered joint: A method for joining copper tube and fittings using capillary action through the heating of metals and filling the space between the tube and fitting with a metal solder alloy that melts at temperatures below 840°F (449°C).

soldering: A copper tube joining method that uses nonferrous filler metal with a melting temperature of less than 840°F (449°C).

spanner: A piece of 2 × 4 lumber or angle iron that spans across at least two trusses and is used to secure a collector mount to a roof using a long threaded rod or bolt, nuts, and washers.

spring check valve: A check valve that prevents backflow through the use of a conical or cylindrical brass disk held in place by a spring. Also known as a silent check valve.

stagnation: A condition in which HTF no longer circulates through a solar collector.

standard: An accepted reference or practice.

standoff mounting method: A solar collector mounting method that involves the use of mounting assemblies to support and raise the collector above a roof surface.

static head: The maximum vertical height a pump can push a liquid; measured in feet or meters.

steamback system: A solar water heating system that uses steam pressure created by high temperatures within the collector to remove the HTF from the exposed areas of the solar loop.

stepladder: A folding ladder with flat rungs built into a supporting frame that stands independently of support.

superheated water: Water under pressure that is heated above 212°F (100°C) without becoming steam.

suspension scaffold: A scaffold supported by overhead wire ropes. Also known as a swinging scaffold.

swimming pool collector: An unglazed solar collector that is specifically designed to use solar radiation to heat swimming pool water.

swing check valve: A check valve that prevents backflow through the use of a hinged disk within the valve body.

swinging scaffold: *See* suspension scaffold.

Sydney collector: A direct flow evacuated-tube collector that has a double-walled glass tube design with a vacuum created between the two glass walls.

T

tagout: The process of attaching a danger tag to a source of power to indicate that the equipment may not be operated until the tag is removed.

tankless water heater: A direct water heater that heats water flowing through it on demand and does not include a hot water storage tank. Also known as a demand-type or instantaneous water heater.

temperature and pressure (T&P) relief valve: An automatic self-closing safety valve installed in the opening of a water heater tank that releases heated water and relieves pressure in a controlled manner.

temperature differential: The difference in temperature between any two points at a given instant.

tempering valve: *See* thermostatic mixing valve (TMV).

thermal expansion of piping material: The expansion or gain in the length of piping when the system is heated.

thermal mass: The capacity of a material to store thermal energy for extended periods. Also known as heat capacity.

thermosiphoning: A method of passive heat exchange based on natural convection in which liquid circulates without the need for a mechanical pump.

thermosiphon system: A passive solar water heating system that consists of a storage tank located above a collector.

thermostatic mixing valve (TMV): A valve that blends hot water with cold water to produce a safe and constant-temperature hot-water flow at the point of use. Also known as a tempering valve.

thermowell: *See* sensor well.

three-way ball valve: A ball valve used in diverter applications suitable for fluid transfer in certain solar water heating systems or as a mixing valve in hydronic systems.

timer controller: A basic operational controller that switches an electrically powered device on and off based on the time of day.

toeboard: A barrier to guard against the falling of tools or other objects.

total global radiation: The total amount of solar radiation that eventually reaches a collector installation site and is available for use in solar thermal systems.

trisodium phosphate (TSP): A white powder used as a cleaning agent or degreaser. Also known as sodium phosphate.

true north: The direction along Earth's surface toward the geographic North Pole, which is 11° from the magnetic North Pole.

true south: The direction directly opposite true north.

tube-and-coupler scaffold: *See* sectional metal-framed scaffold.

tube-in-tube heat exchanger: An external heat exchanger that consists of one tube with HTF flowing in one direction surrounded by another tube containing DHW flowing in the other direction.

two-point suspension scaffold: A suspension scaffold supported by two overhead wire ropes.

U

unglazed collector: A solar collector that does not incorporate an enclosure or glazing to protect the absorber.

V

vacuum breaker: *See* atmospheric vacuum breaker (AVB).

vacuum relief valve: A plumbing valve that is used to automatically allow air into the piping system should a vacuum occur.

valve: A device used to regulate fluid or gas flow within a system.

variable-area flow meter: A flow meter that maintains a constant differential pressure and allows the flow area to change with the flow rate.

W

watt density: The amount of heat generated per surface area unit of a heating element.

wet collector: A solar collector that uses a fluid as the heat transfer medium.

wraparound heating element: A heating element that surrounds the substance to be heated.

Z

zenith: The highest point in the sky reached by the sun directly overhead at a particular location.

Index

Page numbers in italic refer to figures.

USING THE *SOLAR WATER HEATING SYSTEMS* INTERACTIVE DVD-ROM

Before removing the Interactive DVD-ROM from the protective sleeve, please note that the book cannot be returned for refund or credit if the DVD-ROM sleeve seal is broken.

Windows System Requirements

To use this DVD-ROM on a Windows® system, your computer must meet the following minimum system requirements:

- Microsoft® Windows® 7, Windows Vista®, or Windows® XP operating system
- Intel® 1.3 GHz processor (or equivalent)
- 128 MB of available RAM (256 MB recommended)
- 335 MB of available hard disk space
- 1024 × 768 monitor resolution
- DVD-ROM drive (or equivalent optical drive)
- Sound output capability and speakers
- Microsoft® Internet Explorer® 6.0 or Firefox® 2.0 web browser
- Active Internet connection required for Internet links

Macintosh System Requirements

To use this DVD-ROM on a Macintosh® system, your computer must meet the following minimum system requirements:

- Mac OS® X 10.5 (Leopard) or 10.6 (Snow Leopard)
- PowerPC® G4, G5, or Intel® processor
- 128 MB of available RAM (256 MB recommended)
- 335 MB of available hard disk space
- 1024 × 768 monitor resolution
- DVD-ROM drive (or equivalent optical drive)
- Sound output capability and speakers
- Apple® Safari® 2.0 web browser or later
- Active Internet connection required for Internet links

Opening Files

Insert the Interactive DVD-ROM into the computer DVD-ROM drive. Within a few seconds, the home screen will be displayed allowing access to all features of the DVD-ROM. Information about the usage of the DVD-ROM can be accessed by clicking on Using This Interactive DVD-ROM. The Quick Quizzes®, Illustrated Glossary, Solar Radiation Data Sets, Sun Path Charts, Forms and Documents, Flash Cards, Media, and ATPeResources. com can be accessed by clicking on the appropriate button on the home screen. Clicking on the ATP website button (www.go2atp.com) accesses information on related educational products. Unauthorized reproduction of the material on this DVD-ROM is strictly prohibited.